Evolutionary Theory:

THE UNFINISHED SYNTHESIS

ROBERT G.B. REID

Department of Biology,
University of Victoria,
Canada

CORNELL UNIVERSITY PRESS
Ithaca, New York

8   # 11840236
    DLC

10-9-85 rt

© 1985 Robert G.B. Reid

First published 1985 by Cornell University Press

Library of Congress Cataloging in Publication Data

Reid, Robert G.B., 1939-
  Evolutionary theory.

  Bibliography: p.
  Includes index.
  1. Evolution.   I. Title.
QH371.R38   1985   575   85-5747
ISBN 0-8014-1831-3

Printed in Great Britain

EVOLUTIONARY THEORY: The Unfinished Synthesis

Evolutionary Theory:

THE UNFINISHED SYNTHESIS

ROBERT G. B. REID

Department of Biology
University of Victoria
Canada

CORNELL UNIVERSITY PRESS
Ithaca, New York

CONTENTS

FIGURES AND TABLES

Figures

Tables

This is the age of the evolution of Evolution. All thoughts that the Evolutionist works with, all theories and generalizations, have themselves evolved and are now being evolved. Even were his theory perfected, its first lesson would be that it was itself but a phase of the Evolution of other opinion, no more fixed than a species, no more final than the theory which it displaced.

Henry Drummond 1883

To Davie, Mary, Alison and Clio

PREFACE

My favourite literary critique comes from the little girl in James Thurber's *The Darlings at the Top of the Stairs*: 'This book tells me more about penguins than I wanted to know'. My daughter Clio, who learned the alphabet by typing words for the first draft of this book, has similar sentiments about the present work, and she suggests that my next effort 'should be quite short, maybe four pages'. However, this is an epistemological work, in a very general sense. It deals not only with the 'facts' of evolutionary biology and its interpretive logic, but also considers intellectual progress and prejudice, soul-searching and gullibility, and heuristic induction and wishful thinking; which may be more about penguins than students accustomed to cut-and-dried assertions, and teachers who cut and dry, might want to know. Some of my penguins, progenitors of significant epistemological lines, have been detailed down to the last feather, including a number of birds that by conventional wisdom, 'nobody takes seriously'. Some are sketches copied from other naturalists; others are phantoms that cry out for an incarnation denied by a combination of accident and time constraints.

Students in my courses in comparative physiology and the history and philosophy of biology have been the main sounding boards for my ideas. Richard Ring has always listened with polite interest when I have expatiated. Numerous colleagues have suffered their brains to be picked, and many visitors to the University of Victoria have provided useful information, catalytic ideas and valuable reference sources. S. Løvtrup. G. Nelson, A. Packard, J. Ruben, D. Styles, M. Paul, K. Egger and L. Bickell provided some of the reprints I have used. Betsy Sweeny and Ann Stewart assisted with preliminary library work. Dorothy Smith and Barbara Waito processed the early drafts, and Sandy Macmillan effectively threaded the maze of the final revision. Mike and Barbara Ashwood-Smith assisted with the compilation of references.

My departmental chairman, John McInerney, gave intellectual encouragement and provided the resources to facilitate the production of the final draft. My wife, Alison, discussed the central concepts, helped me through the weary, stale, flat and unprofitable moments, and assisted in editing the final draft.

The figures of *Diodon* and *Orthagoriscus* are reproduced from D'Arcy Thompson's *On Growth and Form* by permission of the publishers Cambridge University Press. The first four lines of W.B. Yeat's poem 'The Second Coming' are reproduced by permission of Michael B. Yeats and Macmillan London Ltd.

INTRODUCTION

> When opinions on this issue become more fixed they become
> more questionable, though leading the protected life of the un-
> questioned. No theory should be turned into an institution
> centennially celebrated.
>
> Richard Spilsbury 1974[1]

One hundred and twenty seven years have now elapsed since Darwin
and Wallace (1858) formulated the theory that evolution had
occurred largely as a result of natural selection.[2] Although one
prominent Darwinist found the centenary of the theory an occasion
for remarking that 'a hundred years without Darwin is enough', the
hypothesis of Darwin and Wallace had become the most deeply
entrenched of biological doctrines, and the confidence of Darwinists
had been further stiffened by the revelations of molecular biology:
admiration of the founder reached euphoric levels.[3] My under-
graduate career as a student of zoology began in that centennial year,
and Darwinistic confidence and euphoria saw me through a number
of zoological problems that did not quite seem to fit the Darwinist
mould; but this was probably because of inadequate data requiring
further research. Dissenters who claimed that Darwinism was not
enough, or indeed quite wrong, were not to be taken seriously;
vitalism, Lamarckism, mysticism and senility were recurring conditions in every generation of biology. On occasions when I was
called upon to teach a fragment of evolution theory to first year
biology students I thought my failure to understand selection theory
fully was the result of the specialisation of the subject beyond my
simple comprehension. Confident that every aspect of natural
selection was for the best, I little knew that it had long been criticised
for just that Panglossian felicity.

My views of evolution theory were changed indirectly through an
interest in the history of the universities, stimulated by the turmoil of
the late 1960s. Similar periods of disequilibration had affected both
the universities and the Church in earlier centuries, resulting in some
metamorphoses of educational and doctrinal principles, followed by
long phases of stability and respect for authority. Graduates of those
stable periods, like myself, accepted the normal procedures without

1

question. Proceeding from institutional change to the institutionalisation of ideas, I came to Thomas Kuhn's *The Structure of
Scientific Revolutions* (1962); and a stray remark made by M.
Polanyi in *The Tacit Dimension* (1966) focused my attention on the
institutionalisation of evolution theory: 'A preoccupation with the
way populations of a new kind come into existence has made us lose
sight of the more fundamental question: how any single individual
of a higher species ever came into existence'.[4] Although, in retrospect, there was nothing particularly original about that remark, it
created reverberations in my personal knowledge of 'adaptations'
that seemed inadequate or irrelevant to the organism's needs.
Perhaps the theory of selection was incomprehensible because it was
incomplete, or because its axioms had been institutionalised for
reasons other than their scientific validity. The history of
evolutionism demonstrated that if Kuhn's ideas were of general
applicability then biology was unique. Darwinism was its first and
only universal paradigm, and every crisis had resulted in its deeper
settlement into the substrate of biological thought. Darwin was not
simply a scientific authority; he had undergone a more fundamental
mythification. However, there had been a number of rival paradigms; in addition to Lamarckism there were saltationist theories of
discontinuous evolution, amongst them the emergence doctrine,
which Polanyi had attempted to revive. The saltationist-emergentist
position was that there had been occasional, sudden and unpredictable appearances of macro-evolutionary novelty that had then
settled into slow adaptive adjustment, much in the way that Kuhn had
proposed for the evolution of scientific ideas. Also, while the period
of history occupied by Darwin's own life had been discussed a
number of times, the period of disaffection after his death and
before the resurgence of neo-Darwinism had been dealt with only
fragmentarily. It was a mystery to me how Darwinism could have
been submerged so totally and yet come back so strongly. Although
neo-Darwinism could be experimentally tested, and satisfied the
need for quantification and reduction, characters of a mature
science that had been conspicuously absent from biology, it was by
no means clear that the basic premise of natural selection had been
scientifically justified in the process, and the redefinition of
evolution as frequency changes in the distribution of alleles in
populations seemed to be an attempt to save selection theory at the
expense of the phenomenon it purported to explain. The strength of
polemic and level of invective employed by evolutionists suggested a

skeleton in the cupboard. Certain views expressed in the Darwinian era had become conventional wisdom. A remark to a colleague about Darwin's priority evoked a verbatim quotation from T.H. Huxley. My colleague, believing that he was giving his own considered opinion, had produced an epistemological diamond that had survived generations of mental digestion to emerge as bright as ever. The question that this posed was not the value and longevity of Huxley's ideas, but how much of our total 'knowledge' of evolutionary theory consisted of received aphorisms.

As a student and teacher of comparative physiology, I had been struck by the persistence of additive selectionist hypotheses for the evolution of physiological systems. The concept that biochemistry is evolutionarily conservative ought to have implied that physiological systems must progress by a reorganisation of the conservative biochemical units, and that the key to understanding physiological evolution must be the coordination of interrelated systems. However, adaptive modifications of structural genes, rather than the phenomena of homeostatis, attracted the attention of comparative physiologists.

These suspicions had to be validated by a more thorough search. The matter of Kuhn's view of scientific revolutions in relation to biology had already been considered by Ernst Mayr in a lecture published in 1974, and he had argued that the 'hundred years without Darwin' was due to the many obstacles that both evolutionism and selectionism had had to surmount, concluding that the Darwinian revolution produced the first and only universal paradigm of biology. It seemed to me that some of the 'obstacles' to Darwinism, including the concept of progress in evolution, and saltatory evolution, suggested valid alternative explanations. My comments on this were published in *Systematic Zoology* as part of a correspondence concerning Norman Macbeth's *Darwin Retried* (1973).[5] Macbeth, a lawyer by profession, had found that the evidence and arguments in favour of neo-Darwinism were unsatisfactory, but an anonymous critic of my own opinions wrote that all Macbeth had shown was that the legal rules of evidence were different from, and by implication inferior to, scientific rules of evidence. Søren Løvtrup, in an earlier contribution to the same correspondence, suggested that the logic of neo-Darwinism had more in common with mediaeval scholasticism than with science.[6]

Subsequently, I began a more serious study of the post-Darwinian period to examine non-Darwinist or non-selectionist evolutionary ideas, and to try to comprehend their scientific and philosophical validity, their emotional appeal, and their ultimate failure in the confrontation with neo-Darwinism. The way ahead seemed clear initially. Mivart's *Genesis of Species* (1871), regarded by his contemporaries as the most cogent compilation of criticisms of Darwinism, focused on the causes of variation as the *real agents* of evolution, emphasising the discontinuity of variation. The theme was developed in *Materials for the Study of Variation* (1894) by William Bateson, who later drew support for his belief in discontinuous variation from the rediscovery of Mendelian genetics and De Vries's *Mutation Theory* (1901-03). This line of thought appeared to be a dead end, because De Vries's experimental organism, *Oenothera lamarckiana*, was soon shown to be aberrant, and Mendelian genetics, instead of being used to support discontinuity theories, were grafted on to Darwinism.

Some labyrinthine non-selectionist doctrines germinated even as classical genetics was going from success to success. One was neo-vitalism, whose major proponents Hans Driesch and Henri Bergson were reacting to a perceived inadequacy of natural selection as well as the mechanistic emphasis which it had been given by August Weismann. Driesch, a disillusioned mechanist, invested his faith in a vital force which he called *entelechy,* and Bergson emphasised the autonomous *élan vital*, or evolutionary creativity. Emergence doctrine had been adumbrated by Hegel, J.S. Mill, and G.H. Lewes, developed as a biological and spiritual idea by Henry Drummond, taken as a model for political evolution by Marx and Engels, and independently expounded by Madame Blavatsky and Nietzsche. Its metaphysical aspects were enlarged by Samuel Alexander and C. Lloyd Morgan, both of whom touched on its biological significance. For the neo-vitalists, and Alexander and Morgan, some types of emergence were transcendental; that is, out of a system that had involuted, reached 'critical mass', or had been catalysed into an unpredictable novelty, there could emerge a force that need not conform to known natural laws. Driesch's entelechy, for example, existed in 'ultraspace' and popped out occasionally to take care of its organismic responsibilities, and for Alexander and Morgan, deity was a transcendental emergence. Although there was considerable interest among philosophers, little impact was made on biologists.

Holism, *organicism* when applied to living organisms, had several independent routes of development. In human physiology and medicine Claude Bernard initiated a holistic tradition with the concept of constant life, later called *homeostasis*, namely the maintenance of the mammalian internal environment by physiological responses to disturbance. Some forms of holism began as a reaction to the reductionistic treatment of the cell theory; the belief that if the generalised cell could be understood then the problems of understanding the organism would be solved. Another route to biological holism was the negative reaction to Weismann's premature mechanistic germ plasm theory which proposed that the determinants of heredity were physical bodies borne by the chromosomes.[7] However, holism and emergentism, whether transcendental or realistic, were not scientific explanation. By contrast, neo-Darwinism and neo-Lamarckism were generalisations that suggested experimental tests.

Neo-Lamarckism, in the broadest sense, developed out of the scientific need to find explanations for the origins of variations, and a metaphysical need to believe that the organism was involved in its own evolution, rather than being the product of blind chance. This was combined with a rejection of the all-sufficiency of natural selection. Neo-Lamarckism was an eclectic doctrine consisting of Lamarck's law of use and disuse, inheritance of acquired characteristics, the Buffon/Geoffroy concept of the direct influence of the environment on evolving organisms, together with Darwinian natural selection in a lesser role. Since all attempts to prove the inheritance of acquired characteristics failed, most neo-Lamarckists resorted to induction by elimination: if it was not by natural selection it had to be by the inheritance of acquired characteristics.

Darwin had conceded in *The Origin of Species* that in addition to descent with modification by natural selection there were laws of growth, which participated independently in evolution. Others, unimpressed with natural selection, elaborated perfecting principles and various forms of orthogenesis to explain directed evolution. All of these alternatives to Darwinism could flourish as long as there was confusion about the nature of heredity and the causes of variation, and in the absence of direct proof that natural selection caused evolution. The marriage of Mendelian genetics with the findings of the *Drosophila* experiments by T.H. Morgan and his contemporaries established some foundations of the modern synthetic theory, and the development of neo-Darwinism as theoretical population genetics provided a mathematical base, marking the coming of age of biology as a science. In the confusion of rites of passage two crucial proofs were missing. The nature of heredity and the significance of sexual reproduction had been resolved, the causes of variation located, and a gene theory formulated, and though natural selection had been proved to exist, and 'selection pressure' shown to be more significant than 'mutation pressure' in the cases examined, the molecular nature of the gene was still suppositional. Nor had it been proved that natural selection was the cause of evolution. Allowing that it was the efficient cause of population changes, its position as a cause of evolution was based on the Darwinian supposition that the causes of population variations were also the causes of speciation and phylogenesis. Logically and scientifically this was no sounder than the Lamarckist belief that inheritance of acquired characteristics had to be the cause of evolution because it was the only one that made sense. Re-definition of evolution as frequency changes in the distribution of alleles in populations, and appeals to authority and to the consensus of the majority of biologists covered this omission. Then two events helped to divert attention from the inadequacies of selectionism. The first was that triumph of reductionism, the elucidation of the molecular nature of the gene. The second was the ritual of the Darwin-Wallace centenary, which furthered the mythification of Darwin, and provided an opportunity to pour bile and scorn on his opponents and detractors. However, silently flowing alongside neo-Darwinism was another current of thought developed by a generation of holistic embryologists, geneticists and theoreticians out of whose work came theoretical biology coupled with system theory, and evolutionary epigenetics.

Epigenesis originally implied that all eggs were rather similar and their development in different directions was due to the microenvironment of the embryo as well as larger environmental conditions. The opposite view was preformationism, which held that every detail of the complete organism was contained in the egg and by an *evolutio* — or unfolding — it matured. By the time the Morgan gene theory had developed, a compromise had been reached between epigenesis and preformationism, and the elucidation of DNA brought final acceptance of the molecular nature of the gene. Modern epigenetics therefore accepts the concept of preformation in the form of molecular genetic information, and epigenesis in the form of the control of gene expression in the developing embryo by its intracellular and extracellular micro-environment: Waddington's 'epigenetic landscape'. The pre-Darwinian evolutionist Geoffroy St Hilaire was well aware that the earlier in the development that a change occurred the more drastic would be the change in the mature organism, and many other evolutionists, including Mivart, Bateson and the neo-Lamarckist E.D. Cope, also recognised that early embryonic changes could have greater evolutionary impact than small variations affecting only the mature organism. Richard Goldschmidt suggested the sudden appearance from time to time in evolutionary history of *hopeful monsters* through epigenetic mutations. S. Løvtrup has recently called for a comprehensive theory that would go beyond the inadequacies of selectionism and focus on epigenetic change.

Relatively few of these trends of thought have influenced the consideration of physiological evolution. The holism of pioneering physiologists has been overlooked in the haste to apply the findings of molecular biology to function, resulting in a preoccupation among comparative physiologists with minor biochemical adaptations and their energetic utility, and the neglect of the processes of functional-morphological progress. The concept of progress in biological evolution has been the subject of debate ever since Darwin warned against the careless use of 'higher' and 'lower' in *Origin*. It is sometimes given the progressionistic meaning of the working out of a supernatural purpose, and usually has some implicit value-judgement, for which it is universally condemned. *Progressive evolution*, on the other hand, often means no more than a discernible trend in adaptation. Evolutionary progress implies for me an increase in the complexity and effectiveness of organismic organisation, resulting in increased adaptability.[8] *Organisation* has the connotation of a hierarchical system whose sub-units are linked through feedback to effect a dynamic and self-reproducing whole. *Order* is sometimes used in the same sense. But, since physico-chemical systems are also ordered, for example sand and silt grains are ordered by tides and currents into beaches and mudflats, organisation has become the preferred biological usage. The apperception that humans are intelligent beings who communicate by language, while other primates have limited intelligence and much simpler means of communication, and lower animals have limited consciousness, and plants have none, gives the concept of progress some meaning, although it may be dismissed by some as sheer anthropocentricity. To avoid the undesirable connotations of progress I will use E.S. Russell's equivalent: *grade evolution*, which is similar to Lamarck's *gradation*. The process is distinct from progressive adaptation in that it confers upon organisms the ability to deal with a variety of environments, not simply with one particular environment, and it would seem profitable to consider it in terms of the homeostatic interactions that allow the whole organism to 'persist in being' as Darwin said.[9]

Both developmental and physiological homeostasis imply a steady state and resistance to change. Therefore it is necessary not only to inquire how the homeostatic condition was reached, but also how its resistance to change could be overcome, to allow subsequent evolution. In some cases environmental changes were necessary, effecting a physical alteration of the pre-existing internal milieu as well as the behaviour of the organism. Changes in the biotic environment have also potentiated evolutionary emergences, such as novel associations of prebiotic materials to produce protocells, of prokaryotes to form complex eukaryotes, and subsequent symbiotic associations and integrations. These changes conform to the original meaning of emergence, being large, unpredictable variations with intrinsic qualities that allow them to survive as new, more complex wholes, in a wider range of environments. Looking at evolution from the perspective of emergent change, natural selection, as differential reproduction, appears as the product of these events and not the initiating cause. Faced with this kind of statement selection-ists sometimes refer to natural selection as a 'court of last resort', despite their low opinion of judicial rules of evidence. This is an unfortunate choice of metaphor, since even a complete knowledge of the deliberations of such a court, while it would inform as to the interpretation of the law, would be useless as a means of under-standing the history of actions and behaviour that are the sub-stratum of the law and the court's decisions. The problem is stated more pithily by the account of the three baseball umpires who give their view of the game as follows. The first says, 'I call them the way I see them'; the second, 'I call them the way they is'; the third, 'They ain't nuthin' till I calls 'em'. Natural selection is usually given the creative role claimed by the third umpire. A biological theoretician ought to be the first umpire but usually opts for the role of the second, while admitting that there has to be a play before there is a call, or that natural selection must have variety to act upon. What is needed is a shift of emphasis from the call to the play, that is, from the outcome, selection, to the range of qualities implicit in 'variety', and an elucidation of those that potentiate evolutionary advance.

Other useful thematic metaphors represent recurrent events in the history of evolutionism. 'Throwing out the baby with the bath water' has been used by S.J. Gould in *Ontogeny and Phylogeny*, but it bears repetition since the consequence of the act is that biologists have to keep re-inventing the same wheels by giving them different names. One of the first to warn against this danger was August Weismann: 'We must not on that account "empty out the child with the bath" and conclude that there is no such thing as a "biogenetic law" or recapitulation of the phylogeny in the ontogeny'.[10] Ironically, Weismann was one of the earliest victims of this indiscriminate flushing. His biophore concept, an early molecular gene theory, went down the drain because a particular aspect of it was demonstrably dingy. A similar hypothesis by Nägeli was rejected along with his progressionistic Lamarckism. Uexküll's hypothesis of a supergene that regulated the Mendelian genes was sullied by his perceived idealism. Any theoretical baby with God, deity or 'primordial consciousness' in the bathtub for company has everybody reaching for the plug at the same time. Even if the metaphysics are well-scrubbed materialism, two other factors determine the fate of the baby: time and authority. If an idea has to wait too long before reasonable scientific proof is forthcoming, it becomes a ghost along with its author, and if remembered at all after its vindication is labelled a mere 'lucky guess', as E.B. Wilson called Oken's cell theory and Haeckel's embryology. He also used the expression 'celebrated hypothesis' for the unfounded idea of pangenesis, since it was distinguished by the authority of Darwin, and made enough of a splash to give pause to hygienic positivists.[11] There were plenty of lucky guessers in the two thousand years preceding Darwinism, and Darwin built up just enough epistemological momentum to escape the epithet himself.

A third epistomological gem is Driesch's metaphor of the 'photograph of the problem', in which the problem is simply restated in a sufficiently unfamiliar and interesting way that it lures the observer to think that it has been solved instead of restated.[12] At the turn of the century a photomicrograph of a cell would have been so interesting that its failure to explain anything could be overlooked. The confusion is compounded by the fact that sometimes a photograph can solve a problem by giving an experimeter a longer time to peruse a transient condition, or a more objective view that 'seeing with your own eyes' might provide. Evolution theory is forever being photographed in fashionable modes, whether they be biometrics, quantum mechanics, cybernetics, system theory, thermodynamics or information theory. The process makes the problems more accessible to a wider audience, just as a problem stated in English has more potential solvers if also translated into Russian, but fashionable appeal can mask the absence of a solution.

Ever since Chauncey Wright's review of Mivart's *Genesis of Species* (1871) selectionists have consigned their critics to an abyss of ignorance, a sophism not altogether unjustified. Most critics, as well as most early Darwinists, failed to understand the significance of the Darwinian utilitarian principle — a trait must be useful in some sense to become a general character of the species. On the other hand, selectionists have always tended to identify utility as a simple causal relationship between a single trait and its immediate environment, ignoring useful integrative qualities, adaptability and associations, as well as downplaying latent changes which in the long term of changing environments might prove useful. Darwinism has truly been the only universal paradigm of biology, but it has failed to direct its acolytes' attention to the gaps in their knowledge of the major events of grade evolution. Julian Huxley did not imply a Hegelian sense when he referred to the 'modern synthesis', meaning simply an amalgamation of separate disciplines to produce a united theory. There is no reason to suppose that evolution theory is epistemologically unique, that once the synthesis has been created it is so perfect that there is no necessity for improvement. But because most evolutionists perceive it in this way the synthetic theory has developed an inertia that obstructs dialectical progress, and its survival is largely due to its ability to absorb ideas and data that it did not generate, to 'immunise' itself from criticism, to subsume antitheses under its thesis, or to ignore the antitheses altogether.[13] The absence of a paradigm from a science may indicate its immaturity and deny it the concerted efforts of its practitioners. At the other extreme participants in a science may perceive the current paradigm as being all-sufficient, their science requiring only some finishing touches to reach perfection.

In *The Coming of the Golden Age* (1969) Gunter Stent begins his exploration of biological progress from the same departure point as myself, the university turmoil of the 1960s. He concludes that biology has come close to exhausting the great unknowns, and that the human psyche will have to adapt to this loss of challenge. I take the view that the illusion of a finished synthesis created an intellectual inertia that is unique in the history of modern science, but the prospects for further intellectual progress are in reality unlimited.

The prognosis is not altogether gloomy. After a long period of stasis, dominated by a few evolutionary authorities, a number of crises have developed during the last decade, marked by an explosion of literature. A number of different issues have been involved, including the sudden awareness of the importance of the full life-cycle of the organism to evolution, especially the early epigenetic events; 'hopeful monsters' have raised their lovely heads again; a non-gradualistic view of speciation, as exemplified by the *punctuated equilibrium* hypothesis, has once more been advanced and *molecular drive* may solve some significant problems. [14] A growing impatience with simplistic 'adaptationism' is being expressed, and a variety of flirtations with neo-Lamarckism have been detailed by the scientific gossip columnists. Numerous commentators are talking about an impending paradigm change of greater significance than the settlement of neo-Darwinism into the intellectual substrate of the 1930s. In general there has been a move towards holistic interpretations of evolution: adaptation in the context of the life cycle rather than in the context of the immediate environment; the importance of a well-integrated body plan instead of the external adaptive morphology, and more stress upon behavioural adaptability rather than conformity. These themes have been pursued frequently in the history of evolutionism: their re-emergence is welcome, and has an intrinsic epistemological interest that I will explore in the concluding chapters.

1 LUCKY GUESSES AND CELEBRATED HYPOTHESES

> As to the theory of Natural Selection. . .You had worked it out in
> details I had never thought of, years before I had a ray of light on
> the subject, and my paper would never have convinced anybody
> or been noticed as more than an ingenious speculation, whereas
> your book has revolutionised the study of natural history, and
> carried away captive the best men of the present age.
>
> A.R. Wallace to Charles Darwin, 1864[1]

Evolution was the idea whose time finally came in 1859, with the
publication of Charles Darwin's *On the Origin of Species*. The
individual elements of the Darwinian theory, concepts of the struggle
for existence, adaptation, survival of the fittest, and mutability of
type had drifted in and out of human consciousness for millenia, but
few pre-Darwinian thinkers had projected from these to the broad
sweep of evolutionary progress from a few simple ancestors to their
complex and highly ordered descendants. No other, save Alfred
Russell Wallace, had drawn all of these ideas into a cohesive theory,
and only Darwin had presented the theory along with a
comprehensive body of evidence.

The belief that organisms might be able to change from one type to
another is found in the surviving works of the pre-sophist Greek
philosophers in the period 700 to 400 BC. Thales thought that life
originated in the sea; Anaximander believed that man meta-
morphosed from fish-like forms, and Xenophanes understood that
fossils were the remains of now extinct animals. Empedocles held
that the antagonistic forces of love and hate had acted on the four
elements to produce incomplete organisms, which sought out their
missing parts and emerged as monsters before modern animals
appeared. There was thus a suggestion of progress from the simple
or incomplete to the complex and complete in Empedocles' thought,
together with the inference that some inadequate forms would die
off. Plausibility, and the agreement of other thinkers, were
sufficient reason for acceptance of these ideas. A popular metaphysical universal as flux or change: 'Change is the only reality', said
Hippocrates. This emphasis on change was also contained in the
prehistoric metaphysics of China, and already formalised in the *I*

12

Ching or *Book of Changes* by the beginning of the first millenium BC. However, the Chinese showed no interest in developing theories concerning the origins and evolution of plants and animals, and the changes were cyclic rather than progressive. Similarly, in India the ancient Vedas spoke of diurnal and seasonal cycles which are sometimes taken to symbolise biological and spiritual evolution. To whatever extent mutability and the other separate elements of evolution theory may have hinted at evolution it would be wrong to infer that these early thinkers were evolutionists in the modern sense.

The role of Aristotle in the development of evolutionary ideas has been much debated. Aristotle's *scala natura* placed man at the top and the other animals below in descending order of complexity. At the bottom were the sponges and the plants. The similarity between Aristotle's ladder and later evolutionary trees has suggested to some authors an intuition of progressive evolution. But his scheme may simply have been a ranking by complexity. Aristotle's emphasis on the unchanging entelechy that guided the essential being of an organism, and his aphorism: 'Nature is true to type', make it difficult to conceive of him as an evolutionist. Indeed, his later influence on Arabic and European science may have been an obstacle to the development of the concept of mutability of type.

When Greek and Islamic works became available in Europe in the twelfth and fourteenth centuries, through the efforts of the schools of translators in Spain and Italy, Aristotle was prominent among the classical authors whose lost works could now once more be read, and his acceptance by the Church placed him squarely in the path of proto-evolutionists. Although Islamic alchemy admitted of the mutability of metals, and by association, the mutability or transcendence of the human mind and soul, there was no biological mutation implied. A poem by the Persian mystic Jalaludin Rumi has been taken to suggest a form of evolutionism:

I died from a stone and a plant became,
Died from a plant and took a sentient frame,
Died from a beast and put on human dress,
When by my dying did I ere grow less?

It may be more plausible to suppose that Rumi was referring to advancing grades of reincarnation.[2]

Among the philosophers who hinted at mutability were Giordano Bruno (sixteenth century), Francis Bacon and René Descartes (sixteenth and seventeenth centuries). Leibnitz (seventeenth-eighteenth century) was more explicit than these: 'It is credible that by means of such great changes of habit even the species of animals are often changed.'[3] Leibnitz's speculations on progressive changes of grade in organisms influenced the French evolutionists of the eighteenth century. The speculative evolutionists De Maillet, Maupertuis, Diderot and Robinet were affected by early Greek opinions, and added their own embellishments. The Swedish taxonomist, Carl von Linné, intuitively used phylogenetic relationships between organisms in his taxonomic schemes; and although he was a creationist, he suggested that created species could be 'subsequently multiplied by hybrid generation, that is, by intercrossing with other species'.[4] Georges Buffon (1707-88) speculated about the homologies of vertebrate morphology, and wondered if there could have been a vertebrate archetype. He also mused about the possibility of the evolution of modern species of animals and plants from a few ancestors:

> The naturalists who are so ready to establish families among animals and vegetables, do not seem to have sufficiently considered the consequences which should follow from the premises, for these would limit direct creation to as small a number of forms as anyone might think fit. For if it were once shown that we had right grounds for establishing these families; if the point were once gained that among animals and vegetables there had been, I do not say several species, but even a single one, which had been produced in the course of direct descent from another species; if for example it could be once shown that the ass was but a degeneration from the horse — then there is no further limit to be set to the power of nature, and we should not be wrong in supposing that with sufficient time she could have evolved all other organized forms from one primordial type.[5]

This, he then added, could not be so, since it would be a contradiction of revelation. Buffon further argued that if such evolution had occurred it must have been by hybridisation, and by the gradual accumulation of minor variations brought about by the direct effects of the environment, especially climatic and nutritional conditions.

When the Greek ideas of metamorphosis and flux and progressive spontaneous generation were revived in the eighteenth century, they found themselves in the company of Linnaean taxonomy and

Buffonian natural history, with which they combined to provide a foundation for the more comprehensive evolutionism of Erasmus Darwin and Lamarck. Erasmus Darwin (1731-1802) was impressed with Buffon's ideas concerning the influence of environment on the mutability of species. *Zoonomia, or Origins of Organic Life,* published in 1794, proposed that life was imbued with a vital force and that it changed and evolved by a process analogous to artificial selection, a concept borrowed from Maupertuis. The mechanisms of mutation were: random natural change; the direct influence of the environment; the self-modification of the organism.

Lamarck (1744-1829) had a varied early career as a soldier and writer.[6] During his youth he became an enthusiastic amateur botanist and published a *Flore française* in 1777, which qualified him for membership of the French Academy of Sciences. At the age of 50 Lamarck took the position of Professor of Worms and Insects, the only position available to him after a reorganisation of the Museum of Natural History, and subsequently made his greatest un-disputed contributions to biology in the area of invertebrate zoology. His first evolutionary ideas appeared in a lecture given at the museum in 1800, in which he stressed the importance of relation-ships in taxonomy (as contrasted with differences) and mentioned the reproductive advantage conferred by complexity. His book *Recherches sul l'organisation des corps vivants,* published in 1802, embodied these ideas and developed the concept that an animal's response to its environment induced structural change. He also coined the word *biologie* as did Treviranus in the same year. *Philosophie zoologique,* published in 1809, elaborated his theory of gradation, giving the mechanisms and manifestations of change as: increase in body volume and in the size of the organs up to a limit; the production of new organs out of need and from new movements initiated by the need; the development of organs in direct relation-ship to their use; and the inheritance of acquired characteristics.

The final point is regarded as the epitome of Lamarckism. However, just as important as the genetic fixation of acquired change was the concept of gradual progressive change from imperfect and simple organisms to the complex and perfect. Although he considered the changes brought about by artificial selection, he rejected them as being irrelevant to natural evolution, inferring that response to environmental change would be universal in a species.

Étienne Geoffroy St Hilaire, a member of the Academy of

Sciences who had been a naturalist on Napoleon Bonaparte's Egyptian expedition, supported Lamarck's transformism, although he disagreed with Lamarck's explanations of it. Geoffroy, like Buffon, believed that the environment was the direct cause of change, that such change was not always beneficial, and he denied the inheritance of acquired characteristics. On the basis of congenital abnormalities he concluded that changes in early development would produce the largest transformations in the adult forms and that through such changes the mode of evolution must be saltatory. In the animal kingdom unity of plan was responsible for the phenomena that we now call analogy and homology. This monism induced him to try to unite the invertebrates and vertebrates within a single anatomical plan, and his derivation of the squid from a fish, and similar applications of the principle, drew the fire of Cuvier, resulting in the neglect of the application of the concept to the vertebrates alone.

Goethe was among Geoffroy's supporters, though he was not familiar with Lamarck's opinions. A believer in the harmony of the whole, he expressed his evolutionism in *The Metamorphosis of Animals* (1816).

> All members develop themselves according to external laws,
> And the rarest form mysteriously preserves the primitive type,
> Form therefore determines the animal's way of life,
> And in them the way of life powerfully reacts on form.
> Thus the olderly growth of form is seen to hold,
> Whilst yielding to change from externally acting causes.[7]

Richard Owen, the English comparative anatomist, was also interested in this phenomenon which he called 'homology'. Influenced by the idea of unity of plan, he argued that the archetypes of each major phylum must have been created, because the gaps between them were too large to admit the possibility of mutual ancestry; the original archetypes were nevertheless capable of evolutionary diversification. Owen is commonly presented as a bitter opponent of Darwinism. However, when Chambers' *Vestiges of the Natural History of Creation* was published in 1844 he wrote to the author sympathising with his ideas. He was greatly taken aback by the virulence of the response of the academic establishment to *Vestiges*, and fearful for his own reputation if he were openly to embrace evolutionism.[8] Circumspect in his own publications, his bitterness towards Darwin probably resulted from the fact that this former friend, having finally found the courage to publish and be damned, was now the object of a lionisation that might have been Owen's.

Charles Darwin (1809-82) was born in the year that the *Philosophie zoologique* was published. By the time he entered Edinburgh University as a medical student in 1825, several essays pertaining to evolutionism had been published in England. W.C. Wells read a paper to the Royal Society in 1813 on human skin colour, noting that what breeders do 'by art seems to be done with equal efficiency, though more slowly by nature, in the formation of varieties of mankind, fitted for the country which they inhabit'.[9] This did not pursue the question of species mutability, but did suggest a kind of natural selection. Other British authors advanced the mutability question further. These included Lawrence (1823), Prichard (1826), Blyth (1835, 1837) and Chambers (1844).

Prichard gave his opinions on biological evolution in *Researches Into the Physical History of Mankind* (1826). The following excerpt illustrates the clarity of his concepts of survival of the fittest, adaptation, and speciation.

> The doctrine of variation, or deviation, in the races of animals in general [leads] us to the conclusion, that this is not merely an accidental phenomenon, but a part of the provision of nature for furnishing to each region an appropriate stock of inhabitants, or of modifying the structure and constitution of species, in such a way as to produce races fitted for each mode and condition of existence. A great part of this plan of local adaptation appears to have been accomplished to the original modification of a genus into a variety of species. It has been further continued, and the same end promoted, by the ramification of a species into several varieties. . .Individuals and families and even whole colonies perish and disappear in climates for which they are by peculiarity of constitution, not adapted.[10]

By 1832 von Baer was saying,

> only in a very childish view of nature could organic species be regarded as permanent and unchangeable types, and that in fact they can be only passing series of generations, which have developed by transformation from a common original form.[11]

Darwin is known to have been familiar with Edward Blyth's publications as well as Patrick Matthew's proposal of a form of natural selection (1831), and with Chambers's *Vestiges* (1844). He also knew of Charles Naudin's 1852 paper on species and varieties.[12] Darwin's friend Herbert Spencer coined the expression 'survival of the fittest' in the 'The Development Hypothesis' (1852).[13]

While Darwin began to think seriously about the mutability of species in 1837, shortly after the return of the HMS *Beagle*, he committed nothing to print until after the Darwin and Wallace papers were read in 1858 to the Linnaean Society. Despite his admitted knowledge of Lamarck's work he does not seem to have paid conscious attention to it until his own insights on evolution had developed.[14] These ideas began to incubate during the *Beagle* voyage with his observations of the fauna and fossils of South America. In 1837 his journal hinted at the possibility of mutation of species, and natural selection emerged when in 1838 he read Thomas Malthus's essay on population. Malthus, writing in 1796, had argued that excessive reproduction would result in overpopulation, which was kept in check by natural inhibitors such as famine and disease. Darwin asked himself what qualities would account for survival, and concluded that such qualities must suit the survivors to their conditions, better than those possessed by unsuccessful individuals. If those qualities were heritable they would then be transferred to the new generation. Lyell's remarks in *Principles of Geology* (1830, 1833) concerning the suitability of certain environmental conditions for particular organisms, so that the dominant types changed with climatic change, had prepared Darwin for the concept of adaptation to environment as an aspect of natural selection. In 1842 Darwin wrote his first essay, excerpts from which were read sixteen years later to the Linnaean Society, but which hitherto had been read only by a few friends. It is important to understand that Darwin was connecting two distinct ideas. The first was that species were mutable. This was based on a number of points of evidence. His travels had given him biogeographical reasons to believe in mutability. The South American fauna, for example, included a number of related animals, such as the armadillos, which had fossil relatives including the giant armadillo. Though they occurred there in a number of living and fossil forms they were not found in other parts of the world with similar climates. This distribution was later interpreted by Darwin in terms of an evolutionary diversification of an ancestral type. The Galapagos Island fauna, especially the tortoises, struck him in the same way, and provided the seed that rapidly germinated in his mind long before he set about the accumulation of information, according to 'true Baconian principles'.[15]

Darwin's concept of mutability was further based upon the progressive nature of the fossil record — the simplest organisms occurring in the oldest sedimentary rocks and the most complex in the most recent geological deposits. Embryological evidence was also adduced, phylogenetic relationships being more distinct at the embryonic level. To present these observations as evidence for evolution was an exercise in *petitio principii*, or question-begging. If evolution had occurred then all of the data made sense, but this was not proof of the premise of evolution. Darwin also employed the logical process known as 'induction by elimination', i.e. a plausible theory must be correct if all the alternatives are obviously wrong.[16] This faulty logic brought him and his theory of evolution into disrepute with the philosophers of science of the day, J.S. Mill, John Herschel and William Whewell.[17] However, the development of the idea, and the wealth of observations in *Origin* appealed sufficiently to educated Victorians for the views of the philosophers to be disregarded. The major point of evidence for evolutionism was, in Darwin's view, the experiences of plant and animal breeders, which in recorded history had produced many different varieties which were anatomically and physiologically distinct from the originals. On this evidence, mutability within species was indeed proved, and Darwin contended that species and the higher taxa were produced in the same way. However, even to this day, no artificial selection has ever produced a new species, except in the special case of plant hybrids, when polyploidy has been artificially produced. Despite the circumstantial nature of the case for evolution, the great majority of biologists are content that Darwin was correct and that evolution has occurred. Further supportive evidence continues to accumulate, but it is still indirect. Darwin further elaborated his evolutionism to embrace the grand scheme of progressive evolution from a few simple ancestors to the modern spectrum of complex organisms, a conception which had been absent from the hesitant forays of the other British evolutionists with the exception of Chambers and Wallace.

Descent by modification through natural selection was Darwin's slogan. This linkage of two independent ideas was the core of Darwin's thesis and claim to originality, although it involved the circular argument that evolution occurs because of natural selection, therefore natural selection must operate; natural selection operates, therefore evolution must result. Since the two concepts are independent it does not follow that anyone in history who suggested a natural process of 'survival of the fittest' was an evolutionist. It certainly does not follow that the acceptance of evolutionism implies an acceptance of natural selection: Lucretius understood natural selection but was not an evolutionist; Lamarck was an evolutionist but rejected natural selection. The elaboration of this compound hypothesis was not uniquely Darwinian; A.R. Wallace came to similar conclusions fifteen years later than Darwin. One of the corollaries of Darwin's proposition was that since the mechanism could only be natural selection evolution had to be gradual; change by saltation would make natural selection redundant. Gradual change required a geological time scale, and Charles Lyell whose *Principles of Geology* was taken by Darwin on the *Beagle* voyage had an obvious influence on this aspect of Darwinism. Leibnitz's aphorism, *Natura non facit saltum* (nature does not make leaps) was happily adopted by Darwin, though some of his supporters regarded it as an unnecessary burden.

Darwin had constructed the skeleton of *Origin* by the time Robert Chambers's *Vestiges* was published in 1844. Chambers was a progressionist who believed that evolution proceeded autonomously to a predetermined goal of perfection set by the Creator. 'The simplest and most primitive type, under a law to which that of like-production is subordinate, gave birth to the type next above it'.[18] Chambers's work showed the influence of Buffon, Geoffroy, Lamarck and Erasmus Darwin, but he rejected Lamarck's belief that the organism's volition had anything to do with its adaptation. Evidence for evolution was presented, but he had no scientific suggestions as to the causes. Chambers's book was published anonymously as he feared the consequences, and the hint that he might be the author was enough to make him withdraw his candidacy for the Lord Provostship of Edinburgh, the city in which he conducted his publishing business.[19] *The North British Review* said: 'Prophetic of infidel times, and indicating the unsoundness of our general education, the *Vestiges* has started into public favour with a fair chance of poisoning the fountains of science, and sapping the foundations of religion'.[20] The fascinating controversy fomented by *Vestiges* took it to ten editions before *Origin* came out. As Darwin remarked in his historical sketch in the fifth edition of *Origin*, Chambers had prepared public opinion for evolutionism, and had made some of the scientists all the more ready for a thesis that was supported by scientific observations. One other small part that Chambers played in the evolution drama was to persuade T.H. Huxley to risk an 'episcopal pounding' and stay to confront Bishop Wilberforce at the British Association for the Advancement of Science meeting at Oxford in 1860.[20] Huxley's dialectical victory over Wilberforce in this famous confrontation resolved nothing scientific, but is sometimes taken to signal the collapse of orthodox religious opposition to Darwinism.

The furore over *Vestiges* was one of the factors which contributed to the delay of the publication of *Origin*. The major goad to publish was a letter from Alfred Russell Wallace to Darwin in 1857 enclosing an essay: 'On the Tendency of Varieties to Depart Indefinitely from the Original Type', which outlined a theory of evolution remarkably similar to Darwin's, the catalyst for both having been Malthus's essay on population. Darwin's soul-searching, indecision and lamentation, 'all my originality will be smashed' was compounded by the death of his youngest child.[21] However, an arrangement was worked out that demonstrated Darwin's priority to the theory, while giving credit to its independent formulation by Wallace, and the Darwin and Wallace papers were read to the Linnaean Society in 1858. This stimulated Darwin to edit down his current version of *Origin* and it was published in November of 1859.

Why did Darwin receive so much of the credit after a good century of commitment to and publication of thoughts on evolution by an international array of philosophers and naturalists? Darwin's very hesitation was an important factor. Regardless of his fear of controversy, timidity in making public pronouncements, and the putative malaise, accidie, schizophrenia, Oedipus complex, Fowler's syndrome and Chaga's disease that have been diagnosed by speculators in Darwiniana, he could not have timed it better. Lamarck had planted it, others had nurtured it, Chambers had taken the brunt of outraged reaction; and Darwin took the time to collect the exhaustive information that represented scholarly solidity in nineteenth-century natural history. Therefore Darwin's became 'the theory' of evolution, whose authority elevated his subsequent thought to the level of 'celebrated hypotheses', while his forerunners had merely made 'lucky guesses'. The potential embarrassment with Wallace was resolved in a gentlemanly manner. As A.C. Brackman points out in *A Delicate Arrangement* (1980), Wallace had no say in the plans for the Linnaean Society meeting since he was in the Dutch East Indies. However, there is no doubt that he was satisfied with the arrangement, and pleased to become an instant member of the inner group of Darwinists.

The question of Darwin's claim to originality is, however, interesting. That he felt the right to make such a claim is evident from his initial horrified reaction to Wallace's essay and from his frequent reference in *Origin* to 'my theory', by which he implied not just 'descent with modification' but the concept of evolution in general.[22] This was patently unjustified in view of the earlier claims to priority of Erasmus Darwin, Lamarck, Chambers and the authors who discussed variations of the concept of natural selection. In the German universities the concept of evolution was taught before the publication of *Origin*.[23] But it *was* Darwin's theory in the sense that it was largely his own experience and evidence that forced him to make the intuitive leap that convinced him that evolution was a reality. Some authors who contributed to his development as an evolutionist, such as Malthus and Lyell, were not themselves evolutionists. This personal intuition of evolution is an emotionally affective, almost religious experience, as many authors, including Darwin, have testified, and probably accounts for Darwin's possessiveness, making his claim to originality more comprehensible, if not completely justified. As Brackman also indicates, Wallace was not just a lucky guesser, having set out to explore the causes of speciation almost as early as Darwin, and having spent most of his career as a field naturalist in the tropics, studying the question. Malthus had also been Wallace's inspiration, long before the essay 'On the Tendency of Varieties to Depart Indefinitely From The Original Type' was written in 1858. Brackman's view that Darwin had not developed his own concept of species diversification until he obtained Wallace's essay is questionable, since Darwin was writing about adaptive radiation in 1844. However, the greatest difference between Darwin and Wallace was quantitative, one book manuscript to one essay.

Origin filled a metaphysical void that had been gaping darkly ever since the Enlightenment. The rejection of fundamentalist doctrine led to the search for natural law instead of the interference of the Creator; the emergence of positivism, the idea of progress and its popularisation in England by J.S. Mill, Herbert Spencer and Buckle emphasised the need for natural law in biology.[24] Darwin's was not a deterministic progressionist theory, though it did pay lip-service to the concept in concluding: 'And as natural selection works solely by and for the good of each being, all corporeal and mental endowments will tend to progress towards perfection.'[25] There was a yearning among biologists for their science to mature to the same status as physics or natural philosophy, which could be dealt with empirically and mathematically. Much of biology at that time, along with geology and geography, was mere natural history: stories about nature. If Thomas Kuhn's 1962 theory of scientific revolutions is applied to biology, Darwin can be seen to have given biology its first universal paradigm, the first major step in the maturation of a science.

The response that any modern biologist would give to the question *Why Darwin*? would be similar to Wallace's remarks given in the leading quotation. Huxley held the same view, and like many of his opinions it entered the collective unconscious of biology. Some authors believe that despite the timely arrival of *Origin* Darwinism would not have survived without the support of Huxley, the 'faithful bulldog', ever ready to sharpen his teeth and claws in anticipation of the fray.[26] Huxley dealt with the public relations of Darwinism, with his *Times* review and in oral and written debate. He saw flaws in Darwinism: the burden of gradualism, and Darwin's failure to explain the origins of variation and over-emphasis of natural selection.[27] In trying to come to terms with these difficulties Darwin came round to a quasi-Lamarckist position by suggesting that the environment could have a direct effect on the organism and that these effects might be heritable. His *pangenesis* theory, formulated in *Variation of Animals and Plants Under Domestication* (1868) suggested that hereditary elements called gemmules which could be modified by particular actions of the organism in relation to the environment, or by the direct effect of the environment, could, in their altered state, enter the germ cells and so be passed on to the next generation. This 'celebrated hypothesis' was tested and found wanting in 1871 by Francis Galton, who transfused blood between rabbits to see if it carried gemmules that would alter the recipient in any way. Weismann's germ plasm theory also rejected pangenesis, but it was adopted happily by the neo-Lamarckists, who took the position that since Lamarck had priority and Darwin had reverted to a Lamarckist inheritance theory, credit for the theory of evolution should be returned to Lamarck.

In 1900 Mendel's original work was rediscovered by three workers who had been conducting breeding experiments independently along

Mendelian lines, De Vries in Holland, Karl Correns in Germany and Erick Tschermak in Austria. The modern biology student usually hears an over-simplified history: that the rediscovery of Mendel's laws filled the remaining gaps of Darwinism, by relating the way in which characteristics were inherited, and providing a basis for gene theory. William Sutton's conceptual synthesis of chromosome behaviour and the Mendelian laws, developed in 1902, certainly contributed to that conclusion. However, Mendel's stress on the discrete nature of his genetic factors reinforced the views of the saltationists, who believed that evolution was not always gradual but occurred in discontinuous steps. This interpretation of Mendelian genetics became known as *neo-Mendelism*. Mendel had shown that the genetic factors retained their integrity from generation to generation, even although they might be submerged in a particular individual as an effect of *dominance*, earlier known as *prepotency*. But Darwin's tentative experiments had shown incomplete dominance and had appeared to prove that the characteristics of the parents were blended in the offspring. The outcome was that although blending inheritance was finally laid to rest the immediate beneficiary was not Darwinism but *mutationism*, the form of neo-Mendelism developed by Hugo De Vries, in his *Mutation Theory* (1901-3). The expression *mutation* meant the all-or-nothing change which De Vries thought caused speciation, without the mediation of natural selection. Mutation theory stimulated a flurry of research, including the work of Heribert-Nilsson, Wilhelm Johanssen, William Castle, Jennings and T.H. Morgan and his students. They demonstrated that while the environment can have marked effects on individuals mutations were usually small and imperceptible, the stuff that Darwinian selection needed to work with.

The earliest of the neo-Darwinists, August Weismann, who had rejected all causes of evolution except natural selection, and had to his own satisfaction disproved pangenesis and the somatogenic induction hypothesis, was not taken seriously by all biologists. his materialistic gene theory had obvious inadequacies that made other evolutionists reject it altogether, and contributed to the delay of the acceptance of a general gene theory. However, his staunch selectionism, together with the establishment of the principles of classical genetics by Morgan and his co-workers, paved the way for the neo-Darwinist revival in the early 1930s, despite the period of eclipse during the early decades of the twentieth century.

The mathematical treatment of natural selection by R.A. Fisher and J.B.S. Haldane in England, and by Sewall Wright in America provided the kiss of life for neo-Darwinism, and all but extinguished the rival theories. Starting from the theoretical equilibrium condition formulated by Hardy and Weinberg in 1908, it was shown that the study of populations from one generation to another could determine the relative success, or *fitness*, of particular alleles, and selection coefficients could be assigned to those alleles. The performance of alleles, or their distribution in future populations could then be predicted, and this could be tested experimentally. The hopes of the nineteenth-century materialists that biology would finally become a real science which could be reduced to mathematical principles, appeared to have been realised.

Julian Huxley, the grandson of T.H. Huxley, wrote a definitive text on evolution theory in 1942, calling it *Evolution: The Modern Synthesis* to suggest the multiplicity of his sources, which came not only from Darwin and the population biologists, but also from the founders of genetics. Synthesis, as Huxley used it, lacked the Hegelian implication of dialectical progress and implied only the blending of different ideas. Neo-Darwinism was subsequently referred to as the synthetic theory by some authors. Ernst Mayr believes that the greatest shift in emphasis in evolution theory since the 1930s was the move from *typological thinking* to *demological thinking*, in other words, the recognition that individuals do not evolve, but populations do, and that gene pools of populations are the raw materials of evolution as well as its locus.[28] If the proportions of alleles in the gene pools are altered the individual manifestations of the gene pools, i.e. the organisms, are also altered.

Another important and complementary development was the discovery of the molecular basis of heredity. Before Watson and Crick's *Nature* article on DNA structure was published in 1953, the molecular mechanisms of reproduction, cell division and the translation of genetic information into the physiological activity and structure of the organism had only been guessed at. About twenty years after Weismann's gene theory had been rejected, T.H. Morgan and his collaborators formulated a new gene theory based on their work with chromosome mapping. Even then there was some doubt as to its molecular basis, and a lingering sympathy among biologists for a physical vibrational theory. Neo-Darwinism had already crested in the 1930s before the molecular basis of heredity began to be seriously considered again. Delbrück, a physicist, began to speculate that any molecule that might be at the heart of heredity would have to be large and complicated. Proteins and polynucleic acids (such as DNA) were the most likely candidates, being known to occur invariably in the nuclei which were already known to be the responsible loci.[29] In 1941 Delbrück, along with Luria and Hershey, formed the American Phage Group to study bacteriophages, viruses which were known to contain only the polynucleic acid DNA (deoxyribonucleic acid), and protein. The study of this organism would indicate which molecule was the bearer of genetic information. Prior to the Phage Group's final proof in 1954 that DNA was the only part of the bacteriophage to enter the host cell and participate in the production of the new generation of viruses, there had been other indications that DNA was the molecule being sought, but the evidence was not conclusive.[30]

In parallel developments in the British Isles the physicist Erwin Schrödinger's *What is Life?* (1944) argued that life might have a molecular basis in the form of a 'long, aperiodic crystal'; in other words a molecule sufficiently large and with enough irregularity to contain complex information.[31] Watson and Crick assumed that DNA was the genetic molecule, and guessed it was double stranded, having learned that it was a multi-stranded helix from the X-ray crystallographic work of Wilkins and Franklin at the University of London.[32] From a prior knowledge of the chemical constituents of DNA: thymine, guanine, cytosine, adenine and deoxyribophosphate; and by determining the bond angles at which the sub-units would combine, Watson and Crick established that each strand of the DNA molecule consisted of a backbone of deoxyribophosphate units, with the bases forming side-units arising from the ribose residues. The geometrical configuration allowed the two strands to the helix to form hydrogen bonds between the bases. Adenine could bond with thymine and cytosine with guanine to form the double helix. The exlusivity of the bonding-base pairs explained the analytical results of Chargaff, who had shown that the bonding pairs exist in the same proportions.[33] This molecular model predicted first that reproduction could occur if the strands separated at cell division in the same way that chromosomes had long been known to do. Secondly, the separated strands might act as templates for the formation of new complementary strands to bring the molecule back to its original double-stranded condition. Thirdly, genetic information might exist in a code formed by the sequence of bases on one of the strands to be translated into the form of proteins, a code of triplets being the most parsimonious model. All of these predictions were borne out within a decade. It was further clear that very minor changes in the sequence of bases could bring about changes in protein synthesis, the process of which was concurrently being worked out. This finally explained the kinds of small mutations which had been investigated originally by T.H. Morgan's group, and had been shown by Muller to be induced by X-rays.[34]

The preceding outline of the development of evolution theory has presented the orthodox view, and has largely ignored the criticisms that have been viewed since *Origin* was first published. Doubts are still being expressed not only by biologists but also by philosophers, historians, economists and mathematicians. Such doubts, criticisms, and appeals for more satisfactory alternatives constituted what Darwin called 'Difficulties on Theory', in Chapter 6 of *Origin*. Some of Darwin's problems have been laid to rest, but more have been added.

2 DIFFICULTIES ON THEORY

In some ways evolutionary theories. . .resemble philosophical systems more closely than would generally be admitted. It is rather surprising that they have been largely left alone by logical positivists in search of new demolition work. Perhaps Neo-Darwinism has been saved from this by its essential contribution to the world view that positivists share.

Richard Spilsbury 1974[1]

The centenary of Darwin's death, 1982, brought as much recrimination as adulation, and the chorus of dissent had been growing steadily for the previous decade. Some of the criticism was from the fundamentalist anti-evolutionists, and some from philosophers and mathematicians who were not satisfied that biological evolutionists were meeting their formal logical standards.[2] They ask particularly how single random beneficial mutations, which alone are highly improbable by any standards, could accumulate in an organised manner to bring about evolutionary progress in the time available. Another popular complaint, arising from Karl Popper's philosophy of science, is that selection theory does not meet the criteria of falsifiability required by a rigorous epistemology.[3] Within the ranks of the biological critics are the perennial Lamarckists and transcendentalists who offer the inheritance of acquired characteristics or autonomous forces, or unknown laws of change and growth, as alternatives to natural selection. There are also those who feel that while conventional neo-Darwinism is valid it is not quite enough to cover the contingency that evolutionary progress may be saltatory as well as gradual, and that alterations affecting embryonic development may be the key to the overlooked factor of evolution. Even within the palace-guard of neo-Darwinism there is dissention about the nature of speciation, the possibility of neutral mutations, and the gait of evolutionary progress. Recent critical accounts have been given by Macbeth in *Darwin Retried* (1971), Hitching in *The Neck of the Giraffe* (1982), Taylor in *The Great Evolution Mystery* (1982), and in a series of reviews by Løvtrup (1974 to 1983). These doubts and uncertainties have existed since the publication of *The Origin of Species*. In this chapter I will present in general terms difficulties

with natural selection as the sole mechanism of evolution, problems
with adaptation, a concept whose deceptive simplicity has left much
unexplained, problems with speciation, and with the corollary
philosophical system of reductionism. This will be an epistemological dissertation in the broadest sense of treating not only the
hows and whys of biological and evolutionary explanation, but also
adducing some linguistic, psychological and metaphysical
implications.

Difficulties on Selection

Natural Selection: A common name for a complex of processes of
rather diverse kinds and different biological significance.
                                        Theodosius Dobzhansky 1970[4]

In *Origin*, Darwin, following Herbert Spencer's suggestion, used
survival of the fittest as a synonym for natural selection. The
expression recalls the words of the Preacher in Ecclesiastes 9:11:

I returned and saw under the sun, that the race is not to the swift,
nor the battle to the strong, neither yet bread to the wise, nor yet
riches to men of understanding, not yet favour to men of skill; but
time and chance happeneth to them all.

The paradox is only meaningful if it is contrasted with the conventional wisdom that the swiftest and strongest and wisest and fittest
do find favour and survive under normal circumstances, a concept
that any preliterate society would have discovered as a result of
hunting and agriculture. The Preacher was no more of an evolutionist than Lucretius, who said in *On the Nature of Things,*

Many species of animals must have perished and failed to
propagate and perpetuate their race. For every species that you
see breathing the breath of life has been protected or preserved
from the beginning of its existence either by cunning or by
courage or by speed. There are also many that survive because
their utility has commended them to our care.[5]

These early expressions of survival of the fittest or the struggle for
existence did not generate the corollary of mutability. Lucretius was
aware of such a concept as expressed by Anaximander, and rejected

it: 'For living creatures cannot have dropped from Heaven nor can terrestrial animals have emerged from the briny gulfs of the sea.'[6] Instead, 'the earth has deservedly gained the name of mother, since from the earth all things have been created'; though with time its fecundity had worn out as it would have with a human mother.[7] Similarly, Charles Lyell developed the concept of adaptation as an explanation of how created organisms came to numerically dominate certain geological periods through an accident of fitness, but strenuously resisted the evolutionary implications for many years after the publication of *Origin*.[8]

Darwin recognised that he was risking criticism by using a metaphorical name for the cause of evolution. The selection practised by plant and animal breeders were literal enough, but nature had no such discriminatory power, and natural selection was a metaphor for the utility of certain variations that improved the survival and reproductive chances of their possessors. Darwin tried to preempt criticism by writing:

> In the literal sense of the word, no doubt, natural selection is a false term. . .Everyone knows what is meant by such metaphorical expressions, and they are almost necessary for brevity. So it is difficult to avoid personifying the word Nature, but I mean by Nature, only the aggregate action and product of many natural laws, and by laws the sequence of events as ascertained by us. With a little familiarity such superficial objections will be forgotten.[9]

Despite these qualifications concerning figurative language, Darwin leapt even further into metaphorical space with statements such as, 'Natural Selection. . .is a power incessantly ready for action, and is as immeasurably superior to man's feeble efforts as the works of Nature are to those of Arts',[10] and,

> It may be said that natural selection is daily and hourly scrutinising throughout the world, every variation, even the slightest; rejecting that which is bad, preserving and adding up all that is good, silently and insensibly working, whenever and wherever opportunity offers at the improvement of each organic being in relation to its organic and inorganic conditions of life.[11]

The image of the creator had been replaced by a conscientious schoolmaster, and Herbert Spencer took the public school imagery a step further:

> That organisms which live, thereby prove themselves fit to live, in so far as they have been tried; while organisms which die thereby prove themselves in some respects unfitted for living, are facts no less manifest than is the fact that this self-acting purification of a species must tend ever to insure adaptation between it and its environment. [12]

The Duke of Argyll pounced upon the metaphorical excesses of Darwinism:

> 'Natural Selection' represented no true physical cause, still less the complete set of causes requisite to account for the orderly procession of organic forms in Nature. . .Its only value lay in the convenience with which it groups under one form of words, highly charged with metaphor, an immense variety of causes, some purely mental, some purely vital, and other purely physical or mechanical. [13]

The Duke's real grievance was the lack of an accommodation for a guiding deity, acting through the provision of suitable variations that could then be selected, but his criticism of the metaphorical deception of natural selection remains valid. This debate was not simply the product of Victorian verbal exuberance. Metaphor blooms just as luxuriantly in the modern selectionist hothouse. Selection is an agency 'for generating an exceedingly high degree of probability', or for 'creating meaningful information'. [14] The enviroment is commonly said to exert a 'selection pressure' that causes adaptations to pop out; the organism metaphorically plans its 'evolutionary strategy'. Some modern authorities describe natural selection as being literally a creative force. [15] In *The Descent of Man* (1871) Darwin had used the metaphor of natural selection as an architect who surveyed the rocks that had fallen at random from a cliff, and decided which were suitable for use in construction. Only when the decision had been made did the random qualities of the rocks assume any meaning in an architectural context. Darwin's follower E.B. Poulton preferred the metaphor of a potter who turned amorphous clay into a pot: the design was all in the hands of the artisan, and to say otherwise would imply a ridiculous 'revolt of the clay against the potter'.[16] The argument would favour Paley's argument from design just as well, but the trend continues. Even Ernst Mayr argues: 'We do not hesitate to call a sculptor creative, even though he discards chips of marble. As soon as selection is defined as differential reproduction its creative aspects become evident.'[17]

The need to redefine natural selection as differential reproduction arose when it was realised that natural selection was tautological, a problem raised in 1891 by G.I. Romanes, who suggested that it would be more appropriate to talk about the 'selection value' or 'survival value' of the variants.[18] But R.E. Lloyd later responded,

> To turn the phrase 'survival of the fittest' into the phrase 'survival of that which has the greatest selection value' is to take a wide step which cannot be retraced. After it has been taken others follow as a matter of course, such as 'the survival of that which must survive' or 'the survival of the survivors'[19]

The neo-Darwinist redefinition of natural selection in terms of differential survival or differential reproduction meets these semantic exigencies, but raises the question of how natural selection as differential reproduction can be the agent of evolution. It is an effect of the interactions between varied organisms and their environments. How can it also be the cause? Sexual reproduction is certainly a mechanism of evolution since it preserves novelty and introduces it into the following generation in a variety of combinations of old and new. Differential reproduction cannot be denied: different mating pairs produce different numbers of offspring that in turn survive differentially. However, this is an effect of the life-long interaction between the sum of the qualities of the individual organism and its circumstances, and differential reproduction is the *measure* of quality, but not its cause. Its convenience as a quantitative, empirical measure tends to mask its inadequacy as a casual factor, and one wonders if the redefinition of evolution as, for example, 'changes in the frequency distribution of alleles in populations' is an attempt to disguise the inadequacy of natural selection by semantic tinkering.[20] Whatever the reason, evolutionary progress has been lost in the shuffle. Neo-Darwinists certainly admit that natural selection by any definition is only occasionally the cause of evolution by any definition, and that its most likely outcome is constancy rather than change. Balancing and stabilising selection retain a normal allele distribution; disruptive and even catastrophic selection may result in no more than a well-defined mosaic of distinct sub-populations within the species at large. Selection theory provides what T.A. Goudge has called an *integrating explanation* of the distribution of alles; although the equilibria differ locally according to circumstances, all the changes can be integrated as an overall expression of differential reproduction.[21] Neo-Darwinists use three definitions of natural selection, the formal one: differential reproduction; the rational one: Darwin's 'aggregate action and product of many natural laws', and the working definition that metaphorically adduces an external creative force. To what extent can these alternative usages be justified in terms of the demands of the appropriate level of discourse, and to what extent do they exist to provide sophistical loop-holes? The metalogic is almost as esoteric as that of Lewis Carroll's song 'A-Sitting on a Gate', which was called 'Ways and Means'. While the name of the song was 'The Aged Aged Man', the song was called 'Haddocks' Eyes'.[22] Metalogic is only justified if the symbol is not confused with the reality.

How can evolution be defined to meet the expectations of both specialists and generalists? J.B.S. Haldane used a Darwinian definition of sorts: 'By evolution we mean the descent from living being in the past of other widely different living beings. How wide the difference must be before the process deserves the name of evolution is a doubtful question.'[23] For some neo-Darwinists any point mutation is evolution, but this does nothing to clarify evolutionary changes from the simple to the complex, the process that I have been calling evolutionary progress. At this point it would be as well to introduce the synonym *grade evolution* that was initially used by E.S. Russell.[24] It indicates advance in complexity, implying improvements in mechanisms that allow the maintenance of a constant internal environment in the face of changes in the external environment, and also signifies greater effectiveness in the integration and coordination of the constituent regulatory mechanisms. Since Darwin had difficulties with the concepts of 'higher and lower' in evolution, and since 'progressive evolution' has a number of discordant definitions, and 'progress' is loaded with historical implications, I will employ Russell's expression, grade evolution. It is often argued that the evolution of coordinated complexities is impossible to understand purely in terms of the piecemeal accumulation of single random mutations, since these may have no individual incipient value. For instance, evolution of the eye involved many components that were useless by themselves and could not have appeared all in the right place at the right time in order to have been favoured by natural selection. On the other hand, it is arguable that cytoplasm is light sensitive, that improvements in light sensitivity such as a simple optic cup would be advantageous, and central integration, response to shadow effects, and finally image-formation could all have been added one at a time, each having a certain utility at the time of acquisition. A similar example was presented by the Lamarckist H.G. Cannon concerning the evolution of the reptile egg. Since the aquatic larval stage was given up a heavily yolked egg was required, along with a supply of watery albumen for the developing embryo. Waste products could not diffuse away as in the aquatic amphibian egg, and so the cloacal bladder developed precociously as the allantois, taking on a secondary respiratory function in the later embryo, corresponding functionally with the gills of the amphibian larva. Then this had to be protected with a shell, necessitating, 'a special *ad hoc* structure, a built-in tin-opener, with which to cut its way out of the egg.'[25] Although the yolk and the allantois were modifications of pre-existing structures, the white of egg, the shell and the egg-opener were entirely new. Cannon was using this example as the prologue to a ramshackle Lamarckist explanation which required even more *ad hoc* underpinnings than the Darwinist alternative. The proponents of both of these paradigms have shied away from explaining the emergence of meaningful wholes from individually meaningless parts. Largely ignored by the Lamarckists, they are treated by neo-Darwinists as random fortuities, or epiphenomena of the more fundamental process of the accumulation of selected alleles in the gene pools.

Problems on Adaptation

To buttress the theory of natural selection the same instances of adaptation are used, which in an earlier but not distant age testified to the wisdom of the Creator.

D'Arcy Thompson 1917[26]

The analogy between the function of the human hand and the purpose of human tools was recognised by the classical Greeks, although they argued about whether organs were created for a purpose, or simply found their purpose after they had been created. The relation between form and function was called *adaptation* by the eighteenth-century French philosophers, and the word was used in an evolutionary context by the early nineteenth-century British essayists. 'Adaptation consists in correlation of the structure and habits of animals with the needs of their lives. That this correlation exists is a biological fact that no biologist will deny', said G.S. Carter in *Animal Evolution* (1951).[27] Not even Lamarck would have disagreed with this, but his explanation was that the organism responded to its needs by an appropriate self-modification that became genetically fixed, while the modern inference is that the refinement of the fit between the organism and its needs has been by the action of natural selection. In its loosest sense adaptiveness is equated with *utility*: anything that meets a need is an adaptation, and any feature that exists must be adaptive if natural selection is indeed the true agent of evolution. There is a trace of circularity about this, in addition to a comfortable universality that hints at more semantic tinkering to make the concept meet the need of rhetoric. As is often said about natural selection, any theory that purports to explain everything may actually explain nothing. There are three distinct, though not mutually exclusive, categories of adaptation: internal adaptation, whereby there occurs a structural or physiological refinement to the existing intra-organismic conditions of life. Darwin used the example of improvements to the digestive system, which were adaptive to a particular diet; refinements to enzymes that make them more effective under prevailing physicochemical conditions would also be included here. Secondly, there is adaptation to a specific element of the external environment: if a bird seeks insects under the bark of a tree, then suitable mutations of the beak will be selected, resulting in a more effective wood-pecking mechanism. Thirdly there is adaptation to habit. To say that wings are adaptive to flight is absurd, but there may be adaptive changes that fit a particular mode of flight, such as the habit of catching insects on the wing.

What happens if the needs of the organism change; if, for example, it is transported to an environment different from the one to which it was adapted, or if old habits become unproductive and need to be changed? Darwin described the flycatcher that flew like a kestrel and fished like a kingfisher; and the great tit which might behave like a shrike or a nuthatch. He asked:

Can a more striking instance of adaptation be given than that of a woodpecker for climbing trees and for seizing insects in the chunks of bark? Yet in North America there are woodpeckers which feed largely on fruit, and others with elongated wings which chase insects on the wing, and on the plains of La Plata where not a tree grows, there is a woodpecker, which in every essential part of its organization, even in its colouring, in the harsh tone of its voice, and undulatory flight, told me plainly of its close blood relationship to our common species; yet it is a woodpecker which never climbs a tree. [28]

The last statement was refuted by W.H. Hudson who demonstrated that the woodpecker in question inhabited various woodlands and the large solitary trees of the pampas and fed in the trees as well as on the ground. Hudson concluded that this woodpecker challenged the importance of natural selection since despite its presence in open country,

Its colours are not dimmed, nor its loud notes subdued; but even when it traverses the open country it calls about it the enemies from which it has little chance to escape. Natural selection has not endowed it, for its safety, with the instinct of concealment so common in the true pampas birds. Its peculiar flight also, so admirably adapted for gliding through the forest, here only excited the rapacious birds to pursuit. [29]

A particular bodily structure and flight pattern does not confine birds to a particular habit, or habitat; in Morocco, storks, presumably 'adapted for wading', lead the peck order behind the plough, soar like vultures on the lookout for prey, and rummage in garbage dumps.

Darwin's explanation was

He who believes in the struggle for existence and in the principle of natural selection will acknowledge that every organic being is constantly endeavouring to increase in numbers, and that if any one being vary ever so little, either in habits or structure, and thus gain an advantage over some other inhabitant of the country, he will seize on the place of that inhabitant however different it may be from its own place. [30]

Darwin foresaw no difficulty in 'a race of bears being rendered by natural selection more and more aquatic in their structure and habits with larger and larger mouths, till a creature was produced as monstrous as a whale', and he noted that it was 'conceivable that flying-fish which now glide far enough through the air, slightly rising and turning by the aid of their fluttering fins, might have been modified into perfectly winged animals'.[31] Suppose an extra-terrestrial Darwinist who is unfamiliar with this world's flora and fauna came across the storks of Morocco. Would he not exclaim at how cunningly selection had chosen the longer-legged storks to fit the species for walking in plough furrows, and at the beak so well adapted for garbage-sorting? The Hawaian goose (*Nesochen sandvicensis*) has a running and climbing habit, and neither migrates nor enters water. Julian Huxley noted that its 'adaptations' include reduced webs, elongated flexible toes, strengthened legs and reduced wings.[32] All of these features are trivial compared to its fundamental organismic character, its gooseness, which is assumed to be the product of adaptive accumulation of similar trivial features. To what extent does adaptation exist only in the eye of the beholder? William Bateson cautioned in 1894:

> In dealing with questions of Adaptation more than usual caution is needed. . .Since, at the present time the conclusions arrived at in this field are being allowed to pass unchallenged to a place among the traditional beliefs of Science, it is well to remember that the evidence for the beliefs is far from being of the nature of proof.[33]

J.C. Willis gave the example of a family of water plants, the Podostemaceae, 'one species is obviously adapted to fast water while other species with larger, broader leaves and tenuous attachment are found in water which runs at twice the speed'. The family as a whole showed great diversity of form, but occupied relatively constant conditions, and much of its evolution 'seemed to be completely *de luxe*, for there was no need for the new forms, nor was there any adaptational niche that would suit one form only, and not also many others'.[34]

Darwin's concept of opportunism in evolution, whereby organisms adapted to one enviroment could succeed in a different environment adumbrated Cuénot's concept of pre-adaptation expressed in 1914. Cuénot agreed that Darwinian adaptation made sense as long as it could be shown that the transitional stages of adaptation were useful, but he demurred on the question of the evolution of the electric organs of fishes and the luminous organs of fire-flies. Was it not possible such pre-adaptive or prophetic traits might have spontaneously appeared before the organisms entered an environment or adopted a mode of life that could take advantage of them?[35] Pre-adaptation, a name coined in 1903 by C.B. Davenport, is an exasperating term since it implies that an organism is adapted to an environment which it has never encountered. Carter observed: 'in all adaptations the animal must to some extent be pre-adapted in this sense: it must always possess characters that can be modified to give the adapted character'.[36] But how do modifiable characters arise in the first place; are they adaptations to the previous environment or to the one before that; whence the goose's gooseness? What, moreover, as Willis asked, is the adaptive advantage of the mono-cotyledonous condition in plants compared to the dicotylendonous condition? 'Both grow intermingled almost everywhere and in much the same proportions. There is no monocotyledonous mode of life that suits a Monocotyledon better than a Dicotyledon, yet there are very great structural differences between them.'[37] Perhaps these fundamental characters are reflections of the ability of organisms to *persist in their own being*, an expression that Darwin borrowed from Spinoza.[38] Although Darwin did not elaborate on this self-sufficiency of the organism he understood that it was other than adaptive. Claude Bernard, the physiologist who began to analyse this character of homeostasis, said that it was what gave the organism independence from its environment. This independence is sometimes attributed to *adaptability*. Again Darwin made the distinction, 'adaptation to any special climate may be looked at as a quality readily grafted on an innate wide flexibility of constitution common to most animals'.[39] Adaptation and adaptability are words which are almost identical in form and derivation, but they require entirely distinct definitions. In *What is Adaptation?* (1914) R.E. Lloyd made the distinction between *being adapted*, which was the culmination of selective and genetic processes, and *becoming adapted*, manifest when 'we see that animals and plants are able to adapt themselves, at once, to conditions which are occurring for the first time'.[40] Adaptability is also used by evolutionists in the sense of genetic potential to become specialised, but again this is no explanation of 'flexibility of constitution', within which lurk two important elements. The first, such as gooseness, has something to do with the functional-morphological success of a particular body plan, which may be manifest epigenetically before the goose emerges from the egg. The second pertains to Bernard's concept of independence from the external environment through the regulation of the internal environment. An improvement in this latter physiological adaptability would be useful to an organism regardless of environment, therefore it does not necessarily have to await the nod of natural selection. However, it may not be rapidly spread through a species inhabiting an old stable environment, with highly specialised competitors. Environmental change or the presence of an interface with a new environment may be necessary before population expansion and secondary adaptation can occur. The mammals are believed to have co-existed with the dinosaurs through most of the age of the dinosaurs as small, cryptic animals with comparatively low population densities. But once the conditions of the late Cretaceous began to change, competition disappeared and the potential for great adaptive radiation was realised. The extent of adaptive radiation depends upon the physiological and genetic adaptability of the founding type. The mammals provide the most impressive example, but the dinosaurs themselves may have been more physiologically adaptable than the relict reptiles, having better circulatory systems and some other features indicating a rudimentary homeothermy. By substituting epigenetic and physiological adaptability for pre-adaptation, the concept can be released from semantic confusion. The acquisition of adaptability remains problematical. L.L. Whyte argued that there was a distinct *internal selection* at work that improved the physiological conditions of the organism and eliminated variations that strayed too far from the species 'design'.[41] But while this suggests how internal refinements come about, it does not say how novelties arise.

As organisms become adapted to their environments, and competition for resources increases, their ecological niches shrink. This narrow adaptation is called specialisation.

Specialization may be viewed as the set of adaptations and tolerances of an organism for its adaptive zone in nature, or, more concretely for its normal habitat or niche. If the organism is adapted for a narrow zone, habitat or niche, it is relatively highly specialized. If the organism is adapted for a wider range of environmental conditions its specializations are relatively broad.[42]

Specialisation is a necessary evil. As R. Riedl says, 'the same laws of probability require that the adaptive advantages which they shaped yesterday, must be paid for today by a restriction of adaptability'.[43] It is an almost inevitable effect of evolution in stable environments where competition within and between species is on the increase, although versatility or low specialisation is desirable in fluctuating conditions. According to Carter, 'Specialization is necessarily in the long run a danger to survival, for the conditions to which an animal is specialized may cease to exist, and this is in general more likely the narrower the specialization.'[44] One solution for what to do when the dam bursts is to have to hand a few convenient pre-adaptations like the ability to swim. However, another conceptual problem is how evolution can progress among specialised organisms caught in an environmental bind where any change that is not a refinement of specialisation is disadvantageous. Alister Hardy has a solution:

> However highly specialized a race of animals may have become in its typical adult conditions, provided it has a less or differently adapted specialized young or larval form (which naturally will already be well adapted to its particular mode of living) and has a gene complex which may sooner or later produce neoteny, then given sufficient time it stands a chance of escape from its path to extinction. In the great majority of stocks the end must come before this rare oportunity of paedomorphosis can intervene; but in a very minority the chance comes earlier before it is too late, and such lines are switched by selection to new pathways with fresh possibilities of adaptive radiation. So vast is the time available, that rare as they may be, these escaped from specialization seem likely to have provided some of the more fundamental innovations in the course of evolution.'[45]

Free-living larval animals are often completely different in form, habit and habitat from the adults, and may undergo a drastic metamorphosis to the adult form, for example: tadpole to frog, caterpillar to butterfly, planktonic veliger to sedentary clam. In neoteny the larval form and habit are extended later than usual, and sexual maturation occurs in the larva; metamorphosis does not occur. The most familiar example of neoteny occurs in the axolotl, the sexually mature larva of the tiger salamander. The life-cycle is completed without the metamorphosis into the terrestrial adult. Treatment of the axolotl with low doses of the hormone thyroxine triggers metamorphosis, and hypothetically only one simple epigenetic mutation was needed to produce this change in the original life-cycle — the deletion or strong repression of the structural gene responsible for thyroxine production. Sexual maturation could occur simply as a function of age or seasonal climatic stimuli. On the basis of this model Hardy's invocation of natural selection as a 'switch' is redundant. Most biologists would agree that paedomorphic occurrences in which the adult form is rare, or never appears, have been important stages in the evolution of the chordate line, including the vertebrates and man. On the basis of embryological evidence the echinoderms are believed to be close to the base of this evolutionary branch. The echinoderms, such as the sea urchins and star fish, have adult forms which would make the suggestion of their relationship to fish seem fantastic. Though they are not all highly specialised, their physiological makeup restricts them to clean sea water, with no possibility of invading either fresh water or terrestrial habitats. Their cumbersome anatomy limits locomotive flexibility and the possibility of advances in the nervous system. However, a paedomorphic shift could have brought the echinoderms back to the more plastic form of the larva. The abundance of food in the form of phytoplankton, and the relative ease of collecting it with a simple ciliation mechanism, makes the planktonic habit a relatively easy one for a transitional paedomorphic form to exploit. This transitional form would not have been a reversion to the proto-echinoderm condition. The new planktonic organism would have all of the genetic apparatus acquired during its sojourn as a sedentary bottom-living adult. The redundancy of portions of the genome could afford a significant evolutionary potential as S. Ohno suggested in *Evolution by Gene Duplication* (1970). Redundant genes can mutate in ways which would be lethal at a time when the genes were functional: 'Only the cistron which became redundant was able to escape from the relentless pressure of natural selection, and by escaping, it accumulated formerly *forbidden* mutations to emerge as a new gene locus.'[46] If evolutionary change can be envisaged on the basis of simple gene duplication and redundancy, what might be achieved with large, integrated complexes of many genes made redundant by paedomorphosis? These are not unique events; for example among the urochordates, typically represented by the transluscent sessile sea-squirts, there is a taxon all of whose members are paedomorphic, have a tadpole-like, sexually mature planktonic larval form and no known sessile mature stage. This entire group may have been established as the result of a single mutation occuring in a small number of the urochordates. The Urochordata are also believed close to the base of the evolutionary branch leading to the vertebrates. How many such events have occurred in the evolution of the vertebrates themselves is impossible to say. In human evolution the foetalisation of more massive-boned, hairy ancestors by a similar mechanism of extending the immature anatomy through to sexual maturity, concluded with the former adult stages disappearing altogether.[47] In *The Ghost in the Machine* (1967) Arthur Koestler describes the significance of paedomorphosis thus: 'This *reculer pour mieux sauter* — drawing back to leap — of undoing and redoing — is a favourite gambit in the grand strategy of the evolutionary process; also plays an important part in the progress of science and art.'[48]

Speciation

. . .contributors to the Modern Synthesis, zealous in their defense of natural selection, cast aside not only Goldschmidt's mechanisms but also his idea that most evolution occurs in association with speciation. Almost forgotten was the possibility that established species might be rather stable, in an evolutionary sense, and that natural selection or other agents of change might operate most intensively during the emergence of certain species.

Steven S. Stanley, 1979[49]

If evolution is the same as the origin of species it is desirable to find a universally agreed definition of species, a task that has exercised taxonomists and philosophers for several centuries. A species may be pragmatically defined as what a competent taxonomist asserts it to be, so that if an unidentified specimen is sent to four different authorities four different species may be named. Taxonomists have had the Herculean task of sorting plant and animals into their natural phylogenetic categories, mostly on the basis of fossilised, or preserved specimens, and as a result behavioural and physiological qualities have been largely ignored. For example 'sibling species' are now being discovered which are morphologically identical, but differ functionally so that they cannot interbreed and produce viable offspring. On the other hand, some species which do have superficial morphological difference and do not interbreed in nature due to geographical separation, may be found to interbreed under domestication or in zoos. It is usually assumed that if sub-populations of a species are geographically separated so that gene flow is cut off, allopatric speciation will occur and the new species will be incompatible with the old, due largely to chromosomal changes that make the mechanism of meiosis ineffective in the hybrid offspring. Sympatric speciation, whereby new species appear in the same geographical area as the parent species, is not regarded as a significant phenomenon by many neo-Darwinist systematists, but behavioural or physiological changes could result in sympatric speciation. If, for example, a few members of a nocturnal arboreal species decided that mating on the ground during the day was more comfortable, the results could be the same as in allopatric speciation. There is clearly an arbitrary element in deciding what a good species is, as well as a suspicion of reification; that is taxonomists have been talking about species for so long they assume that the term must have a definite meaning, and represents a real phenomenon.

Another problem relates to the process of speciation. Darwin argued that specialisation was a slow process with linear phyletic series of chronological species being produced in time, and occasional divergences occurring when new or changing environments were encountered. The alternative of saltatory evolution has been debated by every generation of biologists since the publication of *Origin of Species,* the most familiar case being Goldschmidt's *hopeful monster* which by a radical epigenetic change became the forerunner of a new species.[50] Without debating Goldschmidt's hypothesis here, but assuming for the moment that such events might occur, the question is where the emphasis must be put. S.M. Stanley, as is clear from the leading quotation for this section, places the emphasis upon speciation. This is obviously a Darwinian legacy, since if evolution is equated with the origin of species then the product of speciation, the species, is more significant than the process of speciation. The process consists primarily of a fundamental biological change, which might be induced genetically or by the environment, and manifests itself in physiological or behavioural adjustments; this is then followed by genetic adjustments to the original change, and finally reproductive barriers are constructed that allow systematists to recognise that speciation has occurred. Although several steps removed from the effective cause it is the final stage that attracts most attention. Nevertheless, the mode of speciation, and whether it is gradual or sudden, has been a perennial dispute even among orthodox evolutionists. The most recent debate of this kind was triggered by Eldredge's and Gould's punctuated equilibrium hypothesis, which argues that species are stable and relatively unchanging for most of their existence, but may have originated rapidly.[51] As Stanley points out, this is not a new concept, having been mooted by G.G. Simpson and Ernst Mayr several decades ago. Indeed, Darwin himself observed

> that the periods during which species have been undergoing modification through very long as measured by years, have probably been short in comparison with the periods during which these same species remained without undergoing any change.[52]

Neo-Darwinist statements commonly attribute rapid speciation to 'strong selection' operating to speed up the normally gradual processes without altering their quality. The strong selection is associated with new environments with rich resources, and lack of competition and predators. It is logical to say that pre-adaptive variants will be 'strongly selected' in the new environment: 'swim genes' are paramount when the dam bursts. It is just as logical, and more explanatory, to point out that an adaptable organism entering a new environment will have former environmental exigencies relaxed, with the result that random functional-morphological experiments might be tried that would previously have been forbidden, and these might form the basis of a radiative evolution wherein each of the experimental ground plans become more strictly adapted to narrower niches. Stanley's *Macroevolution* (1979) notes that there are a number of palaeological cases where relatively minor speciation events appear to have been responsible for the production of new families early in adaptive radiation.[53] Some modern biologists even agree with Goldschmidt that a single individual with novel features might be responsible for the establishment of a new species. A hopeful monster of this kind would certainly benefit from the hermaphrodite condition, allowing self-fertilisation, as Stanley suggests could have happened when an individual protogastropod mollusc acquired the mechanism of torsion, a developmental process that subsequently dominated gastropod evolution.[54] Some of the antigradualistic interpretations of speciation involve an initial disequilibration followed by re-equilibration and stability, which will be indentified in future chapters with the emergence doctrine, the underwater portion of this epistemological iceberg.

Difficulties on Reduction

So the real problem of life is not that all the structures and molecules in the cell appear to comply with the known laws of physics and chemistry. The real mystery is the origin of the highly improbable constraints which harness these laws to fulfill particular functions.

W.H. Thorpe 1974[55]

The use of reduction as a methodological or philosophical tool creates no immediate problems. Reduction can be used in the loose sense of making the complex simple enough to understand and deal with experimentally. Darwin, with rewarding results, reduced the complexity of evolution to the relative utility of different variants, expressed in terms of their survival. According to Nagel, reduction, in a more formal definition, must meet the *conditions of derivability and connectability*.[56] To be derivable the subjects of investigation of the secondary science, e.g. biology, must be shown to be logical consequences of the theories of the primary science, e.g. physics. To be connectable the terms of the secondary science must be definable in the terms of the primary one. By this definition biology has not been reduced to physics, and much of the debate over this subject concerns the likelihood of biology ever being completely reducible. Ayala, for example, argues that the presence of teleological, or quasi-purposive mechanisms in organisms, such as self-regulating homeostasis or evolutionary specialisation, is likely to preserve the irreducibility of biology, although he applauds the quest for formal statements of unifying principles.[57] Watson and Crick, with spectacular success, reduced the biological phenomena of reproduction and heredity to the molecular structure of DNA, and generated ideas at the physicochemical level concerning the copying of the molecular strands in reproduction and protein synthesis, and the coding of information that could be translated into protein structure. Nevertheless, the reductions of molecular biology began almost at once to reveal the limitations of the method. While viruses and bacteria might well pour their genetic hearts out in praise of molecular biology, genes of the higher organisms are of sterner stuff, spending most of their time in a state of repressed silence. The key to development, form, and function in multicellular organisms is differential gene expression, and the most intimate knowledge of the genetic code reveals nothing about the implementation of its information in space and time. As Watson himself puts it, we have to talk 'in terms of the coordinative interactions of small and large molecules', a statement which has a fine molecular biological ring, but which has smuggled in a secondary regulation, operative at the organismic level.[58]

Reduction is a potent heuristic tool, despite its limitations. However, it is frequently asserted that the expression of biological processes in physicochemical terms is a justifiable end in itself, which shifts us from methodology into the murkier depths of epistemology. To take a chemical photograph of a biological problem opens up the problem to chemists, but is worthless if no answers are forthcoming. Nevertheless, to some people life seems less threateningly complex and scientifically more legitimate at the physicochemical level, and this is a manifestation of the metaphysical domain of materialism that can be called *reductionism;* it does not invariably dilute the value of reduction, but places a restraint on creative imagination. I use the term *selectionism* to describe a form of evolutionary reductionism. Since neo-Darwinism can be relatively unprejudiced at the level of discourse of population biology and its application to the simpler expressions of adaptation, I retain it as a neutral term, even although many neo-Darwinists are also selectionists. Selectionism asserts that selection theory is the all-sufficient explanation of evolution, and is compounded with the belief that all biological phenomena can be reduced to the behaviour of individual genes, and infiltrated with a náive positivism that denies the value of curiosity about phenomena that do not fit the preconceived model. P.-P. Grassé called it *ultra-Darwinism,* and Richard Spilsbury describes it as a faith, generated by 'the general horror of non-materialistic ways of thinking', which is not only speculation disguised by its conformity to a prevailing world view, but also 'confers miraculous power on inappropriate agents'.[59] In contrast to those who are encouraged by the thought that life may be simpler than had been imagined, there are those who like to think that life is more complicated that anyone can imagine.

From a realistic point of view the latter conviction has been guilty of even more abuses that reductionism and selectionism, since it includes the fundamentalist creationists, transcendentalists and vitalists. The two extremes have some qualities in common; as D'Arcy Thompson observed, selection theory used adaptation in the way that creationists used the wisdom of a divine creator:

> When you failed to explain a thing by the ordinary process of causality, you could 'explain' it by reference to some purpose of Nature or of its Creator. This method lent itself with dangerous facility to the well-meant endeavours of the older theologians to expound and emphasize the beneficence of the Divine purpose. *Mutatis mutandis*, the passage carried its plain message to the naturalist. The fate of such arguments and illustrations is always the same. They attract and captivate for a while; they go to the building of a creed, which contemporary orthodoxy defends under its severest penalties; but the time comes when they lose their fascination; they somehow cease to satisfy and convince, their foundations are discovered to be insecure and in the end no man troubles to controvert them.[60]

The 'modern synthesis' allows selectionism to absorb and take credit for independent ideas that it was incapable of generating, while parading a spurious holism. Julian Huxley's original intention in 1942 was that *Evolution: the Modern Synthesis* should show that epigenetic studies were as essential as mutation and population genetics for the comprehension of evolution, and that a synthesis should integrate all of these. But when we read the introduction to the 1974 third edition, compiled by Huxley's associates, we find no section on the effects of genes during development, and only a passing reference to Huxley's interest in allometric evolution. Instead, the discussion is dominated with population genetics, and despite Huxley's original intention the synthesis has become no more than a facade for unreconstructed neo-Darwinism, resistant as ever to the interaction of thesis and antithesis that are supposed to lead to dialectical progress. However, the portents are not all unfavourable: there may be cause for optimism in the chorus of criticism from the evolutionistic critics who are outside the main stream of biology, since it recalls Ernst Mayr's reflection on how some non-scientists were able to welcome the iconoclastic advent of Darwinism:

These well-informed and broadly educated lay people looked at the problem in a 'holistic' way, and thus perceived the truth more readily than did the professionals who were committed to certain well-established doctrines. A view from the distance is sometimes more revealing, for the understanding of broad issues, than the myopic scrutiny of the specialist.[61]

If history is about to be repeated the cycle has been extraordinarily prolonged. T.H. Huxley gave Darwinism only twenty years before hardening of the paradigmatic arteries would set in:

> History warns us, however, that it is the customary fate of new truths to begin as heresies and to end as superstitions; and, as matters now stand it is hardly rash to anticipate that, in another twenty years, the new generation, educated under the influences of the present day, will be in danger of accepting the main doctrines of the *Origin of Species,* with as little reflection and it may be with as little justification, as so many of our contemporaries, twenty years ago, rejected them.[62]

This was just as true one hundred and twenty years after the establishment of Darwinism. The extraordinary resilience of its doctrines may be explicable in terms of an epistemological Red Queen hypothesis: by running hard all the time Darwinism manages to stay in the same place; or by adapting hard all the time to criticisms and new data it has managed to stay in the same relative position of authority.[63] This may be an affront to a normative view of intellectual progress, such as T.H. Huxley embraced:

> The improver of natural knowledge refuses to acknowledge authority, as such. For him, scepticism is the highest of duties; blind faith the one unpardonable sin. And it cannot be otherwise, for every great advance in natural knowledge has involved the absolute rejection of authority, the cherishing of the keenest scepticism, the annihilation of the spirit of blind faith; and the most ardent votary of science holds his firmest convictions, not because the men he most venerates hold them. . .but because. . .whenever he thinks fit to test them by appealing to experiment and to observation. . .Nature will confirm them.[64]

On the other hand, this may be the only way in which biological thought can evolve, a struggle for epistemological existence moulded by the natural selection of ideas. Paul Feyerabend (1978) argues that polemical propaganda are of the essence in the introduction of a new paradigm, and that counterinductive arguments, such as biology has derived from vitalism and even creationism, are necessary for both refining and exposing the inadequacies of prevailing doctrines: 'Ideological ingredients of our knowledge. . .are discovered with the help of theories which are refuted by them.'[65] While keeping this exponent of 'dadaist epistemology' in mind, but at arm's length, I am going to follow a theme of realistic holism, with the intention of restoring to the organism and its internal and external relationships some of the significance that has been culled by selectionism. Fortunately, biologists, regardless of their metaphysics, have the saving grace that by training and experience they are at least tacit holists. I will begin, counterinductively, with the history of bio-logical ideas concerning evolution without selection, or in which the role of natural selection has been diminished, to show that behind the Darwinian paradigm there have been a number of alternative paradigms, at present in disarray, or only surviving as historical relicts. These ideas have contributed to evolutionary thought, and they have the common quality of an emphasis on holistic, or organismic interpretation.

3 EVOLUTION WITHOUT SELECTION

> . . .the Rocky Mountains, and not Natural Selection, have been the agent by means of which species have been modified.
>
> Charles Dixon 1885[1]

In 1864 the German zoologist A. Kölliker challenged Darwinian theory in an essay which was translated into English and reviewed by T.H. Huxley.[2] He observed that there were no known transitional forms between existing species, and similar gaps occurred in the fossil record. Species did not tend to produce useful varieties:

> The varieties which are found arise in consequence of manifold exernal influences, and it is not obvious why they all, or partially, should be particularly useful. Each animal suffices for its own ends, is perfect of its kind, and needs no further development.[3]

He also noted T.H. Huxley's own reservation that no incipient new species, in the form of mutually infertile varieties of a single species were to be found. The weakest aspect of Darwinism was the concept of imperfect types purposefully struggling to perfect themselves.

> Assuredly, every organ has, and every organism fulfils, its end, but its purpose is not the condition of its existence. Every organism is also sufficiently perfect for the purpose it serves, and in that, at least, it is useless to seek for a cause of its improvement.[4]

Huxley responded,

> For the teleologist an organism exists because it was made for the conditions in which it is found; for the Darwinian an organism exists because, out of many of its kind, it is the only one which has been able to persist in the conditions in which it is found.[5]

The task of defining these distinctions continues to plague modern philosophers of biology. Kölliker's final point was that a set of natural laws other than natural selection was needed to explain 'the regular harmonious progress of the complete series of organic forms from the simpler to the more perfect', and he argued that the existence of harmony and order in inanimate nature proved the existence of such laws. As an alternative to natural selection, Kölliker offered the *theory of heterogeneous generations,* whereby new types might arise in both sexually and asexually reproducing organisms, through radical changes in the course of developments, as exemplified by the metamorphosis of echinoderm larvae into adults, the production of medusoid cnidarians from polyps, and the radical sexual dimorphism of some animal species.[6] Huxley rejected the examples of alternation of generation since these were not known to produce new types, and he subsumed the development of some offspring beyond the parental type as an extreme case of Darwinian variation,

> greater in degree than, but perfectly similar in kind to, that which occurred when the well-known Ancon Ram was developed from an ordinary Ewe's ovum. Indeed we have always thought that Mr Darwin has unnecessarily hampered himself by adhering too strictly to his favourite 'Natura non facit saltum'. We greatly suspect that she does make considerable jumps in the way of variation now and then, and these saltations give rise to some of the gaps which appear to exist in the series of known forms.[7]

The Darwinist E.B. Poulton later excused Huxley for this breach of faith on the grounds that he had never been enough of a natural historian to appreciate the full wonder and sway of natural selection.[8]

St George Jackson Mivart (1827-1900) had an ill-fated career as an evolutionist.[9] His father owned Mivart's Hotel, now Claridges, in Grosvenor Square in London, and a number of his guests and friends were scientists and naturalists. At the age of sixteen Mivart, who had become a keen naturalist, was converted to Roman Catholicism and entered St Mary's College at Oscott. Subsequently, he studied Law at Lincoln's Inn, and though called to the bar in 1851 he never practised, since natural history, especially comparative anatomy and Richard Owen's work on homology, began to occupy his attention. In 1858 he attended T.H. Huxley's lectures at the Royal Institute, and was later introduced to Huxley by a member of his father's circle, G.R. Waterhouse, Keeper of Palaeontology and Minerals at the British Museum.[10] By 1861 he had become a personal friend of Huxley and entered the latter's course at the School of Mines as an honorary student. He was then appointed lecturer in comparative anatomy at St Mary's Hospital Medical School, sponsored by Huxley and Owen. His interest in primates and his meticulous anatomical studies of lemurs in the 1860s were stimulated by Huxley's series of essays on 'Man's Place in Nature', at which time Mivart was an enthusiastic Darwinist. The anatomical studies, however, led him to suspect that all was not as straightforward as Darwin and Huxley proposed; problems of parallel evolution and the treatment of the human species as 'just another ape', caused him anxiety. It was at this point, in 1869, that he confessed his mental turmoil to an unsympathetic Huxley. The next year the Vatican Council averred papal infallibility in scientific matters, setting the scene for Mivart's conflict with his Church. In 1871 Mivart published *On the Genesis of Species*.

> My first object was to show that the Darwinian Theory is untenable and that natural selection is not *the* origin of species. My second object was to demonstrate that nothing even in Mr Darwin's theory as then put forth, and *a fortiori* in evolution generally, was necessarily antagonistic to Christianity.[11]

T.H. Huxley's critique of *Genesis* exasperated him because it ignored his biological arguments and took him to task for his religious ones. One of the most serious reservations that the Darwinists had concerning Mivart was that he was a Catholic who was trying to reconcile his religion with their agnostic science. Huxley had performed one of his rhetorical *tours de force* — beating the sophists at their own game, as earlier typified by the debate with Wilberforce. From a study of Suarez, one of Mivart's religious sources, he concluded that Mivart had misread the master, 'Only fancy my vindicating Catholic orthodoxy against the Papishes themselves.'[12] These departures by Huxley from his usual objectivity seem uncharacteristically mean, but his contemporary admirers enjoyed the polemic. Darwin responded: 'How you do smash Mivart's theology. . .Nothing will hurt him so much as this part of your review.'[13] Hooker also applauded, 'What a wonderful man you are to grapple with those old metaphysical divinity books. It quite delights me that you are going to some extent to answer and attack Mivart.'[14] Huxley's distaste for the misuse of the intellect, for which he had criticised Wilberforce, was set aside in Mivart's case, and the patience shown for Kölliker had dried up.

The attack on *Genesis of Species* was joined by the American biologist Chauncey Wright, one of many selectionists who have defended their cause by accusing doubters of ignorance and mis-understanding.

> Such being our author's misconceptions of the principle of Natural Selection, and such their source, it would be useless to follow him in his tests of it by hypothetical illustrations from the history of animals; but we are bound to make good our assertion that the author's difficulties have arisen, not only from his want of a clear mental grasp of principles, but also from an inadequate knowledge of the resources of legitimate hypothesis to supply the unknown incidental causes through which the principle has acted.[15]

Wright also persuaded Darwin that Mivart had dishonestly used quotations from *Origin of Species*.

Mivart cannot be portrayed as an innocent victim of a Darwinist assault. He was at times fulsomely deferential and sycophantic in his personal dealings with Darwin and Huxley. However, given the safety of a wall of anonymity he would snipe with scorn, acrimony and vituperation. It was nevertheless hard to justify the scientific and social treatment that was meted out to him. If 'assault' seems immoderate, note what Hooker had to say of Huxley's letter giving Mivart the final brush-off: '[It] put the screw into the right worm and I could not help admiring the way the turns disappeared till all but the head was sunk in his vitals'.[16]

Since Catholic liberalism was at its peak in England, Mivart's book was acceptable to the Church: Cardinal Manning commended it, and in 1876 Pope Pius IX conferred upon Mivart the degree of Doctor of Philosophy. However, in the same year Mivart's essay *Contemporary Evolution* provoked an attack on his views by the *Dublin Review*, and in 1884 Mivart's ideas were described as heretical by the *Irish Ecclesiastical Record*. Mivart's response, 'Modern Catholics and Scientific Freedom', attacked the historical failures of the Church on scientific matters. 'Happiness in Hell' (1892) attracted more virulent opposition from the Church, including its indexing by the Sacred Congregation, and the publication of the Papal encyclical *Proventissimus Deus*, which upheld a narrow, literal interpretation of scripture. Subsequently, several Catholic evolutionists were called to Rome to recant. Mivart later aggravated the conflict with an attack on the victimisation of Dreyfus by the Church in France, calling him 'the Galileo of the Nineteenth Century'.[17] In 1900 Archbishop Vaughan asked Mivart to sign a declaration of faith, including the statement, 'I reject that it is possible at some time, according to the progress of science, to give to doctrine propounded by the Church a sense different from what the Church has understood.'[18] Mivart refused to sign, was excommunicated, and died a few months later.

Cast out both by his scientific fraternity and his Church, Mivart may have projected his own self-image when he talked about the Galileo of the nineteenth century. Since, from the orthodox scientific viewpoint, Mivart was fundamentally wrong about natural selection, and his position *vis-à-vis* the Church inconsequential, it is not surprising that Mivart is virtually unknown to modern biologists. However, his early rejection of natural selection and his attempts to stress the origins of variations were important, and influential. He clearly pointed out the necessity of separating the 'general theory of evolution' from the 'special Darwinian hypothesis'.[19] He cautioned against the unthinking acceptance of the latter just because it was simple, and because of

> the ready way in which phenomena the most complex appear explicable by a cause for the comprehension of which laborious and persevering efforts are not required, but which may be represented by the simple phrase 'survival of the fittest'. . .At the same time it must be admitted that a similar 'simplicity' — the apparently easy explanation of complex phenomena — constitutes the charm of such matters as hydropathy and phrenology, in the eyes of the unlearned or half-educated public. It is indeed *the* charm of all those seeming 'short cuts' to knowledge by which the labour of mastering scientific details is spared to those who yet believe that without such labour they can attain all the most valuable results of scientific research.[20]

Mivart added defensively,

> It is not, of course, for a moment meant to imply its 'simplicity' tells all against 'Natural Selection' but only the actual or supposed possession of that quality is a strong reason for the wide and somewhat hasty acceptance of the theory, whether it be true or not.

Mivart summarised his objections to the hypothesis that natural selection is the only agent of evolution as follows:

(1)   That natural selection is incompetent to account for the incipient stages of useful structures.

(2)   That it does not harmonise with the co-existence of closely similar structures of diverse origin.

(3)   That there are grounds for thinking that specific differences may be developed suddenly instead of gradually.

(4)   That the opinion that species have definite though very different limits to their variability is still tenable.

(5)   That certain fossil transitional forms are absent, which might have been expected to be present.

(6)   That some facts of geographical distribution supplement other difficulties.

(7)   That the objection drawn from the physiological difference between 'species' and 'races' still exists unrefuted.

(8)   That there are many remarkable phenomena in organic forms upon which 'Natural Selection' throws no light whatever, but the explanation of which if they could be attained might throw light upon specific origination.[21]

The evolution of the giraffe in particular and the ungulates in general illustrated some of these points: how could an incipient increase in the length of the neck be useful; why were there no options such as the development of trunks, or a climbing habit, in order to feed on the higher foliage of bushes and trees? Why among the ungulates did only the giraffes opt for the alternative of growing long necks? (There was no mention of the South American long-necked ungulates.) He also noted that the lengthened neck required coordinated adjustments in the pectoral region, and forelimbs, and doubted that this anatomical burden could be offset by the advantage of increased neck length. This example has significance that Mivart did not read into it, namely the fact that developmental co-ordination in such cases had somehow evolved, burden or not.

Mivart asked how leaf butterflies that mimicked not only the forms of leaves, but also patches of mildew and decay, could have arisen from 'utterly indifferent and indeterminate infinitesimal variations in all conceivable directions'.[22] The siting of the eyes in flatfish was also problematical. Such fishes lie on their sides, and during ontogeny one eye migrates to the uppermost side.

If this condition had appeared at once, if in the hypothetically fortunate common ancestor of the fishes an eye had suddenly become thus transferred, then the perpetuation of such a transformation by the action of 'Natural Selection' is conceivable enough. Such sudden changes, however, are not those favoured by the Darwinian theory, and indeed accidental occurrence of such a spontaneous transformation is hardly conceivable. But if this is not so, if the transit was gradual, then how such transit of one eye a minute fraction of the journey towards the other side of the head could benefit the individual is indeed far from clear.[23]

Darwin himself touched upon a number of these problems. He took the stance that we could not possibly know how the intermediates functioned since we cannot resurrect them. Darwin had particular problems in conceiving the usefulness of the incipient swim bladders of fish and the incipient electric organs of certain fish. These particular problems have been solved by modern evolutionists since the swim bladder is now seen as a modified lung, the incipient stages of which would have been useful to an air breathing fish in stagnant water. The primordial electric organs may have had the function of creating weak electric fields which might be distorted by prey or obstacles; the ability to detect such distortions would mean that in murky waters the sense could be a substitute for sight.[24] But we still have to ask what the incentive was for natural selection to increase electrogenesis from the millivoltages of normal muscle discharge to the extent of creating useful electric field (in an electrically resistant fresh-water environment), in conjunction with a sensory apparatus that would have needed to be most sensitive at its least evolved stage.

Mivart's second objection to the special Darwinian theory of natural selection asked how it explained parallel and convergent evolutionary features if 'on this theory the chances are almost infinitely great against the independent, accidental occurrence and preservation of two similar series of minute variations resulting in the independent development of two closely similar forms'?[25] He could see that analogous structures did not create such a problem, 'The organic world supplies us with multitudes of examples of similar functional results being attained by the most diverse means.'[26] Parallel or convergent evolution between the marsupial and placental mammals was so striking as to have made T. H. Huxley suggest in his Hunterian Lectures of 1866 a direct genetic affinity between the convergent types, i.e. polyphyletic origin of the placentals.[27] Mivart mentioned the placental kangaroo rat, which not only leaped on long hind legs, but also had kangaroo-like dentition. These phenomena must be the expression of 'some other natural law or laws conditioning the similarities and independent evolution of these harmonious and concordant adaptations'.[28]

To support his suggestion that species might suddenly transmute, Mivart cited domesticated species of dogs, rabbits and various plants that changed noticeably within one generation after being introduced to new environment, drawing examples from the work of Darwin, Blyth and Naudin. Greyhounds introduced to the uplands of Mexico

> could not support the fatigues of a long chase in this attenuated atmosphere, and before they could come up with their prey they lay down gasping for breath; but these same animals have produced whelps which have grown up, and are not in the least degree incommoded by the want of density in the air, but run down the hares with as much ease as do the fleetest of their race in this country.[29]

A move to a high altitude is now known to induce increased haemoglobin synthesis, and mammals reared at high altitudes develop deeper lung capacities than their parents. However, Mivart raised other cases of the appearance of new dominant traits in single individuals which permitted the breeding of new strains in domestic animals, and wondered why Darwin had asserted that it was false belief that natural species had often originated in the same abrupt manner. The belief might be false, but it was difficult to see how its falsehood could be positively asserted.

To give credit where it is due Mivart drew heavily on J.J. Murphy's *Habit and Intelligence* (1869) and on Richard Owen's opinions on species mutability, namely that change could be sudden and considerable, and not by 'minute and slow degrees'.[30] Given Owen's own equivocation about evolution and sour-grapes attitude towards Darwin, he was not a dependable witness, and Mivart sought further support from Huxley's conviction that nature does sometimes make leaps. However, although he was clearly aware of Kölliker's essay he made no direct reference to it in *Genesis*. Mivart was particularly taken with Francis Galton's model of species as piles of rock which formed temporary equilibria. Galton had remarked:

Stability is a word taken from the language of mechanics; it is felt to be an apt word; let us see what the conception of types would be, when applied to mechanical conditions. It is shown by Mr Darwin. . .that all forms of organic life are in some sense convertible into one another. . .Yet the changes are not by insensible gradations; there are many but not an infinite number of intermediated links; how is the law of continuity to be satisfied by a series of changes in jerks? The mechanical conception would be that of a rough stone, having, in consequence of its roughness a vast number of stable facets, on any one of which it might rest in 'stable equilibrium'. That is to say, when pushed it would fall back into its first position. But, if by a powerful effort the stone is compelled to overpass the limits of the facet on which it has hitherto found rest, it will tumble over into a new position of stability whence just the same proceedings must be gone through as before, before it can be dislodged and rolled another step onwards. The various positions of stable equilibrium may be looked upon as so many typical attitudes of the stone, the type being more durable as the limits of its stability are wider. We also see clearly that there is no violation of the law of continuity in the movements of the stone, though it can only repose in certain widely separated positions.[31]

What Mivart called 'intermitting conditions of stable equilibrium' pre-dated by a century the *punctuated equilibrium* hypothesis of Eldredge and Gould.[32] Mivart's thinking had been further influenced by Fleeming Jenkin's review of the fifth edition of *Origin of Species*, a contribution to the evolution debate that is usually presented as simply having given pause to Darwin on the implications of blending inheritance.[33] However, as P. Vorzimmer elaborates it, the blending inheritance problem was first raised by Francis Bowen after the appearance of the first edition of *Origin of Species*.[34] Jenkin claimed that there were limitations to the degree of variation of an individual and that variations oscillated around a fixed point. A dog, for example, could not vary beyond the limits of dogness. The alternative was saltation, which Jenkin regarded as being too freakish and likely to be sterile, and that with blending inheritance likely to be homogenised back to the norm. Vorzimmer argued that Jenkin reinforced Darwin's opinion that saltations were impossible, but did not persuade him of limits to variability. Mivart, however, accepted the oscillating variation plus the notion of viable saltation, hence his enthusiasm for Galton's multi-faceted spheroids.

Mivart's recent biographer, J.W. Gruber, judges that it was unpardonable of him to say that by the theory of natural selection a hornbill could be produced from a humming-bird, but it was well within the bounds of absurdity established by Darwin when he suggested a flying fish might evolve true flight, or a penguin regain it.[35] As Mivart observed, Darwin wanted it both ways. While various birds seemed 'to have a singularly inflexible organization', there were also cases of 'a whole organization seeming to have become plastic, and tending to depart from the parental type'.[36]

On the time available for evolution, Mivart allowed himself to be led astray by the conservative calculations of Sir William Thomson, later Lord Kelvin, on the age of the earth, which allowed less than 100 million years, perhaps as little as 40 million years, for life to have appeared and evolved.[37] It was those observations that prompted Huxley's caution:

> this seems to be one of the many cases in which the admitted accuracy of mathematical process is allowed to throw a wholly inadmissible appearance of authority over the results obtained by them. Mathematics may be compared to a mill of exquisite workmanship. . .what you get out depends upon what you put in; and as the grandest mill in the world will not extract wheat flour from peascod, so pages of formulae will not get a definite result out of loose data.[38]

Lord Kelvin later revised his opinions, but too little and too late to console Darwin.

Mivart was revealed as a holistic evolutionist when he said:

> That concrete whole which is spoken of as 'an individual'. . .is formed of a more or less complex aggregation of parts which are actually. . .(from whatever cause or causes) grouped together in a harmonious interdependency, and which have a multitude of complex relations amongst themselves. . .The component parts of each concrete whole have also a relation of resemblance to the parts of other concrete wholes, whether of the same or of different kinds, as the resemblance between the hands of two men, or that between the hand of men and the fore-paw of a cat.[39]

He meant that such homology was not to be explained by natural selection but by

> some innate, internal conditions, power or tendency. . .It is not improbable that, could we arrive at the causes conditioning all the complex inter-relations between the several parts of one animal we should at the same time obtain the key to unlock the secrets of specific originations.

Mivart had adduced no vitalistic nor mystical force up to this point, only an unknown and possibly materialistic cause or causes, somewhat similar to Kölliker's general law of harmonious development. This could have been Ludwig von Bertalanffy speaking sixty years later. Mivart proposed that some manifestations of homology could not be explained by natural selection, and pointed out that Darwin had referred to *ad hoc* 'laws of correlation' to explain them. *Correlation* was Darwin's name for what would now be partly explained by *pleiotropy*, a genetic effect wherein the change in a single allele produces a number of apparently unrelated phenotypic effects. Darwin's correlation also subsumed the necessary morphological adjustments which had to accompany, say, the stretched neck of the giraffe. These are phenomena that the modern epigeneticists would claim are comprehensible only in terms of mutations in the epigenome.

Among the examples of the peculiarities of homology cited by Mivart from Darwin's *Plants and Animals Under Domestication* (1868) were supernumerary digits in humans, primary wing feathers growing from the feet of feather-legged bantams, and skin-webbing between the toes of some pigeons. In endeavouring to explain these phenomena as manifestations of some law of homology, Mivart called Herbert Spencer as a witness,

> We have therefore, no alternative but to say that the living particles composing one of these fragments have an innate tendency to arrange themselves into the shape of the organism to which they belong. We must infer that a plant or animal of any species is made up of special units, in all of which there dwells the intrinsic aptitude to aggregate into the form of that species.[40]

This should come as no surprise to the modern student of molecular biology, but the latter is still not able to explain why the cells of a leg should set about the business that pertains to the wing. And if such large adjustments can occur both in nature and in animal breeding experiments, is this not fuel for the saltationists' arguments?

If natural selection were to have only a subordinate role in evolution the alternative that Mivart proposed was a combination of external and internal causes. From his examples of the effects of environment upon transplanted domestic species he drew a Buffonian conclusion that the environment caused change, which would be reversed if the organisms were returned to the original environment. For Mivart the influence of the environment was effective only as long as the organism was exposed to it, rather than an inducer of change that would be fixed. The analogy to which he referred on this matter was the influence of a violin bow drawn across the edge of a metal plate covered with sand grains. The variety of patterns in the sand which the vibrations caused changed according to the physical conditions, but Mivart supposed further that there might be autonomous vibrations emanating from the plate, analogous to the 'innate tendencies of the organism'.[41] Cumulative modification of the external stimuli could produce distinctly new patterns of innate vibration. Mivart denied that 'volition' was in the nature of the innate force, preferring the 'physical' or materialistic alternative. Supernatural causation was also ruled out at this level. As to the locus of specific change: 'It seems probable. . .that new species may arise from some constitutional affection of parental forms — and affection mainly, if not exclusively, of their generative system.'[42] Darwin's *Plants and Animals under Domestication* was quoted as an authority for this view. However, Mivart's materialistic pretensions were vitiated by his agreement with Murphy that 'organic nature. . .speakes clearly to many minds of the action of an intelligence resulting. . .in order, harmony and beauty, yet of an intelligence the ways of which are not ours'.[43]

Mivart believed that his explanation of the genesis of species lay between Owen's derivation theory, which was a mixture of the special creation of types and their subsequent evolution by natural selection, and Darwin's theory. Perhaps he wanted to draw upon the remaining reserves of the old guard, represented by Owen and Sedgwick, while at the same time trying to curry favour with Huxley, to whose authority he appealed many times. Simultaneously, he was attempting to stiffen the backbone of Catholic liberalism. His failure in these aspirations, and indeed the dire straits that he got into with Huxley and the Church made it impossible for him to produce a more mature exposition.

Joseph John Murphy was the author of *Habit and Intelligence* (1869), which was frequently cited by Mivart in *Genesis of Species*. Murphy set out to investigate the mechanical and vital principles of both unconscious and conscious life: environmental effects; the 'border land' where life came into contact with inorganic matter and force, which was the domain of physiological laws such as those of nutrition and respiration, and the inner laws of organisation and mind.[43] By *habit* Murphy meant what an animal does repeatedly in its way of life, and what is perpetuated in the species. Vital energy was initially treated as a distinct form of physical energy, like heat, electricity or magnetism, but different in being unique to organisms:

> It is not improbable that every muscular action is accompanied by some chemical change in the very complex and unstable compounds that constitute muscular substance, and it is very probable that these compounds, which are at once organic compounds and nitrogenous compounds, give out energy during such changes.[44]

But he insisted on supplementing this process with special vital energy.

Murphy believed that

> the highest organic development is the most complete physiological division of labour, and the most perfect physiological centralization. In the lowest organic species, and in the germs of the highest, the parts are all alike, and all independent of one another; in the mature forms of the highest species the parts are all different, and the whole organism is bound together into a system, with all its parts mutually dependent. To speak technically organic progress consists in increasing *differentiation* and increasing *integration*.[45]

Murphy admitted a subordinate role for natural selection, but not as the agent of evolutionary progress, which was due to some primordial formative influence.

Kölliker, Murphy and Mivart were among the first to enunciate cogent criticisms of selectionism. A.W. Bennett remarked in his *Nature* review of *Genesis of Species*,

In order to save defeat, the next move must be made by the advocates of Natural Selection, a *prima facie* case against them having at all events been made out. Mr Mivart has no counter theory to propose, beyond a belief that there exists in all organic life an innate power analogous to intelligence, which controls their actions as reason does those of men. Should the inquiries which are now being energetically pursued on every side result in our acquiring more accurate knowledge of such a force, it will be safe to predict that to it will then be ascribed a more easy and natural solution of many phenomena we are now forced to attribute to Natural Selection.[46]

Bennett was himself highly sceptical of the adequacy of natural selection, whose discovery

marked an era in the history of natural science, and gave a wonderful impulse to original research. The danger now is that the law will be pressed into services which have no claim upon it; and that, in the hands of injudicious partisans, it will become a hindrance rather than an aid to science, by closing the door against further investigations into other laws which lie behind it.[47]

Mivart and Murphy recognised that evolutionary progress was a phenomenon apart from adaptation, and that the integration and interdependence of the complex parts of highly evolved wholes was an important clue to the comprehension of evolution. They focused also on the inability of the Darwinian theory to explain spontaneous variation. Like Owen they took too little account of developments in embryology, perhaps out of ignorance or chauvinism. Mivart, for example, felt that

in the first half of the present century the speculative dreams of our Teutonic neighbours were trite subjects of ridicule. We laughed, not without reason at the farthingsworth of fact on which such an unconscionable amount of theory was too often based.[48]

His opinions of the German theoreticians of the second half of the nineteenth century, Haeckel and Weismann, were almost as low, and his critique of Weismann's germ plasm theory illustrates his failure to grasp the need for some kind of materialistic basis for heredity.[49] He first of all implied that Weismann's theory was pre-formationistic in the old-fashioned sense. If every detail of the whole organism was represented in the germ plasm by its own special and peculiar arrangement of the groups of molecules, the germ plasm would have to be complex enough to contain determinants for the precise number of all the succeeding cells of each lineage: such a 'collocation of particles' was impossible.[50] Herbert Spencer had similar objections:

> There must be a determinant for each scale on a butterfly's wing: the number on the four wings being over two hundred thousand. And then each cluster of biophores composing a determinant had to find its way to the place where there is to be formed the part it represents.

Spencer's *reductio ad absurdum* was the calculation that the peacock's tail would require 480,000 biophores per feather, and all this had to be packed into a sperm cell.[51] The problem is solved when we realise that all the scales on a butterfly's wing are determined by a few cistrons of DNA, their form and colour patterns being modified by a small number of regulators and colour-determining structural genes, and that every cell has all the information contained in the original zygote. While all the obstacles that Mivart placed in the way of a molecular theory of heredity have been removed, there remains the challenge to epigeneticists to discover how the particulars of molecular inheritance operate to produce organismic diversity and evolutionary progress.

The title of this chapter was borrowed from *Evolution Without Natural Selection* (1885), by the ornithologist Charles Dixon, whose choice of title was curious since he was a thoroughgoing Darwinist. He had been struck by apparently meaningless minor variations in the colouration of birds of the same species, e.g. the coal-tit, in different geographical areas. There had appeared to be no causal relationship between the characteristic and the environment. Citing the example of the thrush-robin (*Catharus*) of the mountains of Central and South America, Dixon observed that the twelve species had specific characters that differed so slightly that only geographical separation justified their taxonomic separation. Postulating the disappearance of a homogeneously distributed ancestral population from the interconnecting lowlands, he concluded that the relict populations in the mountains developed new characters, 'accidental variations. . .being preserved simply because they were isolated'.[52] Dixon proposed that similar effects were to be found among island populations of birds.

Dixon also noted that 'a perusal of Baird, Brewer and Ridgeway's elaborate work on North American birds. . .will reveal numerous cases where the Rocky Mountains, and not Natural Selection, have been the agent by means of which species have been modified'.[53] Dixon imagined a preglacial period in which a North Polar land-mass, with a sub-tropical climate, was homogenously populated with the ancestors of modern organisms. The advance of the glacials caused southerly migrations, and isolation resulted, for example, by divergence on each side of the Rockies. Non-adaptive accidental variations then occurred in the separated populations and were preserved by isolation. Dixon was writing long before the concept of continental drift had been suggested, but clearly felt that some drastic geographical change must be adduced to explain the curiosities of biogeography. His ideas were not so far from the modern concept of the snow-free Pleistocene Beringean land bridges.

G.J. Romanes, who reviewed Dixon's book for *Nature* in 1885, was critical of the book's misleading title and of its author's contention that geological upheavals, together with natural selection, were the most potent agents of evolution: 'Not in chaos or in cataclysm is the influence of natural selection to be sought, but in field, in river, lake and sea, where all may seem most orderly and eloquent of peace.' Although Romanes adhered to the 'universally-accepted teaching of uniformitarianism' together with the Darwinian interpretation of it, he proposed, a year later, the hypothesis of *physiological selection*, to meet some of the 'cardinal difficulties' of natural selection.[54] Pertinent to these difficulties was Dixon's 'admirable collection of facts'. Romanes claimed that physiological selection was distinct from Darwinian selection because it explained the true origin of species rather than 'cumulative development — of *adaptations*'. The gist of physiological selection was the sympatric construction of reproductive barriers: 'Some individuals living on the same geographical area as the rest of their species, have varied in their reproductive systems so that they are perfectly fertile *inter se*, while absolutely sterile with all other members of their species.'[55] To further emphasise the distinction from natural selection, he cited Darwin's contention that the sterility of species must be due to some principle quite independent of natural selection. The consensus of neo-Darwinists is that most if not all speciation is allopatric, and that sterility barriers arise only after the event of geographical separation. This does not, however, take into account the possibility of spontaneous behavioural changes in some members of populations, which could fit Romanes' model, the reproductive barrier being behavioural rather than physiological.

But however wrong Dixon might have been about geological time and place, the essentials remain relevant to biogeography. His isolation theory had the elements of the modern concepts of allo-patric speciation, founder principle and genetic drift, and even modern selectionists would admit that natural selection is not the cause of the latter two phenomena. Dixon's hypotheses seem remarkably close to the *vicariance* principle developed later by Leon Croizat, who also believes that geological events are major causes of evolution. Croizat's exhaustive detailing of biogeographical evidence for his theses, developed in *Panbiogeography* (1958) and in *Space, Time, Form: The Biological Synthesis* (1962), conform much more closely to the current understanding of continental drift, but, like Dixon's they were developed primarily on the basis of distri-bution, ignoring modern geography, and at a time before continental drift was accepted as orthodox.

Dixon believed that climate would directly affect colouration. The hot sun of a desert environment could, for example, fade the bright colours of the exposed feathers, but leave the shaded parts un-affected. He stressed that this was not due to any volition but to a physicochemical cause. Dixon evaded the problem of how cold climates might cause albinism, but he did note that white birds, such as ptarmigans, would seek out snowy mountain tops. This example was so similar to one given by Moritz Wagner that it is hard to believe that Dixon was being completely honest when he noted that although he had heard of Wagner's essays he had not read them until *Evolution Without Natural Selection* was complete.[56]

Dixon believed like Darwin that the use or disuse of organs was a direct cause of variation, and could furnish natural selection with abundance of material to work on. He also believed that occasional interbreeding between isolated related species might be responsible for the colour variants in which he was interested, though such hybridisation would not be responsible for the evolution of new species. In conclusion, Dixon's most interesting criticism of selection theory was that some variations are so slight that natural selection would be unable to discriminate between them. These variants might now be considered *neutral mutations*, a concept which has been at the centre of some recent squabbling in the ranks, since the strict application of the neo-Darwinist utilitarian principle demands that only the fittest characters are spread by natural selection.[57] His hypotheses concerning the necessity of migration and separation for the preservation of such variations resembled those of Mortiz Wagner and to some extent those of Gulick, which will be discussed in Chapter 11, 'The Struggle for Existence of Selectionism'.

Some of the evolutionists mentioned in this chapter might loosely be described as neo-Larmarckists, eclectic theorists who embraced the Lamarckian laws, the direct effect of environment proposed by Buffon and Geoffroy, and natural selection, the latter being afforded a minor role chiefly as an eliminator of the unfit. Mivart and the others did not utterly reject the agency of natural selection, and they were interested in the direct effects of environment, but with the exception of Spencer they did not adopt the inheritance of acquired characteristics (the central doctrine of Lamarckian and neo-Lamarckist thought) as the major alternative agent of evolution.

They did question Darwin's failure to explain the causes of variation and their implications for evolution theory. At one end of the critical spectrum were Dixon's variations that were too insignificant for natural selection to notice; at the other end were the variations which were large enough to bring about an evolutionary saltation, establishing new centres of specific stability without the participation of natural selection. Chapter 4, while continuing the theme of evolution without natural selection, will focus on the importance of understanding the origins and quality of variation for evolution.

4 THE UNKNOWN FACTOR

> . . .for Buffon's and Lamarck's factors we have no theory of
> Heredity, while the original Darwin factor, or Neo-Darwinism,
> offers an inadequate explanation of Evolution. If acquired
> variations are transmitted, there must be, therefore, some
> unknown principle in Heredity; if they are not transmitted, there
> must be some unknown factor in Evolution.
>
> H.F. Osborn 1895[1]

Darwin's own attempt to find the causes of the variations that were
the necessary fuel for selection produced the *pangenesis theory*,
based on the argument that the environment caused somatic
changes, which were then incorporated in the germ plasm in the form
of gemmules. The fact that Weismann had condemned pangenesis
for its revisionism brought even Mivart to Darwin's defence.[2]
Francis Galton believed that pangenesis was a testable hypothesis
with a simple mathematical basis, which he then disproved experimentally; but the new wave of Lamarckists, together with many
Darwinists, felt that it was too good an idea to let drop for mere
empirical reasons.[3] So it is from a literature laced with opinions on
pangenesis and similar speculative models that ideas about
variations have to be extricated.

Herbert Spencer (1820-1903) has been called a Darwinist, a
Lamarckist and a neo-Lamarckist.[4] He is not easy to categorise as a
follower of a particular evolutionary doctrine, and only his opinions
relating to variation will be considered here. In his early career he
was an assistant teacher, and a civil engineer and draughtsman with
companies of railway engineers, but by the age of twenty he had
written to his father that he would like 'to make public some of my
ideas upon the state of the world and religion, together with a few
remarks on education'.[5] A year later he took his savings to try to
patent some engineering inventions, and in the meantime study
natural history and phrenology, and collect fossils, and eventually
he became a sub-editor to *The Economist* in 1848. Through the
soirées of the publisher John Chapman he became acquainted with
G.H. Lewes, with whom he discussed the 'development hypothesis',
Thomas Carlyle, Marian Evans (George Eliot), whose social

companion he became for several years, T.H. Huxley and Thomas Tyndall. In 1852 Spencer had various articles published anonymously in the *Leader*, whose editor was G.H. Lewes. One of those articles was 'The Development Hypothesis', a Lamarckist essay which coined the expression 'survival of the fittest'. An essay for the *Westminster Review* on 'Progress: its Law and Cause' (1857) gave him the idea for a synthetic system of philosophy based on progress, or evolution, as applied to various departments of science and philosophy. *First Principles*, published in serial form, was completed in 1862, and he than began to write *Principles of Biology*. Volume I was published in 1864, the year that the X Club was established, with Huxley, Hooker, Tyndall and others as Spencer's fellow members. These associates and J.S. Mill gave him encouragement at a time when he was in mental and financial difficulties. Volume II appeared in 1866, by which time the serialised version had brought adequate financial support in America. 'The Factors of Organic Evolution', a series of articles published in *Nineteenth Century* in 1886, affirmed Spencer's acceptance of the inheritance of acquired characteristics and claimed that natural selection was inadequate to explain evolution. Spencer's confrontation with Weismann on the subject of the 'all-sufficiency' of natural selection, occupied a series of articles for the *Contemporary Review* in 1893 and 1894.

It might be asked how and why Spencer set out to write books on subjects in which he had no formal training, such as psychology, biology, sociology and philosophy. As one biographer wrote:

> There can be no doubt that he lost much from ignorance of the work of his predecessors. There can be equally little doubt that he gained more by coming on the ground unbiased and untrammelled by older methods of looking at things. Most of the sciences on which he wrote were in their infancy; and it was possible at that time to write upon them with very little previous knowledge. Moreover, the commercial success which ultimately attended his works is doubtless due in part to the fact that he started from ground that was common to most educated people and could therefore be appreciated by the more intelligent of the general public.[6]

The second edition of *Principles of Biology* (1898) is an interesting insight into pre-Mendelian speculative biology. What little was known of prepotency (dominance) confused rather than enlightened. Variation that resulted from meiotic chromosome assortment and recombination had been detected, but its significance was not appreciated outside Weismann's circle. Spencer was a severe critic of the germ plasm theory, and was possibly influential in delaying the acceptance of a molecular theory of inheritance. Some biologists had already suggested that hereditary traits were represented by discrete portions of the germs, but these were regarded as exceptional cases, and blending inheritance still dominated genetical discussions.

For Spencer life was 'the definite combination of heterogenous changes, both simultaneous and successive, in correspondence with external co-existences and sequences.[7] It followed then that

> an organism exposed to a permanent change in the arrangement of outer forces must undergo a permanent change in the arrangement of inner forces. The old equilibrium has been destroyed; and a new equilibrium must be established. There must be functional perturbations, ending in a re-adjusted balance of functions.

He then argued from this that the differences in the offspring from a pair of parents were due to minor functional changes in the adults, and so the longer the offspring were apart in time the greater the differences were likely to be, due to the cumulative effects of change in the parents. Like Mivart, Murphy and Lewes, Spencer was attracted to chemical analogies. Noting that chemical units of the same kind tended to form more stable aggregates than those of different kinds, and applying this observation to biology, he wrote

> it will follow that by uniting a group of units from the one organism with a group of slightly different units from the other, the tendency towards equilibrium will be diminished, and the mixed units will be rendered more modifiable in their arrangements by the forces acting on them: they will be so far freed as to become again capable of that re-distribution which constitutes evolution.[8]

There is a suggestion here of evolution by hybridisation. Hybrid vigour was already known to Darwin, and discussed elsewhere by Spencer, who, like Galton and Mivart, was fascinated by what it took to destabilise types. Some extraordinary force, or odd combination, was needed to shake up the stable system and produce a new order of biological novelty.

While rejecting the ideas of 'inherent tendencies', for which he criticised Erasmus Darwin and Lamarck, Spencer himself espoused the 'principle of equilibrium' or 'law of the instability of the homogenous', which 'must be recognized as an ever-acting cause of organic evolution'.[9] The first form of life, homogeneous organic matter, was unstable, because of different and variable external forces being applied to it. Since the internal homogeneity was disrupted, new interactions resulted in a new heterogeneous equilibrium, more complex than the original. At the organism level individual variations acted as destabilising forces, and as the biosphere itself evolved, becoming more complex, its increasingly varied impact on the resident species produced greater heterogeneity. 'Clearly the process, ever-advancing towards a temporary limit, but ever having its limit removed, must go on unceasingly.'[10] This comes close to recent proposals by some population ecologists that the subtle complexities of the environment are the 'motor' of adaptive and progressive changes in organisms. It also has the flavour of punctuated equilibria. Spencer's agent of evolution was thus initially entirely external, and in later stages remained largely external.

Although all of the biologists mentioned in this chapter were seeking unknown factors beyond Darwinism, none of them abandoned natural selection entirely, since it was always retained in the subordinate role of eliminator. As Spencer pointed out, Darwin had contributed to doubts about the adequacy of natural selection not only by the adoption of a Lamarckist mechanism, but also with the proposal of corollary laws of correlation, compensation and growth. He proposed that his 'law of the instability of the homogenous' superseded natural selection and all the rest. Natural selection was a form of indirect equilibration. Spencer is interesting not only for ideas 'unbiassed and untrammelled by older methods of looking at things', but also for his influence as an educator and populariser of evolutionism. Therefore his biological hypotheses, and especially his acceptance of some Lamarckist principles, must be treated as significant factors in the epistemology of biology.

E.D. Cope wrote that papers published independently in 1866 by himself and Hyatt were the

first attempt to show by concrete examples of natural taxonomy, that the variations that result in evolution are not multifarious or promiscuous, but definite and direct, contrary to the method which seeks no origin for variations other than natural selection.[11]

In *The Origin of Genera* (1868) Cope proposed the *law of retardation and acceleration*. This consisted of

> a continual crowding backward of the successive steps of individual development, so that the period of reproduction, while occurring periodically with the change of the year, falls later and later in the *life history* of the species, conferring upon its offspring features in advance of those possessed by its predecessors. . .This progressive crowding back of stages is not, however, supposed to have progressed regularly. . .There are well-known periods when the most important transitions are accompanied in an incredibly short period of time.[12]

Metamorphoses exemplified such rapid changes. When the metamorphosis was placed early in the life-cycle, the results could be dramatic, capable even of creating new genera. 'Fitness' did not enter into this scheme except as a possible result of the law of acceleration and retardation. This doctrine touches upon the problem of escaping from the trap of specialisation; neoteny would be one effect of acceleration and retardation. Cope attributed some aspects of this law to Geoffroy, Owen, Agassiz and Duméril, in addition to his contemporary, Hyatt. He suggested that many genera were polyphyletic and did not arise by adaptive radiation from a single archetypal species. Cope's *law of repetitive addition*, described in his 1871 essay, 'The Method of Creation of Organic Types', proposed that some animal structures originated from simple repetitions of identical elements, e.g. segmentation in Annelida, Arthropoda and Vertebrata. The agent of these changes was a vital *growth force*, which could be differentially expressed to produce anything from a simple allometric change to segmentation, and the further development of these ideas carried Cope into the mainstream of vitalistic Lamarckism.

Conwy Lloyd Morgan's early views on evolution appeared in *Animal Life and Intelligence* (1891), when he argued that 'selection proper' should be distinguished from 'natural elimination' which was by the action of surrounding physical or climatic conditions; by parasites and diseases, and by competition.[13] Elimination by enemies was productive of mimicry and camouflage, swiftness and cunning. 'Selection proper' was behavourial and included sexual selection, the choice of attractively fragrant or coloured flowers by insects, and the choice of the more succulent fruits as food. Morgan excluded 'neutral variations' from natural elimination, hinting that although not directly favoured by selection proper, they might provide some of the raw material for 'new experiments in the combination of variations, occasionally. . .with happy results'.[14] Still confounded with the problems of blending inheritance, Morgan stressed the need for isolation for evolutionary experiments in novelty. Both occurred in the organic, social and intellectual sphere. The organism could adapt in an orderly progression, becoming more and more in harmony with surrounding conditions, and if the conditions became more and more complex, the organism would progress in complexity.[15] The influence of Spencer, who corresponded with Morgan, is evident here. The question of *natural revolution*, which became his later concept of emergent evolution, was mentioned but not developed at this point.

Morgan equivocated over the causes of variation. He could see that the inheritance of acquired characteristics would be distinctly advantageous to the species which possessed it, but was drawn at that time to the position of the neo-Darwinists, as he called Weismann and his supporters, on the grounds that they proposed the all-sufficiency of natural selection. Morgan had three classes of variation: *superficial variations* included those of colour and form; *organic variations* included those of the size, complexity and efficiency of the organs of the body; *reproductive and development variations* included the modes of reproduction and fertilisation, variations in the time taken by development, in the time at which reproduction commenced, and in the period at which secondary sexual characters and the maximum efficiency of the several organs was reached. Morgan mentioned known cases of neoteny such as the axolotl and premetamorphic sexually mature insects, and also gave the discussion an orthogenic flavour by saying that the 'condensed development which is familiar in the embryos of so many higher animals may be regarded as the result of variations constantly tending in the same direction.[16] He agreed with Francis Galton that

the theory of natural selection might dispense with a restriction for which it is difficult to see either the need or the justification, namely, that the course of evolution always proceeds by steps that are severally minute, and that become effective only through accumulation. That the steps may be small and that they must be small, are very different views. It is only to the latter that I object, and only when the indefinite word 'small' is used in the sense of 'barely discernible', or small as compared with such large sports as are known to have been the origin of new races.[17]

As an example of the 'large sports' Morgan cited the ancon sheep, which seems to have been a case of an epigenetic mutation affecting growth hormone production, i.e. the product of a single point mutation. Whatever the mechanism, the sheep, which appeared spontaneously in 1791 in a Massachusetts herd, was popular with breeders because its short legs made it easier to confine.[18]

Morgan's interest in the evolution of mind was stimulated by A.R. Wallace's opinions.

> If a material element, or a combination of a thousand material elements in a molecule are alike unconscious, it is impossible for us to believe that the mere addition of one, two or a thousand other material elements to form a more complex molecule could in any way tend to produce a self-conscious existence. The things are radically distinct. To say that a mind is a product or function or protoplasm, or its molecular changes, is to use words to which we can attach no clear conception. You cannot have in the whole what does not exist in any of the parts; and those who argue thus should put forth a definite conception of matter with clearly enunciated properties, and show that the necessary result of a certain complex arrangement of the elements or atoms of that matter will be the production of self-consciousness.[19]

Wallace, while arguing on behalf of Divine intervention, through the creation of mind, was tacitly illustrating the difficulty of any gradualistic or hylozoic theory of mental evolution. More recently, Bernhard Rensch, arguing that the series of phylogenetic transformations are gapless, says that

> the gradual evolutionary transformation traceable in the phylogenetic tree, in particular, implies that one has to attribute at least primitive and not highly differentiated sensations accompanied by positive or negative feelings and also simple images (retentive ability, memory) to all those invertebrates

which are provided with the 'psychological structure' of nerve and sense cells.[20]

But the phenomena need not in Rensch's view originate with nerve cells, since parallel functions exist in unicellular organisms. Therefore, it is legitimate to ask whether the origin of psychic phenomena coincided with the origin of life.

> We have seen that it is possible to assume parallel 'psychic' components even in unicellular organisms and that the laws of parallel correspondence may possibly be as 'eternal' as causality. Hence, molecules and atoms should also be credited with basic parallel components of some kind.[21]

This is a *reductio ad absurdum*; but proponents of a particular world view always seem to lack the sense of humour necessary for the recognition of absurdity within that mind set.

For Morgan, Wallace's dilemma had only one horn, the precept that consciousness is distinct and separate from matter. He disagreed that the only alternative was that all matter is conscious, arguing that neural anatomy and consciousness were both manifestations of the same phenomena. Using the analogy of the outer surface of a sphere and the inner concave surface he noted that each could be viewed separately, from inside or outside, but they were attributes of the sphere which could not exist independently of the sphere or of each other. Complex psyches evolved with complex nervous systems, but should not be separated off intellectually as products of complex nervous systems. He agreed with Wallace that the mathematical faculty of the brain was unexplained by natural selection: there existed certain self-sufficient neural kineses not subject to the law of elimination; and these were conceptual thoughts, emotions and ideas.[22] The development of Morgan's later emergentistic interpretation of consciousness was diverted by his *law of congruity* whereby ideas evolve and survive if they are in harmony with pre-existing ideas, and analogously neural novelties must be in harmony with the pre-existing nervous system. Morgan had to transcend this comfortable platitude before he could conceive of an emergent quality that might instead of conforming to pre-existing patterns make those patterns change to fit it.

H.F. Osborn was an early historian of evolutionism, a perspective evident in his 1895 essay 'The Hereditary Mechanism and the Search for the Unknown Factors of Evolution', which discussed the early concepts of variation of Buffon, Lamarck and Geoffroy, together with those of contemporary workers. Like the other authors cited in this chapter he did not believe that natural selection explained the origin of the fittest, nor did he embrace the inheritance of acquired characteristics. It followed that there must be other causes, and that 'chief among the unknown factors of evolution are the relations which subsist between the various stages of development, and the environment'.[23] It was futile to study variations, *en masse*, according to the Baconian procedures employed by William Bateson in his *Materials for the Study of Variation* (1894). It required a more incisive formal approach.

He divided heredity into its two aspects of *repetition* and *variation*. The first involved repetition of parental type and regression to present or past race types, i.e. atavistic or throw-back variations, which he called *palingenic*. The second aspect, variation, included neutral anomalies and abnormalities, which did not have a role in evolution, and progressive variations, which comprised ontogenic variations from parental type or from race type, and phylogenic variations towards new types, or *cenogenic variations*. Osborn agreed with Galton's reasoning that 'there must be some strong progressive variational tendency in organisms to offset the strongly retrogressive principle of Repetition wherever the neutralizing or swamping effect of natural inter-breeding is in force', and that this was illustrated by 'a strong undercurrent of phylogenic variations'[24] in natural interbreeding populations. This argument demonstrates how great an obstacle blending inheritance was to understanding evolution. Twenty years later the steam had gone out of the problem; the full implications of Mendelian inheritance had been grasped, and it had become clear that 'selection pressure' was more important than 'mutation pressure' in determining allele distribution.

Osborn suggested that much ontogenic variation was what we now call phenotypic — partly induced by external influences and not heritable as such. Immersed among these variations, and disguised by them, there had to be phylogenic, heritable changes. As to the time and place of origins of these variations Osborn felt that they could arise during maturation or fertilisation of the gametes as Weismann had proposed, in early embryogenesis, or in later development. That Osborn should have been ambivalent on this question illustrates the power of the idea of the direct effect of environment or direct response of the organism to the environment, at that time in biological history. As possible causes of variation Osborn accepted Kölliker's and others' physicochemical, neurological, nutritive and pathological agents, together with Weismann's *amphimixis*, which in modern language includes chromosomal assortment and recombination. There remained the matter of 'definite' variation, by which he meant the autonomous process of orthogenesis.[25] He regarded Bateson's 'continuous' and 'discontinuous' variations as useful descriptive terms which distinguished gradual from sudden ontogenic variations, and he gave priority to Geoffroy for the concept of saltatory evolution caused by early embryonic changes.

Of all the publications which dealt with variations in the 1890s William Bateson's *Materials for the study of Variation, Treated with Especial Regard to Discontinuity in the Origin of Species*, (1894) was the most exhaustive. Bateson (1860-1926) studied natural sciences at Cambridge where he was a friend and fellow student of W.F.R. Weldon, who later became his greatest antagonist. Bateson conducted graduate research under W.K. Brooks in America on the subject of *Balanoglossus* and the evolution of the protochordates. Brooks's *The Law of Heredity: A Study of the Cause of Variation, and the Origin of Living Organisms*, published in 1883 while Bateson was associated with him, stimulated Bateson's interest in the mechanisms of variation. Brooks's study, which had been influenced by Mivart and Spencer, accounted for apparent saltations in evolutions with the *theory of correlated variation*. An initial variation caused by an environmental change would be inherited by a kind of pangenesis, the male germ cells having 'a peculiar power to gather and store up germs'.[26] This conclusion arose from his study of insect parthenogenesis. Such variations would have a cumulative effect, and rapidly cause new variations to occur, like the ripple effect of a stone thrown into water. In later works he reverted to a more orthodox Darwinism.

Bateson's expeditions to Russia and Egypt to study variations correlated with different limnological environments were unproductive. However, over the next ten years his studies on variation strengthened his convictions that evolution was discontinuous, rather than gradual, and he regarded embryology as a fruitful subject for such studies. A letter to his sister Anna in 1888 attempted to explain his obsession with variation and shows the influence of Brooks:

If then, it is true no variation could occur if it were not arranged that other variations should occur in correlation with it, in all parts, all these correlated variations are dictated by the initial variation acting as an environmental change.[27]

He added that the primary variation to which other correlated variations accommodated themselves might itself be very small. If broad correlations did not occur, 'a long time must elapse before the whole organism is again a system'. He concluded, 'The accommodatory mechanism is the thing to go for. I don't believe it is generally recognised as existing, though when stated it seems obvious.' This accommodatory mechanism, (which allowed a viable equilibrium to be sustained or attained in the face of disequilibrating changes) was to intrigue Driesch and the later systems theoreticians: it remains a mystery, and there is still no general recognition of its existence.

Bateson was an active controversialist, engaging the neo-Lamarckists, biometricians and neo-Darwinists; Weismann's theories were, 'evolution for amateurs'[28]. Sometimes his criticisms misfired; near the end of his career his address to the American Association for the Advancement of Science in Toronto in 1921, 'Evolutionary Faith and Modern Doubts', was interpreted by some American journalists as an attack on evolutionism in general, and William Jennings Bryan used this in his political campaign.[29] 'The critic's critic' as J.H. Woodger called him, followed his own aphorism, 'righteous anger is the very salt of good work'.[30]

Materials for the Study of Variation fits the Kuhnian model as a book written by a 'normal scientist' at a time of scholarly crisis and confusion, pending the introduction of a new paradigm. It criticised the old paradigm (Darwinism) and provided a Baconian compendium of data which Bateson believed ought to be considered, without making a final commitment to wild ideas. 'To collect and codify the facts of variation is. . .the first duty of the naturalist. Whatever be our views of Descent, Variation is the common basis of them all'.[31] Bateson felt that the 'representative type' method of teaching zoology which took an intimate look at the dogfish as a representative of the Pisces, and the frog for the Amphibia, was responsible for the neglect of variation within such groups. He believed that the student sees in a specimen 'what he has been told to see and no more, rarely learning the habit of spontaneous observation, the one lesson that the study of natural history is best fitted to teach'.[32] This method of teaching zoology was popularised by T.H. Huxley so successfully that many university courses still employ it.

For Bateson the two main factors requiring explanation by evolutionists were the discontinuity of variation and adaptation. He credited Lamarck for attempting to explain adaptive variation and noted that Darwin had neglected to do so. He cautioned,

> In dealing with the Question of Adaptation more than usual caution is needed. . .since, at the present time the conclusions arrived at in this field are being allowed to pass unchallenged to a place among the traditional beliefs of Science, it is well to remember that the evidence for these beliefs is far from being of the nature of proof. . .We have no right to consider the utility of a structure demonstrated in the sense that we may use this demonstration as evidence of the causes which have led to the existence of the structure, until we have this quantitative knowledge of the causes of its utility and are able to set off against it the cost of the production of the structure and all the difficulties which its presence entails on the organism.[33]

This statement was a warning against simplistic and teleological interpretations of nature. It is easy to see a character of an organism as useful if we believe that to exist a character must be adaptive, and even if a character is obviously more useful than the possible alternatives, this cannot be used to explain how the organism acquired the character. Bateson stressed the difficulty of understanding how the intermediate stages in the evolution of complex organs might be useful, but he never quite came to terms with the accommodatory mechanism and the importance of internal organisational harmony in evolutionary progress, concentrating as he did on outward appearances. The closest he came was on the subject of 'meristic variation', or variation by multiplication of parts, such as digits, segments and petals. He said,

> this may be a strictly *mechanical* phenomenon, and the perfection and symmetry of the process, whether in type or in variety may be an expression of the fact that the forms of the type or of the variety represent positions in which the forces of Division are in a condition of Mechanical Stability.[34]

Meristic variation, like Cope's 'repetitive addition', is an early morphological statement of the Ohno concept of evolution by the redundancy made available by gene duplication. The original segments go on with the old biological functions while the new ones are free to change and potentiate different functions.

Bateson also picked out what Gertrude Himmelfarb has called 'the logical difficulty that was to plague Darwin and all later evolutionary thought, the confusion between naturalism and gradualism, and assumption that the method of naturalism necessarily involved a theory of gradualism'.[35] Bateson said, 'There is in the minds of some persons an inherent conviction that all natural processes are continuous'.[36] Even the discontinuity of chemical combination was regarded as the epiphenomenon of an underlying continuous process. Among his 'substantive variation', i.e. differences in the parts as opposed to numbers of the parts, Bateson discussed distinct size discontinuities in insects, as Løvtrup, Rahemtulla and Hoglund have recently done, giving vertebrate examples.[37]

His final conclusions, based upon exhaustive descriptions of over eight hundred cases of what he thought to be relevant examples of variation, was that

> the existence of Discontinuity in Variation is therefore a final proof that the accepted hypothesis is inadequate. If the evidence went no further than this the result would be rather to destroy than to build up. But besides this negative result there is a positive result too and the same Discontinuity which in the old structure had no place, may be made the framework round which a new structure may be built. For if distinct and 'perfect' varieties may come into existence discontinuously, may not the Discontinuity of Species have had a similar Origin?[38]

Bateson's emphasis on the discontinuous emergence of 'perfect' varieties helps us to understand why the rediscovery of Mendel's work, and Bateson's own Mendelian experiments were initially treated as an alternative to Darwinism, and why *neo-Mendelism* and *mutationism* offered a temporary paradigm of genetics and evolution. Mendel's experiments showed that phenotypic traits were determined by discrete pairs of genetic 'elements' which segregated when germ cells were being formed, and come together again when a zygote is formed from the fusion of male and female gametes.[39]

Bateson and De Vries were so biased in favour of the discontinuity of evolution that they over-emphasised the discrete nature of the genes and neglected Mendel's observation that multiple factor inheritance could produce phenotypically continuous variation. The tangled controversy that developed between Bateson as champion of neo-Mendelism, and the biometricians, who were the forerunners of the quantitative Darwinists, has been carefully disentangled by William Provine in *Origins of Theoretical Population Genetics* (1971), and will be touched upon again in Chapter 11 on 'The Struggle for Existence of Selectionism'.

Hugo De Vries (1848-1935) was Professor of Botany and Director of the Botanical Gardens at Amsterdam. Between 1886 and 1899 he carried out breeding experiments with the evening primrose *Oenothera lamarckiana*, finding that up to 3 per cent of a given generation were 'sports', i.e. had some distinct features unlike previous generations, not simply of recessive character. Many of these sports bred true. These apparent sudden changes he called *mutations*, and believing that he had discovered a species in an active state of evolution, he developed a hypothesis of *intracellular pangenesis* to account for these mutations.[40] *Pangens*, not to be confused with the gemmules of Darwin's pangenesis hypothesis, were definite material particles in the organism which were capable of independent variations. This came close to Mendel's concept of hereditary elements, so De Vries had the 'prepared mind' that the chance discovery of Mendel's paper favoured. He also rejected that aspect of Darwin's pangenesis that accounted for the inheritance of acquired characteristics. De Vries met Bateson at a conference in 1899, and they discussed their mutual interests in discontinuity and their dislike of the biometrics school.[41] In 1900 De Vries sent Bateson an advance copy of the first volume of *The Mutation Theory*, and Bateson consequently took up cudgels on De Vries's behalf when the book was criticised by the biometrician Weldon. De Vries soon began to downplay the importance of Mendel's principles in understanding saltatory evolution, but Bateson continued to enlarge upon neo-Mendelism. Both volumes of *The Mutation Theory* became available in English in 1910.

De Vries distinguished between environmentally induced variation, which could not be sustained by natural selection and become hereditary (what his contemporary Johanssen called phenotypic quality), and *mutability*.[42] He pointed out that the horticulturalists had never produced a new species, and that even varieties created themselves somehow before being selected. Concerning Darwinism he wrote:

> The several propositions and hypotheses which Darwin employed as supports for this theory should be regarded now only as such, since their interest is mainly historical. They have served their purpose and are thereby fully justified. Whether they contain in part what is unproven or what is incorrect matters not. But they contain over and above that, a large mass of important facts which can be made use of to build further on the foundations laid by Darwin. This is especially true of the theory of selection, which now has served its time as an argument for the theory of descent; happily this theory no longer stands in need of such support. We are now concerned to bring the origin of species into the field of experimental investigation.[43]

While he thought that *O. lamarckiana* was in a 'condition of mutability' he qualified this with

> But not all plants and animals are mutable at the present time; on the contrary, mutability is a very rate phenomenon. This circumstance can only be brought into harmony with the theory of the ever mutable main lines of the pedigree by assuming that they have produced lateral branches, in which the capacity for mutation has been lost.[44]

He had to make this qualification because of his failure to produce similar results with other species of *Oenothera*. He also observed that his mutation theory did not require the disappearance of the paternal stock of a new species. *Oenothera* is now recognised as a genetically unusual plant, not because it is in a 'condition of mutability', but because of its unusual chromosome makeup. None of the mutations discussed by De Vries were the now-familiar DNA point mutations.

The belief that new species could arise by single mutations, first mooted by Geoffroy, was very popular during the first decade of the twentieth century. There were multifold reasons for this, primarily the epistemological need for variations with sufficient impact to offset blending inheritance. T.H. Huxley never enlarged upon saltationism, but his overt support of other authors on this matter was influential. Bateson became a skilled polemicist and outmanoeuvred the biometricians with his power play in the Royal Society Committee on Evolution, as will be demonstrated in Chapter 11. Neo-Mendelism and mutationism were unhampered by metaphysics and had empirical substance.

Some conceptions of variation had a romantic appeal, restoring to the organism the apparent control of its evolutionary destiny, instead of being at the mercy of a random, capricious and soulless external cause. There was consequently a cross-fertilisation between these ideas and neo-vitalism, which was then in the ascendant phase.

5 TRANSCENDENTAL EVOLUTION

> Entelechy is affected by and acts upon spatial causality as if it
> came out of ultra-spatial dimension; it does not act in space, it
> acts into space; it is not in space, it only has points of
> manifestation in space.
>
> Hans Driesch 1908[1]

> The vitalist is more impressed by man's ignorance than are the
> others. It is scarcely too much to say that he gets from his
> ignorance that very satisfaction which the others get from their
> knowledge of it.
>
> R.E. Lloyd 1914[2]

Reaction to the simplistic positivism of nineteenth-century science,
the reductionism of the cell theory and biophore theory, and the
mechanicism of embryology contributed to a wave of romantic
biology and evolutionism. The 'formative intelligence' and 'inherent
tendencies' of Murphy and Mivart were symptomatic of the early
stages of this movement. The neo-vitalists talked about the
evolutionary role of organising energies such as entelechy and *élan
vital*, and early expressions of emergence doctrine conferred transcendental qualities upon newly evolved levels of complexity, so that
new wholes were not only more than the sum of their parts in terms
of novelty of arrangement and organisation, but their wholeness
acquired an independent metaphysical existence with the supernatural ability to escape the restraints of matter: deity was the highest
emergent. The exploration of these epistemological eddies has an
intrinsic fascination, but they are also counterinductively significant
since they came close to exhausting the possibilities of the actions of
vital forces, and sometimes focused on biological phenomena that
had evaded reduction.

Vital airs and fluids appeared in ancient myths; for example
creation was supposed to be effected by the transfer of divine energy
such as the breath of God into his inanimate creations, and vital
elixirs could bring the dead to life. In Aristotle's time the soul
was believed to be composed of vital pneuma obtained from the air
and purified by the heart. With the advance of anatomy the

pneumatic spirits became the vital fluids residing in the blood and in the nervous system. In the seventeenth century Descartes proposed that humans consisted of corporeal substance and spiritual substance, the latter subsuming soul, mind and its function of reason. In his scheme animals were purely corporeal automata, and some contemporary iatromechanists quickly discarded the spiritual aspect of humans, embracing a monistic materialism or mechanicism in which all functions, including mental ones were to be regarded as mechanical, based upon physical matter and the laws of motion. However, the role of biochemical, and particularly enzymological functions in organisms were unknown, and the progress of mechanicism was hindered by the inability of mechanical models to explain functions such as digestion, and metabolism.

By 1828 Wöhler had synthesised urea, an organic substance whose production had been previously thought possible only with the mediation of a living organism. Vital force as an exclusive property of living organisms was thus shown to be non-essential, at least for the synthesis of some organic compounds typically found in live organisms. A premature flowering of chemical reductionism followed Wöhler's findings, retarding the understanding of the role of micro-organisms in nature. This was rectified by Louis Pasteur. By the late nineteenth century the enzymes responsible for fermentation had been extracted from yeast and had been demonstrated to carry out the process *in vitro*, in the absence of living yeast cells. By the same time the role of electricity in the nervous system was well enough understood to obviate the idea that a special vital energy was required for neural activity.

Materialism and *mechanism* have been used interchangeably by biologists. Both imply that all of the phenomena exhibited by living organisms can be explained naturalistically in terms of their constituent molecules, atoms or sub-atomic particles and their activities. Mechanicism is sometimes used in the special sense that machines can be adequate models for living beings. In the history of this particular usage, clocks, sluice gates, ships' rigging, heat-engines and computers have been used as mechanical models. A curious kind of mechanistic romanticism is common in science fiction: the apparent wish to see the present unsatisfactory blood-and-guts model replaced by intelligent machines. Stanislav Lem has satirised this vision of machine evolution:

Know then that that race of the Galaxy originated in a manner as mysterious as it was obscene, for it resulted from the general pollution of a certain heavenly body. There arose noxious exhalations and putrid excrescences, and out of these was spawned the species known as paleface — though not all at once. First they were creeping molds that slithered forth from the ocean onto land, and lived by devouring one another, and the more they devoured themselves, the more of them there were, and then they stood upright, supporting their globby substance by means of calcareous scaffolding, and finally they built machines. From these protomachines came sentient machines, which begat intelligent machines, which in turn conceived perfect machines, for it is written that All is Machine, from atom to Galaxy, and the machine is one and eternal and thou shalt have no other things before thee.[3]

Some biologists are dualists, who believe in the applicability of both mechanistic and supernatural concepts to nature; but this kind of dualism often requires a form of hysterical dissocation of beliefs because some phenomena that might be believed possible in the idealistic or supernatural mode appear absurd in the materialistic or natural mode. Consider, for example, a physiologist who believes in transubstantiation, the conversion of bread into the flesh of Christ. As a materialist he knows that the wafer is hydrolysed by carbohydrases in his gut, producing glucose; as a Catholic, obedient to the Council of Trent, he would know that the wafer turns into the flesh of Christ and is then digested by proteinases which convert it into amino acids. Descartes, who agonised over transubstantiation, was spared the heartburn of enzymological knowledge. J.J. Murphy believed that while chemical and electrical energy must have something to do with muscular contraction, room had to be made for a special vital energy. As long as he did not understand the energetics of muscle, there was room for the vital spark, and no matter how much of a materialistic smoke screen was emitted, the vitalists usually confessed that there was no difference between the vital spark and Divine breath. In 1912 E.G. Spaulding classified vitalistic theories as follows:

(1) Traditional vague vitalism, which attributed an undiscovered vital energy to the organism. This, Spaulding argued, would not make the organism non-mechanical since energy was subject to mechanical principles. The chemical energy that fires exothermic

reactions in the organism is usually contained in adenosine triphosphate (ATP), which does not occur naturally outside living organisms. At the time when Spaulding was writing this was an undiscovered mechanical 'vital energy'.

(2) Vitalism which made 'a mental factor universally parallel with the physiological factors'. Spaulding argued that if the physiological factors were mechanical then any parallel vital quality would also be founded on mechanical laws.

(3) Vitalism which gave a 'psychical entity' control of the discharge of potential energy, which it was one of the distinctive features of the organism to store up. This was the only version that Spaulding would admit as distinctly non-mechanistic. His criticism of it was that the organism was 'found to do the same thing under the same circumstances', with respect to 'vital forces' such as regeneration and morphogenesis. A vital principle might be expected to have the demonstrable ability to respond *differently* under the same circumstances, due to the exertion of Lamarckian psychic force for example. Spaulding regarded Driesch's entelechy as a related category, being a principle which produced the *same* end under *different* circumstances.[4]

This is writing the review before the play has been staged. The neo-vitalists are important protagonists in the epistemological drama.

Hans Driesch (1867-1841) was a zoologist who studied briefly with Weismann and completed his doctorate at Jena under Haeckel. His biological research, which included the growth patterns of hydroid colonies and invertebrate embryology, was supported by private means, and his co-workers included Curt Herbst and T.H. Morgan. When he formally re-entered academic circles it was as a philospher, first at Heidelberg, and then at Cologne and Leipzig. Driesch's autobiographical account of his research and theories is given in *The Science and Philosophy of the Organism* (1908), based on his Gifford lectures. He began in the late 1880s to investigate embryology following the mechanistic approach taken by Wilhelm Roux: the study of 'the mechanics of development' (Entwickelungs-mechanik).[5] In one experiment he had destroyed one of the two cells formed from the fertilised egg of a frog after the first cell division, and discovered that only half an embryo developed. Like Weismann, Roux had believed that there was an organising substance in the nucleus of the zygote which was sub-divided at each cell division in the embryo. Cell lineages leading from particular embryonic cells to particular organs in the mature organism had already been traced, and Roux's experiment had appeared to further confirm the disintegration of the hereditary material.

When Driesch separated the cells in the two-celled stage of the sea urchin *Echinus,* he found that instead of developing into the expected half-embryos the surviving cells became small, whole embryos.[6] This was clear refutation of Weismann's proposal that the hereditary material was sub-divided during develoment. After obtaining similar results with the four-celled stage of *Echinus,* Driesch went on to conduct the experiments which were the foundation of his vitalism. First of all he found that heat-treatment of *Echinus* embryos interfered with the normal progress of cell cleavage. Nevertheless, the abnormal embryo would develop into a normal larva. Mechanical distortion of the embryo, achieved by pressing the zygote between glass plates, caused the embryo to develop up to the sixteen-cell stage as a round plate of cells, one cell thick. On removal of the pressure the development of the distorted embryo was gradually restored to normal. There was some quality in the embryo that restored development to normal if there had been some external interference, a quality that Driesch later called *entelechy,* and what he attributed to it at first was responsibility for the orientation and symmetry of the embryo. The pressure experiments were repeated on frog embryos by Hertwig and on annelid worm embryos by E.B. Wilson, with similar results. Driesch thought that the organising factor resided in the cytoplasm, and not in the nucleus as Roux and Weismann had proposed. He was determined to prove the importance of the cytoplasm and how he went wrong is a cautionary tale; scientific method is no protection from the experimenter's desire to find what he thinks is necessary. It was later decided that there must be two kinds of egg, the mosaic or determinate egg, in which the organisation of the cytoplasm of the egg strongly affects embryonic development, and the regulative or indeterminate egg in which the organisation of the cytoplasm is relatively unimportant. The distinction between the two types is not hard and fast, and much of the apparent difference was due to the way in which experiments were conducted. For example, T.H. Morgan made a frog embryo develop normally from one cell of the two-cell stage, in contrast to Roux's result of a half-embryo.[7] To prove that the cytoplasm possessed that crucial entelechal ingredient Driesch, in collaboration with Morgan, chose the eggs of the ctenophore *Beroë,* a primitive jellyfish-like organism which traditionally was included in the mosaic category. They knew from earlier experiments by Chun that this organism was likely to give the required results. Removal of a portion of the cytoplasm, leaving the nucleus unharmed, resulted in an incomplete embryo.[8]

In his determination to prove Weismann wrong Driesch had jumped to the conclusion that the nuclear material was relatively unimportant, ignoring the possibility that the hereditary determinants might be nuclear, but not shared out among the daughter cells following cell division. These inductive leaps in the dark sometimes pay off and sometimes do not. Darwin, for example, had already 'flown up' to his hypothesis of evolution and natural selection before he had spent much time following Baconian principles of data-collecting.[9] On the other hand, he dismissed Naudin's 'living mosaic' theory of heredity as a special case, and persisted with his own concept of blending inheritance on meagre experimental evidence, despite the problem it caused for selection theory.[10] Even experimental rigour connives at false premisses. Although Driesch's inductive errors led him to interesting conclusions, the abandonment of his empirical approach resulted in the conventional wisdom that,

> Although Driesch had a long-standing record of sound scientific work, he never fully recovered from the awesome implications of this isolated blastomere experiment. Eventually he retreated to a monastry with several of his protégés and devoted the rest of his life to mysticism.[11]

It is difficult to pin down Driesch's exact meaning of entelechy, since he said more about what it was not than what it was. Entelechy was, 'an "intensive manifoldness" realising itself extensively'.[12] The determinants in Weismann's concept of heredity could easily have fitted this definition, not to mention the modern concept of DNA function. 'Entelechy was *order* of relation and absolutely nothing else.'[13] The 'order of relation' was hierarchical; for example, the large category of *morphogenetic entelechy* subsumed the *psychoid* (neural) category, and psychoids subsumed instinctive and voluntary actions.[14] Entelechy was therefore thought to govern the development of the embryo which had a *'prospective potency'* which could be realised even when normal development was disturbed.[15] It could be restored to a *causal harmony* which expressed 'the unfailing relative condition of formative causes and cause recipients'.[16] In other words, entelechy ensured that during development the right bits were in the right place at the right time, perhaps by 'permitting' enzymes to appear, which then operated 'along purely chemical lines'.[17]

> Entelechy suspends *all* possible reactions at first, and then, if differentiation sets in, allows only *one* particular reaction in each cell to occur, according to its plan. . .In this way a *sum of possibilities* is transformed into an *ordered whole of actualities,* as it is characteristic of harmonious differentiation.[18]

Entelechy could restitute missing parts in the mature organism to some extent and was responsible for functional harmony.

In many ways Driesch came close to the modern epigeneticist's view of homeorhesis, and the molecular biologist's understanding of gene expression. Entelechy desuppressed the potential of the embryo, without any vital energy being expended, but it did not arise from a 'living chemical substance', nor was it a form of energy; yet so many attributes of entelechy are actually those of DNA, including the 'intensive manifoldness expressing itself extensively', the 'prospective potency' of the cell, and the 'suspending of possible becoming'. Because of the association of entelechy with a cytoplasmic organising force, the grand scheme was bound to fail eventually. But Driesch made an error of logic, identical to the one made by selectionists. Entelechy would have been a useful catch-all expression for all of the checks and balances and regulations of gene activity involved in development, just as natural selection is a useful catch-all for organismic interactions with the environment which affect reproduction. The mistake was to treat entelechy as the agent of harmonious development, just as it is a mistake to treat natural selection as the general agent of evolution.

Although he explored the physical and chemical possibilities, his aversion to preformationism blinded Driesch to the possibility that a 'living' chemical substance might act as a code for the production of the 'ferments' and indeed, might carry the instructions for its own differential expression. Having taken away from entelechy all of those attributes which now belong to DNA, there still remains that aspect of it that Waddington called *homeorhesis.*[19] This is the process of keeping the development of the organism on the right tracks and compensating for physical insults. Waddington did not repeat Driesch's mistake of regarding homeorhesis as an agent of embryogenesis, but rather as the logical analogue of homeostasis.

Both Driesch's and Waddington's concern with the way in which the developing organism remains 'true to type' should focus our attention on a most important evolutionary question. How is the organism released from its embryonic governors to emerge to a new grade, a more highly ordered state? Or as Bateson asked: what is the accommodatory mechanism? Driesch had hardly anything to say about this problem except that entelechy used material means in each individual morphogenesis, handed down by the material continuity in inheritance, and the material means were harmonious and resistant to change and evolution.[20] Therefore, the entelechy itself and not the material morphogenetic expression of it must be transformed: it was an 'immanent evolutionary force'.[21] As an actual cause entelechy defied definition and could not be pinned down experimentally: 'entelechy is affected by and acts upon causality as if it came out of ultra-spatial dimension; *it does not act in space, it acts into space;* it is not in space, it only has points of manifestation in space'.[1] If you are wondering why this has a familiar ring try substituting 'flying saucer' for entelechy. Entelechy was also self-actualising, being 'an agent that has the "idea" of the end in its imagination'. It was trying to evolve: 'Super-entelechy wants to know what it has done in a quasi-instinctive way, suffering from its own products. Finally it does know, and it now has the means of a conscious rectification of all its mistakes.'[21]

Driesch believed that natural selection was unable to create diversities, and was particularly critical of selectionists who confused the elimination of the unfit with the creation of the fit, agreeing with Nägeli,

> To say that a man had explained some organic character by natural selection is the same as if some one who is asked the question, 'Why is this tree covered with these leaves?' were to answer, 'Because the gardener did not cut them away.'[22]

That individual organisms could modify their functions according to environmental conditions gave Lamarckism some substance in Driesch's view, but he was neutral to the 'psychological vitalism' advanced by E.D. Cope and August Pauly, and to the whole apparatus of the inheritance of acquired characteristics. Neither Lamarckism nor Darwinism could properly explain adaptability, nor organisation as due to 'contingent variations, which accidentally have been found to satisfy some needs of the individual and therefore have been maintained and handed down'.[23]

Driesch was no more inclined to accept the mutation theory than Darwinism or Lamarckism. De Vries had, in his opinion, shown only that variation was discontinuous, and, mutations or no, neither De Vries nor the army of horticulturalists and breeders had produced a new species. Many evolutionists had inferred, 'that a sort of organisatory law must be at the base of all transformism'; but none had explained how it worked.[24]

With the hindsight of the molecular biological revolution entelechy is seen to retreat back into the ultraspatial dimension. But what remains, as *homeorhesis,* represents an area of our own ignorance of both ontogeny and phylogeny that should impress us sufficiently to want to overcome it. Jane Oppenheimer asks,

> Why do we place so little emphasis on Roux's strong emphasis on the interrelationship of parts, and on Driesch's remarkable materialistic analysis? Perhaps in part because we no longer read what they wrote, but in part also because these were early, not late concepts in the minds of Roux and Driesch themselves. Roux had a defect of character that led him to overdogmatism, and after he performed the separation of the blastomeres his dogma was self-differentiation; he could brook no dispute, and those who read only his later works have only a bare glimmering of the richness of his earlier thought. But his early words may well have found their way into conscious or subconscious memories. Driesch's case was different. In becoming a philosopher, and a vitalist at that, he emphasized the nonchemical and the nonmechanical in all but the earliest of his writings; but his early words too may have had their influence on the great minds that must have been exposed to them in the happy days when there seems to us to have been more time for reading and contemplation'.[25]

Driesch helped to create the intellectual diversion of neo-vitalism, and in so doing broadcast an odour of unreliability over any form of evolutionism that had vitalistic characteristics. However, he also addressed the problem of how the equipotential of embryonic cells might be harmoniously expressed to develop into a differentiated but integrated whole. Entelechy, taken as a metaphor, in the way that later organicists used 'fields' and 'landscapes', had great heuristic potential, but prejudice against his later ideas denied its realisation.

Richard Mocek sees Driesch's later life as a model of commitment to organicism or holism, and in it a justification for his rehabilitation. Donna Haraway argues that Driesch's thought cannot be reduced to a vitalist theory of the organism; his biology was not vitalist but was the ground for philosophic vitalism. Moreover, his theory of organism was rationally articulated with his overall thinking.[26] Entelechy was the wrong answer to the right questions: how did the germ contain the directions for making the organism; how were the directions expressed; how was interference circumvented, and how did evolution transcend developmental stability?

Henri Bergson (1859-1941) was educated in classics and philosophy, and between 1883 and 1888 he made an informal study of biology and its metaphysical implications while holding a professorship at Clermont-Ferrand.[27] In 1908 *L'Évolution créatrice* was published, and by 1911 the English translation of this book, together with Bergsons's Huxley Lecture on 'Life and Consciousness' at the University of Birmingham, and Balfour's essay on 'Creative Evolution and Philosophic Doubt', brought Bergson to a wide audience, further enlarged by his lecture tour to the United States in 1913. In 1913 he was also President of the Society for Psychical Research, his interest in hypnotism and psychical phenomena having paralleled his interest in evolution and metaphysics. By 1914 *L'Évolution créatrice* had been indexed by the Roman Catholic Church.

Bergson viewed evolutionary progress not as a finalistic but as an open-ended process. Mechanistic interpretations of organic processes could be useful, and the process of adaptation to environment was almost mechanical, but, 'adaptation explains the sinuosities of the movement of evolution but not its general directions.'[28] It was necessary to understand how the *élan vital*, which is usually translated as 'vital spark', affected mechanical events. He thought that aspects of neo-Darwinism and neo-Lamarckism might be true, but disagreed with the neo-Darwinist position that variations occurred in the germ plasm fortuitously. They were instead, 'due to the passage of the *élan vital* from germ to germ, and they might well appear at the same time, in the same form, in some or all the representatives of a species.'[29] This is also implied by Lamarckian theory, and Bergson believed that De Vries's mutation theory supported this position.

The trials and tribulations of the creative life-impulse were given a near-poetic expression by Bergson, and much of his success may have been due to his ability to confer mythic dimensions upon evolutionary progress. As Lovejoy wrote:

> How exciting and how welcome is the sense of initiation into hidden mysteries! And how effectively have certain philosophers — notably Schelling and Hegel a century ago, and Bergson in our own generation — satisfied the human craving for this experience, by representing the central insight of their philosophy as a thing to be reached not through a consecutive progress of thought guided by the ordinary logic available to every man, but through a sudden leap whereby one rises to a plane of insight wholly different in its principles from the level of mere understanding'.[30]

Bergson's view of the evolutionary process retains a strong subjective allure.

> In the evolution of life. . .the disproportion is striking between the work and the result. From the bottom to the top of the organized world we do indeed find one great effort; but most often this effort turns short, sometimes paralyzed by contrary forces, sometimes diverted from what it should do by what it does, absorbed by the form it is engaged in taking, hypnotized by it as by a mirror. Even in its most perfect works, though it seems to have triumphed over external resistances and also over its own, it is at the mercy of the materiality which it has had to assume. It is what each of us may experience in himself.
>
> Life in general is mobility itself; particular manifestations of life accept this mobility reluctantly and constantly lag behind. It is always going ahead; they want to mark time. Evolution in general would fain go in a straight line; each special evolution is a kind of circle. Like eddies of dust raised by the wind as it passes, the living turn upon themselves borne up by the great blast of life. They are therefore relatively stable, and counterfeit immobility so well that we treat each of them as a *thing* rather than as a *progress,* forgetting that the very permanence of their form is only an outline of a movement. . .Regarded in what constitutes its true essence, namely as a transition from species to species, life is a continually growing action. But each of the species, through which life passes, aims only at its own convenience. It goes for that which demands the least labour. Absorbed in the form it is

about to take it falls into a partial sleep in which it ignores almost all the rest of life; it fashions itself so as to take the greatest possible advantage of its immediate environment with the least possible trouble. Accordingly, the act by which life goes forward to the creation of a new form, and the act by which this form is shaped, are two different and often antagonistic movements. The first is continuous with the second, but cannot continue in it without being drawn aside from its direction, as would happen to a man leaping, if, in order to clear the obstacle, he had to turn his eyes from it and look at himself all the time'.[31]

Bergson turned the process of evolution into a heroic progression. Evolution was confronted with obstacles, lured down the primrose path of adaptation and specialisation, and the desire to linger, eat lotus and forget the quest, all of which could only be surmounted by self-transcendance. However successful the poetic vision of the evolutionary process, Bergson left us with little clear perception of how we might deal intellectually with its content.

The current of consciousness that had penetrated matter had led to its organisation, but its flow was retarded and divided among divergent series of organisms, dissipating its energy to the point of torpor or unconsciousness. However, the origin of the intitial consciousness which imposed itself on matter is as unclear as the origin of Driesch's entelechy. Of the primordial consciousness he said that it had to *detend* in order to *extend*. It was not the equivalent of human consciousness, which was obliged to analyse past experience. In order to coincide with primordial consciousness 'it must detach itself from the *already-made* and attach itself to the *being-made'*.[32] This is a popular sentiment among modern transcendentalists, who agree with Bergson that the rate moment of transcendent consciousness is a real experience of the force that drives evolution. Bergson also noted that the transcience of these experiences allowed a variety of dialectical interpretations. 'Intuition, if it could be prolonged beyond a few instants, would not only make the philosopher agree with his own thought, but also all philosophers with each other'.[33] Most of the manifestations of evolution, the set-backs, arrests, adaptive divergence and form were regarded as contingencies. 'Two things only are necessary: 1. a gradual accumulation of energy; 2. an elastic canalization of this energy in variable and indeterminable directions, at an end of which are free acts'.[34] Bergson's vitalism captured the interest of people who might otherwise have been remote from evolutionism. He helped to compound the general distaste for vitalism among scientists in general, but he marks a crossroads on the epistemological road map. From that intersection one route led on to the modern transcendentalists.

The transcendental evolutionism of the Jesuit palaeontologist Père Teilhard de Chardin was strongly in the Bergson tradition. His major work, *Le Phénomène humain* was completed in 1938, but its publication was opposed by his order and it did not appear until 1955, after his death. Teilhard's early ideas concerning the relationship between original sin and evolution had been too unorthodox, and he had lost his teaching position at the Catholic Institute of Paris in 1924, whereupon he went to China to work with the Geological Survey. He was a member of the expedition that discovered Peking Man, and he also studied *Australopithecus* in South Africa.

Teilhard viewed evolutionary progress as a series of *complexifications,* which in the prebiotic history of the earth involved polymerisation, followed by the origin of life as an 'unprecedented and unrepeated. . .chemical transformation' — or 'primordial emission'.[35] The organised association of cells made possible 'sufficient bulk to escape innumerable external obstacles' — fluctuations in physical and chemical environmental conditions, and to begin to allow tissue differentiation.[36] Julian Huxley's introduction to the English edition, published in 1959, reminds the reader that *The Phenomenon of Man* was a pro-evolutionistic polemic with the intention of persuading conservative sceptics that man was a natural part of the evoluticnary scheme, and he was not interested in debating the various explanatory hypotheses. His inclusion of man in the evolutionary scheme was not the reductionistic 'naked ape' approach. Man was to be seen as a step beyond the ape towards the pre-determined *Omega Points,* which represented a confluence of consciousnesses transcending the individual mind. Though not a Lamarckist, Teilhard appreciated the fact that the evolution of intelligence, which led to the reduction of chance as a significant factor in life, was more Lamarckian than Darwinian.[37] The emergence that characterised man separated the species from other animals by a chasm, or threshold, contrary to the conventional notion that there was no sharp line to be drawn between instinct and thought. Teilhard saw this emergence as a kind of quantum jump, followed by the growth of a collective unconscience, or of a culture tradition that augmented consciousness and led to further evolution of mind.

The new human environment of reflection, consciousness, tradition and education Teilhard called the *noosphere*. Its acquisition involved a complexifying disequilibration and re-equilibration. Just as strikes disequilibrated industry, Teilhard expected 'strikes' in the noosphere to demand for individuals the full opportunity to progress towards the limits of consciousness, and he predicted that the holding together of a number of cells to make a metazoan, or the symbiosis of different organisms, would have a physical equivalent. By a mega-synthesis of thinking elements the Omega Point might be reached. Progress was not inevitable; at the time of writing of *The Phenomenon of Man* a perversion of noogenesis was ascending: the mass movements of Communism and National-Socialism, 'the crystal instead of the cell; the anthill instead of brotherhood. Instead of the upsurge of consciousness which we expected, it is mechanisation that seems to emerge inevitably from totalisation.'[38] Omega was not an absolute whole in which all individuals merged like grains of salt dissolving in the ocean; *union differentiated.*[39]

> In every organised whole the parts perfect themselves and fulfill themselves. . .following the confluent orbits of their centres, the grains of consciousness do not tend to lose their outlines and blend, but on the contrary, to accentuate the depth and incom-municability of their egos. The more 'other' they become in con-junction the more they find themselves as 'self'.[40]

The Omega Point was not simply 'a centre born of the fusion of elements which it collects or annihilating them in itself. By its structure Omega, in its ultimate principle, can only be a *distinct centre radiating at the core of a system of centres*'.[41] It is in the dis-tinctness of the centre and of the 'supremely autonomous focus of union', that the transcendental resides, although Teilhard tried to disguise it as a materialism arrived at by the remorseless logic. To some minds the transcendental is a necessary route to understanding the central metaphysical problems of what are we doing and where are we going. Moreover, transcendental agencies have often intui-tively defined particular sectors of the unknown. To the positivist it comes as a surprise when molecular biology reveals the answers to problems that were ignored or regarded as spurious. The latter position is the more anti-intellectual. The areas of our metaphysical and scientific ignorance or malaise have to be explored and exhausted, not simply set aside, and the physical signposts that writers such as Bergson and Teilhard have left us from their own journeys are to be cherished. Teilhard remarked that

> Religion and science are the two conjugated faces of phases of one and the same complete act of knowledge — the only one which can embrace the past and future of evolution so as to contemplate, measure and fulfill them.[42]

There is no shortage of piety in Teilhard; and perhaps the only 'bunk' is his admonition that *The Phenomenon of Man* should be read as a 'scientific treatise'.[43]

6 HOLISM AND BIOLOGY

> All collective life depends on the separation of offices and the concurrence of efforts.
>
> Aristotle

Looking for clues to the origins of biological holism is like looking for a needle in a needle-stack. As W.E. Ritter indicated in *The Unity of the Organism* in 1919, there have been as many versions of its history as there have been interested scholars, and the literature has expanded considerably since then. I will detail only those aspects that have been under-emphasised by other writers, and those which are particularly relevant to the subsequent development of my arguments.

The first and most general aspect of holism is one that other biological authors have omitted; either because it seems so self-evident as to be a truism, or, more likely, because of a distaste for anything that smacks of mysticism. This is the ontological aspect; that unitary view of the self and cosmos that Leibnitz called the *perennial philosophy*. In Aldous Huxley's words it 'is primarily concerned with the one, divine Reality substantial to the manifold world of things and lives and minds'.[1] One of the common aspects of such experience is a sense of the unity of nature and fusion of the observer with nature, which mythification may transduce from personal experience though some metaphysical universal such as light or time into a humanoid personification, i.e. into a god. Or the sense of oneness may be intellectualised into metaphysics as an explanation of nature. The early Greek metaphysicians transduced their visionary experiences into elemental categories such as fire, flux or motion; this may also have been a conscious attempt to reduce the confusing diversity of nature; the atomism of Democritus was the epitome of reductionism: 'nothing but atoms and the void'. Whichever interpretation we like to put on it we are dealing with the need to give intellectual expression to the underlying sense of oneness. A number of biologists and psychologists have taken the view that the sense of oneness and other aspects of religious experience have a materialistic basis, much in the way that Scrooge took Marley's ghost to be the manifestation of a piece of undigested cheese.

99

Timothy Leary, in an experiment on 'the psychodynamics of religion' before the proscription of LSD-25 and his subsequent notoriety, administered a hallucinogen, psilocybin, to a number of divinity students and theologians, who emerged from the experiment with pledges to the effect that the natural and drug-induced states of altered consciousness were identical.[2] Aldous Huxley observed that the natural prerequisites of religious experience were that the subject be 'loving, pure in heart, and poor in spirit', and reading between the lines one might conclude that the physiological state induced by partial starvation and lack of sleep might have something to do with it, hence the difficulty for rich, well-fed men to enter the Kingdom of Heaven.[3] Various authors have noted that the psychedelic drugs interfere with serotonin function in the brain, and so it could be that the sense of oneness is no more than an inhibition of neural centres that deal with the ability to distinguish between self and not-self, so that all sensory input is consciously unified. Whether we can establish materialistic terms for these experiences or not, a holistic world view satisfies metaphysical needs that are intermittently felt by many people. These needs and experiences are ineffable, but that does not make them illusory. It is impossible to convey the feeling of orgasm in the printed word, but not even a positivist would infer that orgasm is an illusion. A holistic view has developed spontaneously and independently in many minds during the course of history, and it may be a fruitless task to trace the historical development of holistic ideas purely in terms of intellectual tradition.

Holism, as applied to biology, stresses the need to deal with wholes, whether organs, organisms, or ecosystems and the relationships of their constituent parts, as opposed to inferring the nature of wholes from a knowledge of their parts. It cautions against the oversimplification of complex systems, and acts as a baby-filter for analytical bathtubs. The aphorism 'The whole is greater than the sum of its parts' suggests that only a holistic approach is likely to provide an adequate understanding of complex systems. Henri Bergson went as far as suggesting that analysis was a futile, destructive procedure, since it only provided knowledge of damaged parts.

One of the earliest and most illuminating attempts to place biology in the context of the history of ideas was L.J. Henderson's *The Order of Nature* (1917). He attributed the development of the importance of organisation as a physiological principle to Johannes Muller and Von Baer, and regarded the latter and the French histologist Bichat as the first *organicists*. However, he accorded Claude Bernard the greatest role in the promulgation of physiological holism. Organicism in its broad sense refers simply to the holistic treatment of the organisation of organisms, but some authors regard pure organicism or *organismalism* as a twentieth-century phenomenon, with W.E. Ritter as its chief elaborator.[4]

Claude Bernard, a major pioneer of experimental medicine, taught that the organism must be regarded as a whole. His contributions to biology are particularly relevant to this work because he investigated the manner in which *constant life*, or free life, was organised, i.e. the form of life whose internal milieu was kept constant in its physicochemical quantities, thus allowing it independence from the fluctuations of the external environment: 'La fixité du milieu intérieur c'est la condition de la vie libre.'[5] It is the acquisition of Bernard's *constant life* (homeostasis) that I have been equating with evolutionary progress or grade evolution, and it is a subject that evolutionary biologists have largely ignored.

Bernard (1813-78) was born in the Province of Beaujolais in France.[6] Science and philosophy were neglected in his early education and he became a druggist's assistant when he left school, devoting his spare time to writing plays, one of which was performed locally in Lyons. A literary critic commented to Bernard that he was no dramatist, and that since, as a druggist, he was already on the road to medicine, he should consider that as a future profession, which advice Bernard took, thus demonstrating the positive value of literary criticism. Later Bernard became a model for characters invented by Zola and Dostoyevski. In 1841, having interned in Paris under François Magendie, Professor of Physiology, he became his teacher's laboratory assistant. Magendie's approach to science was Baconian: non-speculative with emphasis on experiment and data collection, but the student was not only more imaginative but also able to obtain more consistent experimental results than the master. Moreover, as P. Medawar has remarked, he was the only biologist of the period to formalise scientific method as well as to practise it.[7] In 1854 Bernard took the Chair of Physiology, succeeding Magendie, and eleven years later (after a prolonged illness which gave him the opportunity to ponder and formalise his experimental methods) he published *An Introduction to the Study of Experimental Medicine*. Bernard's research successes were too numerous to detail here, but included original contributions to the study of mammalian blood-glucose levels, the role of glycogen in the liver, the physiology of digestion and starvation, neuromuscular physiology, haemodynamics, kidney function, osmoregulation, ionoregulation and acid-base balance, as well as plant physiology and physiological and embryological studies on lower animals. Bernard was therefore in a unique position to understand the integration and control of diverse physiological functions. His holism was a realistic one which started out with analysis and followed through with synthesis:

> Physiologists, finding themselves. . .outside the animal organism which they see as a whole, must take account of the harmony of this whole, even while trying to get inside, so as to understand the mechanism of its every part. . .Physicists and chemists can reject all idea of final causes for the facts that they observe; while physiologists are inclined to acknowledge an harmonious and pre-established unity in an organized body, all of whose partial actions are interdependent and mutually generative. We really must learn, then, that if we break up a living organism by isolating its different parts, it is only for the sake of ease in experimental analysis, and by no means in order to conceive them separately. Indeed when we wish to ascribe to a physiological quality its value and true significance, we must always refer it to this whole, and draw our final conclusion only in relation to its effects in the whole.[8]

He added that vitalists were mistaken in objecting to analysis and experiment in medicine, and that this hampered scientific progress because analysis was necessary to 'grasp hidden conditions of the phenomena, so as to follow them later inside the organism and to interpret their vital role'.

However, Bernard was a dualist who criticised his contemporary Léon Rostan's organicism for the opinion that

> the Creator did not add an additional vital force to the organized being, He has placed in this being along with its organization the molecular disposition appropriate for its development. . .This ability is nothing other than that which results from its structure; it is not a separate property, a super-added quality, it is the assembled machine.[9]

Bernard felt there had to be some other cause: 'One must be a materialist in form and vitalist at heart'.[10] He argued a mechanical experimental approach to the study of organs, while believing that an organising force was imbued in the organism at its moment of creation. 'Physics and chemistry explain only the execution of a physiological phenomenon but not its directing cause, which is vital, as a consequence of the starting point created by development and maintained by nutrition.'[11] Bernard declared himself a physiological determinist, but this had no performationist implication, and he seems to have meant no more than that the organism must obey natural laws, both physicochemical and vital, in its development and life. He agreed with C.C. Wolff that development was epigenetic, and gave as an example of epigenetic holism an experiment on the removal of metatarsal bone from a young rabbit and its placement under the skin of the back. The bone, while continuing to live and partially ossify, was ultimately resorbed. But in the metatarsal space a new bone appeared and persisted. Thus, he concluded that the place of the element in the total plan exerted a strong influence, and that there was another condition that did not relate to the element itself, but related to the morphological plan, to the whole organism.[12]

Bernard was an indifferent evolutionist:

> Whether one is Cuvierist or Darwinist is of little importance; these are two different ways of understanding the history of the past. . .The limits between which morphology is fixed, if they are not absolute (there is nothing absolute in the living animal), are at least very circumscribed. If one tries to make a being deviate from its path, as happens by the creation of artificial varieties, one must constantly maintain it in the new path. Varieties tend ceaselessly to return to their starting point.[13]

He did suggest, however, that changes in environment could cause the appearance of new types, and ruminated on the possibility of 'remaking living things. . .by the repetition of organic phenomena of which nature would retain the memory'.[14] The *Cahier Rouge* noted the possibility of such an experiment on young rabbits, by cutting off their ears for several generations.[15] It was not until 1929 that Sir Joseph Barcroft suggested the link between evolution and physiological holism by pointing out that evolutionary progress was manifested by the degree of sophistication of the homeostatic balance.[16]

Claude Bernard's work was quoted frequently in G.H. Lewes's *Problems of Life and Mind* (1874, 1875). Lewes (1817-78) was the author of a number of popular works on philosophy and science.[17] Through Herbert Spencer he met John Chapman's editorial assistant Marian Evans (George Eliot), with whom he formed a life-long association. His popular science works included *Seaside Studies* (1858) and *Physiology of Common Life* (1859). *Problems of Life and Mind*, published as two volumes, was an attempt to establish a materialistic foundation for life and mind, following the positivistic approach of Comte. The *vital force* or *vitality* of an organism was for Lewes an abstraction analogous to the mechanism of a machine.

> No one tries to reach and modify the mechanical force. . .he only tries to reach and modify the mechanical conditions (which are reals), certain that if he lessen the friction of the parts he will increase the mechanical product. In like manner, no philosophic biologist now tries to reach and modify a vital force, but only to reach and modify those biostatical conditions which, when considering them as causes and condensing them all into a single expression, he calls Vitality or the Vital Forces.[18]

The same was held true of abstractions like sensibility, life, mind, intelligence and thought. Lewes also noted that the organism is unlike the machine in that 'the organism's parts are all identical in fundamental characters, and diverse only in their superadded differentiations: each has its independence, although all co-operate'.[19] While Lewes urged that the synthetic view must be sustained, he admitted that the necessities of investigation would be largely analytical. Bernard's influence was obvious in Lewes' emphasis on the importance of its internal medium to the organism — a medium which was a reservoir of materials transformed from the food of the organism and available for the regeneration of organs — a medium which could fluctuate in nutritive and waste-product concentrations, but which provided the essential internal environment for the organs — a medium which in concert with the organism produced every vital phenomenon.[20] Thought constituted an analogous *psychostatic medium*, which provided man with a unique buffer between external stimulus and automatic response, and assimilated the knowledge and traditions of the larger social organism.[21] The sentient process, although analytically assigned to the molecular changes in the nerve centres, was synthetically the reaction of the whole organism.[22] Lewes agreed with Comte's position that in order to understand the individual it is necessary to understand the collective evolution of the society, and inferred that the attributes of intelligence and consciousness must have evolved as a result of interactions between the individual and society.[23]

Lewes's Law of Correlated Development was thoroughly holistic:

> There is a marked tendency in organic substance to vary under varying excitations, which results in the individualisation of the parts, so that growth is accompanied by a greater or less differentiation of structure. Were this tendency uncontrolled, there would be no organic unity: the organism would then be simply an assemblage of organs. But owing to the solidarity which underlies all differentiation, the parts are not only individualised into tissues and organs, but all connected. Thus each new modification of structure is secured, each organ is independent yet subordinated to the whole; and instead of being an obstacle, this independence becomes, through the consensus and co-operation of independent agents — the Freedom which is subordinated to Law, and the Law which secures freedom.[24]

The holistic outlook of Mivart and Murphy was well considered and realistic, and not simply dropped into their discourses to induce a disarming note of piety. This rhetorical use of 'motherhood' statements as they are called in the modern idiom, is used by many self-styled holists, resulting in the loss of incisiveness from the holistic approach. Herbert Spencer's definition of life as 'the continuous adjustment of internal relations to external relations' had the flavour of Bernard's physiological holism, and he also had a holistic view of evolution, believing that if the parts of a whole organism were brought somehow into an unstable equilibrium the result was greater multiformity of the parts, together with a new stability.

Hans Driesch's pioneering steps in experimental embryology and development physiology were also holistic. *Harmony of constellation* expressed the harmonious development of the parts, and *functional harmony* expressed the unity of organic function.[25] Driesch also talked about *systems*, which were equivalent to the concept of *holons* developed by A. Koestler in *The Ghost in the Machine* (1967). Driesch broke down the complexity of morphgenesis into *elements*, such as elementary processes, formative stimuli, and potency, and then rearranged the elements into *systems*, such as the equipotential system. Then he ranked the systems in a hierarchy resembling a military organisation, and claimed that such a formalisation of the components of living organisms had universal validity. This was an adumbration of the form of holism that has come to be called the *systems approach*, developed and popularised by L. Von Bertalanffy, Paul Weiss and others (see Chapter 12).

In reaction to the anti-intellectual 'pragmatism' and trans-cendentalism of James, Driesch and Bergson, and the over-simplicity of naturalism, a group of American philosophers compiled *The New Realism. Cooperative Studies in Philosophy* (1912). New realism was to be distinct from *naive realism* which conceived of objects 'as directly presented to consciousness and being precisely what they appear', with consciousness taken as analogous to a light which illuminated the world, and which encountered serious paradoxes with dreams, memory and soul.[26] The Realistic Programme of Reform was to consist firstly of the scrupulous use of words and careful definition of terms. Analysis was to be regarded as a valid procedure which did not destroy its object. However, it was equally necessary to take into account the 'combining relations' of the parts, if it was found that the objects of study were lacking something when the products of analysis were put together again. It was also important

> not to confuse analysis and synthesis with the physical operation that often accompanies them. For the purposes of knowledge it is not necessary to put Humpty Dumpty together again, but only to recognize that Humpty Dumpty is not himself unless the pieces are together.[27]

They pointed out that confusion about analysis and synthesis was due partly to a confusion between things and words, that as soon as a word obtained currency (such as 'entelechy' or 'natural selection') it began to pose as a thing in its own right, by reification. This had to be guarded against, 'if words are to be kept in working order, like sign-posts kept up to date with their inscriptions legible and their pointing true'.[28] Next, there had to be regard for logical form, with theories kept consistent and simple. Problems had to be dealt with one at a time with intensive application of investigation, as Francis Bacon recommended in *Novum Organum*. It was 'of more importance in theoretical procedure that two or three should agree, than that all should sympathize'. Implicit agreement and disagreement should be made explicit: 'If agreement is to be based on tradition then tradition, with all its ambiguity, its admixture of irrelevant associations, and its unlawful authority, is made the arbiter of philosophical disputes.'[29] This is a double-edged sword that cuts at both holism in its 'motherhood' form, and at reductionism. The whole programme had as its stated aim the provision of 'assistance and clarification to biology and psychology'.[30] Gifts to biology of this nature are rarely acted upon; which may be just as well: if Darwin had been confronted with the modern critique of his logic and the explanatory validity of his ideas, *Origin* would never have seen the light of day.

Together with a programme of reform the introduction to the new realism provided a nominalistic summary of particularly inimical fallacies, which have not lost their relevance, even if they have been ignored for the last half-century,[31] including:

Pseudo-simplicity (or false parsimony). A good example of simplicity is the creationist appeal: the Bible says God created the world and the plants and animals in six days; how much simpler this is than the random process of biogenesis in primordial ooze over a billion years, followed by more billions of years of complicated evolution, especially since we know that the world is only six thousand years old! The pseudo-simplicity is in 'God did it'. The statement 'It evolved' is just as simple. *How* God or evolution did it is the unknown. Fortunately biologists regard Occam's razor (*entia non multiplicanda praeter necessitatem*), which can be paraphrased as 'the simplest explanation is usually the best one' as a guide and not an absolute.

Exclusive Particularity was the assumption that a particular term of any system belongs to such a system *exclusively*. In the selectionist tradition naturalistic evolution is exclusively gradual, and the product of saltatory evolution would be assumed simply to be a gradualistic phenomenon with an incomplete fossil record.

Definition by initial prediction, the definition of any subject of discourse might be construed in terms of its initial expression. Then, owing to the error of exclusive particularity, assumed to have no other aspect, or belong to no other relational set, 'Thus the initial characterization becomes definitive and final.' So the initial Darwinian characterisation of evolution as *by natural selection* has become definitive and final.

The Speculative Dogma: the assumption that there is, 'an all-sufficient, all-general principle, a single fundamental proposition that adquately determines or explains everything.' The real epistemological curiosity of biology is its possession of only one dominant paradigm, the Darwinian one, unlike the Kuhnian progressions of the other sciences.

The Error of Verbal Suggestion: 'Words which do not possess a clear and unambiguous meaning, but which nevertheless have a rhetorical effect owning to their associations, lend themselves to a specious discourse having no cognitive value in intself and standing in the way of attainment of genuine knowledge.' This is Bacon's 'Idol of the Forum', or what Koestler castigated as 'reification': if a term exists it must represent something real.

Finally there were verbal abuses which made it possible to invent utterly fictitious concepts simply by combining words. Natural selection is a classical example of such portmanteau concepts. It is not fictitious but it does not in its Darwinian interpretation possess clear and unambiguous meaning, and has led to a proliferation of other combination metaphors which contribute to specious discourse; e.g. *selection pressure* and *evolutionary strategy*, whose usefulness rests upon the validity of an already suspect *natural selection*.

The Fallacy of Illicit Importance: the inference that, because a proposition is self-evident or unchallengeable, it is important. Sure, confident action in human affairs is more effective than dithering, but the certainty on which that action is based is not necessarily validated by the results. 'If one person is certain that a distant object is a tree while his companion is equally certain that the same object is an automobile, is it not obvious that certainty is a negligible factor in the problem of deciding what the object really is?' This criticism was directed particularly at the idealists. The new realists explicitly rejected all mystical philosophies and warned against mystical terms which possessed simplicity only because analysis had not been applied to them. A step in the opposite direction brings us to the positivistic extremity: if an experience cannot be analysed then it is unreal. The non-analysability has to be carefully examined. Is it due to 'illicit importance', or the lack of a suitable analytical technique, or the disinclination of the positivist to approach something that would mar the simplicity of his cosmology?

Realistic analysis had to reveal not only parts but also *relations* between the parts, and also to organise the parts into wholes. Analysis had also to discover those characteristics that the whole might have, over and above the characteristics of the parts, i.e. those characteristics that Spaulding called the *specific properties of the whole*.[32] New realism rejected the Bergsonian assumption of a transcendental unity which mediated between the parts, and which was lost when the parts were taken apart.

The major achievement of the new realists with regard to holism was their elucidation of the relationship of part to whole, establishing to their own satisfaction that there was nothing holy about the whole. Parts of organisms could be interfered with or switched surgically without the integrity of the whole being seriously affected. A whole could profitably be taken apart, the parts examined, and meaningful conclusions drawn about the specificity of the whole. They were holists in the sense that they recognised that wholes do have specific qualities contributed to by the parts but not determined by the parts. Their realistic holism contrasts with *transcendental holism* in which the quality of the whole is regarded as an agency which is almost independent of the parts, such as entelechy or *élan vital*. Unfortunately, the pseudo-simplicity of the expression *specific properties of the whole* makes it easy to ignore altogether. The history of biology of the twentieth century is largely a history of analytical biology, and the applications of realistic holism to evolution theory have been few and far between. Once the idea of the specific quality of the whole has been separated from the value of analytical knowledge of the parts, analysis, being the accessible procedure, submerges synthesis. Cartesian dualism allowed the same fate for the mind-soul.

A curious mixture of realism, idealism and holism is found in the writing of J.S. Haldane, the respiratory physiologist and philosopher, whose son J.B.S. Haldane helped to roll the stone from the crypt in which Darwinism rested. Haldane rejected the concept of entelechy as vital force, writing in *Mechanism Life and Personality* (1913),

> There is no evidence at all that each cell, in growing and dividing in the one particular manner which constitutes normal development is not determined by special physical and chemical stimuli peculiar to its position relatively to other cells, and to the external environment. We do not yet know what their stimuli are; but probably no physiologist would doubt that they exist.[33]

None the less, he would have no truck with Weismann's molecular theory of heredity, recoiling from it for the same reasons as Mivart and Spencer, and ultimately rejecting a mechanistic hypothesis of heredity. He would as soon have gone back, 'to the mythology of our Saxon forefathers as to the mechanistic physiology'.[34] He conceded, however, that the mechanistic approach had been practically useful in the development of physiology, and had performed the service of driving out the vital spirits from conceptions of function. His own organicism was equivocal:

> Life manifests itself in two ways — as structure and as activity. But we also recognise — a biologist feels it in his very bones — that this is a *living* structure and *living* activity. Each part of the structure not only bears a more or less definite spatial relation to the other parts, but it is actively maintained in that relation. . .it follows that the metabolic activity of the living body is also organised, every aspect of it bearing a definite relation to every other aspect. . .The relation of the living organism to its environment is no less peculiar and specific than the relationship of the internal parts and activities of the organism itself. . .The living body and its physiological environment form an organic whole, the parts of which cannot be understood in separation from one another.[35]

What a biologist might feel in his bones is not admissible evidence, but Haldane cannot easily be filed and forgotten, since he was as enthusiastic and effective an analyst as Bernard, and fully appreciated the mechanistic tools of physiology. The living organism was Haldane's holistic 'ground conception', and specificity of the whole organism could be approached in terms of *organic unity*, or what we now call homeostasis.

> To leave it out of account in physiology, or treat it as a mere 'heuristic principle' of a very uncertain value, seems to be about as foolish as it would be to reject the idea of mass in chemistry, and retain the phlogiston theory.[36]

Twenty years after the publication of *Mechanism, Life and Personality* Haldane had compromised little with the current advances in biology. The essays collected in *Materialism* (1932), were still anti-mechanistic. The discovery of viruses had done nothing to turn him around to accept a molecular basis for heredity. Instead, he believed that such minute organisms might possibly furnish the link in the extension of biological interpretation into the inorganic world of molecules and atoms. The discovery of quantum phenomena suggested to him the possibility of such an extension, and Planck had contended that 'the conception of wholeness must. . .be introduced into physics, as in biology, to make the orderliness of Nature intelligible and capable of formulation;' suggesting that mathematical physics could be regarded as a backward branch of biology.[37]

The holists did not have it all their own way during this period, since there was a strong mechanistic tradition continuing on from Huxley, Weismann and Roux. Jacques Loeb's *The Organism as a Whole: from the Physico-chemical Viewpoint* (1916) argued that the whole had no special significance, but was the aggregate of physico-chemical reactions and interactions. The book was chiefly a response to Hans Driesch and Henri Bergson, and Claude Bernard and von Uexküll were added to the roster of villains. The integrity of the organism, Loeb contended, was the integrity of the cytoplasm, and the egg was an 'embryo-in-the-rough' with the independent Mendelian characters superimposed upon it.[38] Species specificity was determined by specific, self-replicating hormones and enzymes, a belief built upon many instances of complex biological effects, which were mediated by simple physicochemical stimuli. Loeb believed that it eliminated the need for inherent design such as was proposed by Bernard to account for the harmony of physiological homeostasis. From the modern materialistic viewpoint Loeb was correct in his general conclusion about the physicochemical basis of life, though wrong in his choice of molecular models. But he did not try to explain what the new realists called with disarming simplicity, 'the specific properties of the whole'.

Uexküll came far closer both to the problem of the organisation of biological molecules and to the specificity of the whole, with a proposal that the Mendelian genes were as Loeb paraphrased it, 'the foreman for the different types of work to be done in a building. But there must be something that makes of the work of the single genes a harmonious whole, and for this purpose, he assumes the existence of "super-genes"'.[39] This Loeb dismissed as an exercise in idealistic question-begging. But that is what scientific modelling is: temporary question-begging, with mathematical or empirical testing used to validate the assumptions. Uexküll's 'lucky guess' was not validated until the success of Jacob and Monod's lac-operon work forty-five years later. Even now biologists are only gradually coming to feel more comfortable with the concept of *regulators*, the modern synonym for Uexküll's supergenes, and the technically difficult access route for their elucidation has barely been explored.

J. von Uexküll's work was not widely known until the translation of *Theoretical Biology* in 1926. Condemned by Loeb as a purveyor of Platonic idealism and anti-Darwinist propaganda, he deserves some further mention here as an evolutionist and transcendental holist, not to mention the influence of his *umwelt* concept in the field of ethology.[40] Uexküll argued that speciation could only be the precursor of specialisation, resulting in reduced genetic variability or increased morphological rigidity. The important question of increase in complexity could be solved in terms of the enrichment of the genotype through new genes arising, together with the activity of a supergene which was to regulate the Mendelian genes. But the appearance of new genes was to be by 'conformity with plan', which he saw as a transcendental 'environmental holism' or the 'great universal law'.[41]

The concept of homeostasis is a good example of realistic holism in biology. The word was coined by the American physiologist Walter B. Cannon in 1929, to express the regulated stability of the higher vertebrate internal environment, or what Bernard had called the condition of constant life.[42] The word means by its Greek derivation 'same state' and is usually loosely defined as 'the maintenance of a steady state', but neither Bernard nor Cannon meant to imply that the internal environment was always constant in its physico-chemical parameters. It was understood to be in a condition of equilibrium within viable limits, which the homeostatic or physiological control mechanisms worked to maintain, or as Cannon put it, 'a condition which may vary, but is relatively constant'.[43] Neural and vascular receptors convey information about pressure, temperature, and chemical composition of the internal environment, which consists of the blood and the other body fluids. Changes in those parameters bring neural, muscular and hormonal responses which tend to return the internal environment to its normal condition. Subsequent physiological research has revealed that there are also negative feedback mechanisms which control the responses to changes by stopping them when the appropriate conditions have been restored. These control mechanisms operate at the molecular level, controlling protein synthesis, enzyme activity, and hormone synthesis and release, as well as at the organ level. In a sense the behaviour of the organism can also be described as homeostatic. The concept can be even further stretched, as by some geneticists who regard reproduction as a mechanism of population homeostasis. Genetic homeostasis is also used to describe stable equilibria within gene pools.

Homeostasis is most commonly used in its narrowest sense of the maintenance of the steady state of the mammalian blood system. But for both didactic and theoretical reasons it is a useful concept when applied both to single cells and the physiology of organisms lower on the physiological scale than the mammals, despite the fact that these are all much more at the mercy of their environment and possess what Bernard called *oscillating life*, i.e. their internal environment changes along with changes in the external environment, and they either accommodate to those changes or succumb.[44] However, from the beginning of their history cells have had individual homeostatic mechanisms including the selectively permeable membrane which controls the physicochemical constitution of the cytosol, and which prevents the random dissipation of the cell solutes to the external environment. In animal evolution the emergence of an aqueous buffer between the cell and the external environment — Bernard's *milieu intérieur* — meant that the cell's homeostasis was not overworked. Cellular specialisation could begin because the cells could exchange their special products via the internal milieu without losing them to the external environment.

By the time Cannon's *The Wisdom of the Body* was published in 1932, great advances had been made in endocrinology. However, it is only in the last ten years that a broad understanding of the operation of hormones at the molecular level has been gained, and knowledge of how these and other molecules affect gene expression is still very vague. Gene expression is not only significant for the homeostasis of the mature organism, but at the developmental level is the basis of both normal development and evolutionary change. With the notable exceptions of Waddington's *Strategy of the Genes* (1956), S. Løvtrup's *Epigenetics* (1974) and R. Reidl's *Order in Living Organisms* (1978), remarkably little attention has been paid to the theoretical aspects of this central and fundamental problem. Our ignorance of these matters is partly due to the technical difficulties involved. But there are also epistemological reasons, including the dominance of positivistic reductionism and a relative over-investment in population biology as the most important area for evolutionary study.

Yet another stream of holistic biological thought arose from the confluence of reactions against the elementalism of cell theory, which held that the cell was the fundamental unit of life and that the organism as a whole was of lesser importance: as Virchow said, 'Every animal appears as a sum of vital units, each of which bears in itself the complete characteristics of life.'[44] This has been discussed by T.S. Hall in *Ideas of Life and Matter* (1969), and I will mention only one of the early contributors to it. Edmund Montgomery was particularly excoriating about the popular scientific attitude of the late nineteenth century that regarded human consciousness of personality as 'nothing but the resultant of the activities of myriads of such separate beings, of which your so-called body is a mere aggregate'.[45] If this were true then it necessitated a supernatural force to take care of the harmonious co-operation of the separate cells. Montgomery was quite happy with molecular analysis, though he wondered how molecules could be stable enough to survive from one generation to the next. The interrelation of cellular activities was, however, the most important clue to life, and a theory of organic unity was required to deal with this. Montgomery preferred a 'theory of specification' which signified 'the specification of one single protoplasmic unit into definite areas of disparate stimulation, not association of a number of elementary organisms for the purpose of directing among themselves an hypostatised physiological labour'.[46]

A later proponent of this anti-elementalistic holism was William E. Ritter, a marine biologist with a special interest in unicellular marine organisms. Ritter's holistic review of current physiological advances, set out in the second volume of *The Unity of the Organism* (1919), not only antedated Cannon's *Wisdom of the Body* by thirteen years, but though less analytical, was just as comprehensive, and more philosophically profound. An updated abstract of his 1919 work was published as 'The Organismal Conception' in 1928, with Edna W. Bailey as co-author. They observed that holism had made its greatest impact in the social sciences, crediting Herbert Spencer with this development, noting also that it was being extensively used in current clinical medicine and education theory. But in biology there were problems of specialised applications and strange nomenclature. Ritter preferred *organismal* as an adjective for holism as applied to individual organisms, and subsequent commentators have called him an *organismalist*. *Organismalism* was coined by L.W. Sharp (1926) who expressed it thus:

ontogenesis is a function primarily of the organism as a whole and consists in the growth and progressive internal differentiation of a single protoplasmic individual, this differentiation often, but not always, involving the septation of the living mass into subordinate semi-independent parts, the cells. Since the septation is rarely complete, all parts remain in connection and the whole continues to act as a unit. . .Thus development is not primarily the establishment of an association of multiplying elementary units to form a new whole, but rather the resolution of one whole into newly formed parts; it should be thought of not as a multiplication and co-operation of cells, but rather as a *differentiation of protoplasm*.[47]

This shift from regarding the organism as an expression of the aggregation of cells, to regarding the cells as an expression of the whole organism epitomised organismalism. In Ritter's own words, the 'organism seems as much a causal explanation of the cells as the cells are a causal explanation of the organism'.[48] Ritter judged that the 'obstreperous controversy' between preformationism and epigeneticism had been a major obstacle to the acceptance of organismalism. The view of the cell as an expression of organism and as a living system in its own right, not simply as a manifestation of its physicochemical components, was a holistic refinement of cell theory, which had been undergoing the same fluctuations of emphasis and interpretation as had evolution theory.

Ritter and Bailey named C.S. Sherrington as a major proponent of organismalism for his *The Integrative Action of the Nervous System* (1906). However, J.S. Haldane was given pride of place, his physiological studies were pronouncedly organismal, and his vitalism had been an over-reaction to the epistemological difficulties of elementalism.

Although Ritter was much in sympathy with A.N. Whitehead's rebuke that biologists were 'aping physicists' too much, he drew the line at accepting the generalisation that all kinds of existence, including atoms and molecules, should be called 'organisms' because of distinction in levels of organisation between the inorganic and living forms.[49] On the whole, however, he was impressed by Whitehead's *theory of organic mechanism*, which held that, 'The concrete enduring entities are organisms, so that the plan of the *whole* influences the very characters of the various subordinate organisms which enter into it.'[50] Whitehead elaborated the entities-as-organisms argument thus:

> In the case of an animal the mental states enter into the plan of the total organism and thus modify the plans of the successive subordinate organisms until the smallest organisms, such as electrons, are reached. Thus an electron within a living body is different from an electron outside it, by reason of the plan of the body. The electron blindly runs either within the body or without the body; but it runs within the body in accordance with its character within the body; that is to say in accordance with the general plan of the body, and this plan includes the mental state. But this principle of modification is perfectly general throughout nature and represents no property peculiar to living bodies.[50]

Ritter and Bailey concluded that the organismal concept was essential for a complete theory of knowledge; epistemology was a psychobiological problem. Moreover, the organismal concept in its contribution to the knowledge of reality looked out towards a true synthetic philosophy. Although they made no reference to Bergson there was a distinct flavour of his participatory ontology:

> The mystery we feel concerning existence — if mystery is what one decides to call his feeling about what extends beyond his positive knowledge — is our consciousness of the fact that we respond sensuously to vastly more of nature than we can know through our rational and analytical processes. Limitations on our positive knowledge of the natural order appear to be a necessary consequence of our being only parts and not the whole of that order. In a very literal sense we *know* things immediately only as we live them.
>
> Although we are only exceedingly small parts of the natural order our connection with that order is so close and peculiar that our existence depends on our ability continually to transform parts of nature external to us into our very selves. Since we can become immediate participants in only a small part of the whole of reality, while we can respond sensuously to much larger parts, a good deal of this sensuous response puzzles us (expressing it mildly) as rational beings; and produces in us feelings which we name wonder, mystery, adoration, terror, dread, loathing, depending on the particular conditions, external and internal, existing at particular times.[51]

I began my discussion of holism and biology with the observation that such concepts might arise from a spontaneous feeling of unity, whatever the causes of such a feeling. Ritter and Bailey were brought to conclusions concerning the sense of wonder at our participation in reality. Such conceptions run through the bulk of literature on biological holism, and I am bound to ask if reductionism, as a view of reality, is not simply a refuge from a sense of terror or loathing that to holistic minds is mystery and wonder. As J.B.S. Haldane remarked,

> The fact about science is that everyone who had made a serious contribution to it is aware, or very strongly suspects, that the world is not only queerer than anyone has imagined, but queerer than any one can imagine. This is a most disturbing thought, and one flees from it by stating the exact opposite.[52]

7 EMERGENT EVOLUTION

> The higher quality emerges from the lower level of existence and
> has its roots therein, but it emerges therefrom and it does not
> belong to that lower level, but constitutes its possessor a new
> order of existent with its special laws of behaviour.
>
> S.A. Alexander 1920[1]

Evolutionary emergence signifies the appearance of a novelty which
might affect the behaviour, physiology or anatomy of whole
organisms, with the associated sense that novelty is unpredictable
from a knowledge of the pre-emergent forms. Implicit in
emergentism is the belief that such novelties usually represent
progressive improvements in complexity, or in organisational
integration. Emergence has traditionally been seen as an intrinsically
valuable or self-sufficient phenomenon, and natural selection thus
has a lesser impact. The concept of emergence has sometimes been
equated with saltatory evolution, but obvious leaps need not be its
exclusive gait.

Hegel's philosophy was one of the catalysts of emergentism.
Although *The Phenomenology of Spirit* (1807) predated modern
evolutionism, it portrayed a self-actualising universe which came to
understand itself through the minds of conscious beings. Within the
human species mind evolved through ascending perceptive levels,
and the discontinuity of those levels was inferred from the subjective
ontogenic experiences of those individuals who had attained higher
consciousness.

During the nineteenth century there developed three independent
currents of emergentistic thought. One was realistic and dealt with
the emergence of unpredictable inorganic and biological wholes,
which were believed to be greater than the sums of their parts.
Another was inspired by concepts of psychological and spiritual
emergence, and had strong transcendental and vitalistic
components. R.W. Sellars argued that a naturalistic world view
produced a gradualistic conception of evolution, and that a super-
naturalistic philosophy generated a conception of discontinuity or
emergence in evolution.[2] The two are not however mutually
exclusive, since there are cases of saltation being proposed by

materialists such as T.H. Huxley, and the provenance of emergentistic thought from supernaturalism does not exclude its realistic applications. A third philosophical trend with Hegelian origins, dialectical materialism, occasionally touched on the emergentistic nature of biological evolution as well as social evolution:

> Nature is the test of dialectics, and it has furnished this test with very rich and daily increasing materials, and thus has shown that in the last resort nature works dialectically and not metaphysically, that she does not move in an eternally uniform and perpetually recurring circle, but goes through a genuine historical evolution.

So wrote Frederick Engels in 1878.[3] He gave Darwin the credit for driving out the metaphysicals, but noted that so few biologists had learned to think dialectically there was a conflict between an evolutionism that ought to be revolutionary, and traditional uniformitarian modes of thought, resulting in the endless confusion of theoretical natural science. Engels developed these concepts further in *Dialectics of Nature,* which presented evolution as the dialectical resolution of the conflict between heredity and environment. However, since Engels's notes on this topic were not published until 1925, his influence on early emergentism was insignificant.

John Stuart Mill in *A System of Logic, Ratiocinative and Inductive* (1843) described novelty of combination in chemical terms, believing that mechanical interactions would be entirely predictable if the properties of the interacting bodies were known. Using the term *composition of causes* for wholes that were identical with the sum of their separate parts, Mill pointed out that there were other kinds of wholes, with properties different from those of their parts taken separately, or taken together.

> Not a trace of the properties of hydrogen or of oxygen is observable in those of their compound water. . .We are not, at least in the present state of our knowledge, able to see what result will follow from any new combination until we have tried the specific experiment.

This was still more true of more complex combinations of elements which constituted organised bodies;

in which those extraordinary new uniformities arise which are called the laws of life. All organized bodies are composed of parts similar to those composing inorganic nature, and which have even themselves existed in an inorganic state; but the phenomena of life, which result from the juxtaposition of those parts in a certain manner, bear no analogy to any of the effects which would be produced by the action of the component substances considered as mere physical agents. To whatever degree we might imagine our knowledge of the properties of the several ingredients of living body to be extended and perfected, it is certain that no mere summing up of the separate actions of those elements will ever amount to the action of the living body itself. The tongue, for instance, is, like all other parts of the animal frame, comprised of gelatine, fibrin, and other products of the chemistry of digestion, but from no knowledge of the properties of those substances could we ever predict that it could taste, unless gelatine or fibrin could themselves taste, for no elementary fact can be in the conclusion, which was not in the premises.[4]

Neither could a molecular biologist with a perfect knowledge of the genetic code for all proteins predict the nature of a given organism. The picture is clouded by the fact that nature, or evolution, has already tried most of the experiments, and since we already know all those answers it makes it easy to forget that the powers of prediction of the primary sciences are limited. Mill himself did not postulate any vitalistic hypotheses to account for the combination of organic complexes, feeling that such laws as might govern these combinations must be related to the nature of the combining entities.

The application of these concepts to biological and evolutionary situations was developed by G.H. Lewes, whose essay *Mr Darwin's Hypothesis* supported the gist of Darwinian evolution, with the qualification that it was one thing to accept the development hypothesis, and another to accept natural selection as the last word on the subject. He noted,

We shall have to seek our explanation by enlarging the hypothesis of Natural Selection, subordinating it to the laws of organic combination. . .Similarity in the laws and conditions must produce similarity in organisms independently of relationship, just as similarity in combination will produce identity in chemical species. . .the link which unites all organisms is not always the common bond of heritage, but the uniformity of organic laws acting under uniform conditions.[5]

Lewes began to use the expression *emergence* synonymously with 'laws of organic combination' in the first volume of *Problems of Life and Mind* (1874).

> We do not suppose that when what is called the physical motions of molecules are grouped into what is called chemical actions, and surprisingly novel phenomena emerge, there has been anything essentially superadded to the primitive molecules and their forces. Nor do biologists now suppose that when physical and chemical action are specially grouped and vital phenomena emerge, anything essential has been superadded to the primitive threads of objective existence. The chemical phenomenon is new, the vital phenomenon is new; but the novelty is one of a special grouping of the old material and the old energy. In like manner, when the psychical phenomenon emerges from the vital, and the social phenomenon emerges from the psychical, there is a regrouping, not the introduction of new material, above all not a casting away of the old.[6]

In the second volume of *Problems of Life and Mind* (1875) Lewes used the term *emergent* more formally, as an effect which should be distinguished from *resultant*. It was not always possible to trace the steps of the process of emergence, so as to see in the product the mode of operation of each factor. Lewes made no further reference to the relevance of emergents to evolution.

However, his contemporary the Rev. Henry Drummond (1854-1896), a lecturer in natural science at the Free Church of Scotland College in Glasgow, was developing a dualistic scheme of evolutionary emergentism. Despite his religious circumstances, which might have been expected to be hostile to evolutionism, he was an enthusiastic evolutionist, regarding it as 'the last romance of science', and 'A Vision — which is revolutionizing the world of Nature and of thought, and within living memory has opened up avenues into the past and vistas into the future such as science has never witnessed before.'[7] However, Darwin had done too little to incorporate all the manifestations of human nature into the general scheme. Like Lewes, Drummond felt that 'all prudent men can but hold their judgement in suspense both as to that specific theory of one department of Evolution which is called Darwinism, and as to the factors and causes of Evolution itself.'[8] Like T.H. Huxley he felt that specific theories would prove to be ephemeral:

> This is the age of the evolution of Evolution. All thoughts that the Evolutionist works with, all theories and generalizations, have been themselves evolved and are now being evolved. Even were his theory perfected, its first lesson would be that it was itself a phase of the Evolution of further opinion, no more fixed than a species, no more final than the theory which it displaced. Of all men the Evolutionist, by the very nature of his calling, the mere tools of his craft, his understanding of his hourly shifting place in this always moving and ever more mysterious world, must be humble, tolerant, and undogmatic.

The inadequacy of evolution theory, for Drummond, was its failure to account for the development of moral and social forces, in particular altruism, or what he called 'the struggle for the life of others'.[9]

In *Natural Law in the Spiritual World* (1883) Drummond proposed his 'law of continuity', which held that the universe consisted of a hierarchy of levels of complexity and organisation. Evolution developed in tiers like a pyramid, or in successions of kingdoms,

> at the bottom the inorganic kingdom, which, though dead, furnished the physical basis of life to the next kingdom containing plant and animal and man, the organic kingdom. The organic kingdom was in turn the preparation for, and the prophecy of, the Spiritual.[10]

The moves from the inorganic to the organic to the spiritual might be *per saltum,* but Drummond regarded them as part of a continuum of evolution from the simple to the complex, with new orders of complexity, or emergents, arising at the apparent barriers.

> When we pass from the inorganic to the organic we come upon a new set of laws — but the reason why the lower set do not seem to act in the higher sphere is not that they are annihilated but that they are over-ruled.[11]

The continum of evolution, in Drummond, was analogous to the continuum of mental progress in Hegel. Neither implied nor discounted a gradual continuity; the process might continue without pause, but it was seen to be composed of distinct phases. He was aware of the experiments of his contemporary Robert Tyndall, which, for his generation, finally 'disproved' the possibility of spontaneous generation, but was sure that the Creator used natural laws. Ergo, the appearance of a new order of complexity, life, must have occurred as part of a natural progress, and not as a special creation; similarly with spirituality. Accused of *petitio principii* and worse malfeasances by his theological and scientific contemporaries, Drummond tempered his more exuberant speculations in *The Ascent of Man* (1894). Observations of nature indicated that on the whole evolution had certainly been a rise: 'but whether a rise without leap or break or pause or — what is more unlikely — a cataclysmic ascent by steps abrupt and steep — may possibly never be proved'.[12] He continued to hold ground against the creationists:

> There are reverent minds who ceaselessly scan the fields of Nature and the books of Science in search of gaps — gaps which they will fill up with God. As if God lived in gaps? What view of Nature or of Truth is theirs whose interest in Science is not in what it can explain but in what it cannot, whose quest is ignorance not knowledge, whose daily dread is that the cloud may lift, and who as darkness melts from this field or from that begin to tremble for the place of His abode.[13]

In organismic evolution Drummond agreed with Spencer's stress on the evolution of complexity and adaptability, 'The organism with the most perfect set of correspondences [with the environment], that is, the highest and most complex organism, has an obvious advantage over less complex forms. It can adjust itself more perfectly and frequently.'[14] Drummond studied philosophy, classics, physics, mathematics, botany, zoology and geology at Edinburgh University between 1866 and 1872, and was especially influenced by Mill, Comte and Spencer.[15] His blend of natural science, analogy, allegory and religion was received enthusiastically by a large readership. *Natural Law in the Spiritual World* sold more than one hundred and twenty thousand copies in Britain, and he had a sizeable following in North America. The book's greatest appeal, however, was inspirational, reaching the same kind of audience that he had met as a preacher associated with the Moody and Sankey evangelical mission to Britain (1873-5). Incidentally, the newly founded YMCA enjoyed a great growth spurt as a result of Drummond's ministrations during the mission. His books were chiefly received and reviewed from the religious point of view, and in this respect it may be said that he helped to do for presbyterians and the evangelical nonconformists what Murphy and Mivart did for the Roman Catholics, persuade them that their religion and the new findings of biology were not in conflict.[16]

Interest in the evolution of spirituality was not unique. His contemporaries Madame Helena Petrovna Blavatsky and Friedrich Nietzsche were also affected by Darwinism to make pronouncements on these matters. Some of Nietzsche's thinking paralled Mivart's, although he made his *will to power* work harder than Mivart's *innate tendencies.* Critical of Darwin's teleological emphasis on utility, Nietzsche argued that apparent purposes were merely signs of the expression of will to power. Adapted organs might have to regress or be destroyed during evolutionary progress:

> The magnitude of an 'advance' can even be measured by the mass of things that had to be sacrificed to it; mankind in the mass sacrificed to the prosperity of a single *stronger* species of man — that *would* be an advance.[17]

Nietzsche believed that biology and evolution theory had already been adulterated by a faint-hearted democratic world view, and that his own concept of evolutionary progress by exertion of will to power was closer to reality.

Nietzsche's concept of evolution advancing by simplification, regression and extinction has echoes of the occult evolutionism preached by Madame Blavatsky, who, according to T. Roszak was

> the first person to aggressively argue the case for a transphysical element in evolution against the rising Darwinian consensus. . .Her effort was not to reject Darwin's work but to insist that it had, by its focus on the purely physical, wholly omitted the mental, creative and visionary life of the human race.[18]

H.P. Blavatsky described the cyclic evolution of man, whom she supposed to have gone through a number of progressive stages, each one reaching a peak and degenerating back to a simplified form from which the new cycle might emerge. 'The Occultists believe in an *inherent law* of progressive *development*. Mr Darwin never did, and says so himself.'[19] She gave as examples of pre-human occult evolutionary cycles the ages of the amphibians and reptiles, each group now represented by what she regarded as degenerate forms. The 'Occultists' were supposed to be part of an esoteric continuum going back to pre-Vedic Hinduism. An account of the early Hindu concepts of evolution, divested of Blavatsky's theosophical mystification, is found in R.S. Srivastava's *Sri Aurobindo and The Theories of Evolution* (1968). The emphasis in the Vedas and Upanishads was on spiritual and cosmogonic evolution, and some recent Indian philosophers have inferred a transcendental, emergent biological evolution from their ancient traditions. S.K. Maitra wrote,

> The sweep and range of the principle of emergence are infinitely greater than that of the principle of continuity. Evolution on the lines of continuity is a very tame affair, compared with that based on the principle of emergence. Its dance is a marionette dance not at all comparable to the world-shaking and world-shaping dance of Shiva, which is envisaged by emergent evolution.[20]

Modern Indian emergentists retain the idealistic concepts of evolution found in Bergson, but criticise him for his intrusive materialism and naturalism. These examples of emergentism illustrate how a universal ground-swell of spirituality has always been ready to surge around the metaphysical pilings provided by various philosophers. However, the only emergentist with strong spiritualistic leanings who became generally known was Henri Bergson. His emergentism was implicit in the nisus of the *élan vital*, but he did not formalise it biologically. A 'realistic' emergentism was developed by Samuel Alexander and Conwy Lloyd Morgan.

Samuel Alexander (1859-1938), an Australian who studied mathematics and classics at Oxford, was elected to a Fellowship at Lincoln College, teaching mainly metaphysics and ethics.[21] A period of practical post-graduate study of psychology at Freiburg and of biology at London occupied him from 1889-1891, and during the last two years of his Fellowship he returned to Oxford to give a course on psychology, in addition to philosophy. His first evolutionary ideas appeared in *Moral Order and Progress* (1899), a study in ethics that brought him into contact with C. Lloyd Morgan who reviewed it for *Nature.* In 1893 Alexander was elected to the chair of philosophy at the University of Manchester, where he became an enthusiastic supporter of the Women's Suffrage movement. Alexander also travelled round the Universities, particularly in Scotland, as a propagandist for psychological realism, and this extract from a 1910 Glasgow Herald article summarises his philosophy.

> If we describe faithfully what happens when we perceive, we recognise that the object is. . .entirely distinct from us and that consciousness is only another thing along with it, and, let us add, excited by it. The true philosophical method consisted 'in recognising that you yourself are but a thing among other things' and in accepting the fact that in perception imagery and ideation a physical object is 'revealed'.[22]

Alexander was struggling against a strong current of idealism in British academic philosophy, and it was this realistic view of consciousness that made him want 'to order man and mind to their proper place among finite things'.[23] A biologist might point out that T.H. Huxley had set out to do exactly that in *Man's Place in Nature* in 1862. But Huxley had shown no inclination to devise a treaty of reconciliation between biology and academic philosophy, this had to wait until Bergson, Alexander and their colleagues came along to persuade other philosophers to catch up. The dichotomy between biology and philosophy allowed evolutionistic philosophers to develop an independent and creative view of evolution. At the same time it virtually guaranteed that they would be ignored by most biologists.

Alexander's reputation as the polemicist of realism brought him an invitation to give the Gifford Lectures in 1917 and 1918, a series to which Driesch had been an earlier contributor. Lord Gifford had made the benefaction for 'Promoting, Advancing, Teaching and Diffusing the study of Natural Theology, in the widest sense of the term. . .without reference to or reliance upon any supposed special, exceptional or so-called miraculous revelation'.[24] Despite his conditions that the lecturers did not have to be members of an ecclesiastical denomination, but could be 'so-called sceptics, or agnostics or free-thinkers', he clearly thought that the lecturers would set out with an *a priori* conception of God or the infinite, and that their study of nature would rationally lead to a classification of these metaphysical categories.

Space, Time and Form (1920), which was based on Alexander's Gifford Lectures, was, in the words of his literary executor Laird, 'metaphysical in the grand manner'.[25]

> It builds on. . .space-time and comes to deity, beyond which human conjecture is dumb. It reviews the pervasive. . .features of reality. It describes the evolution of the grander and more novel complexities of movement (for all is movement, it says) when life and mind 'emerge'.[26]

Alexander's philosophical maturation from 1906 took the route of epistemological realism, a theory of knowledge that was commonly called 'the new realism' in England at the time.[27] But Alexander, unlike the American New Realists, was also working towards an ontological theory of existence. His metaphysical categories resembled Spinoza's, and the pervasive complexities of motion in the oceans of reality were to be the primary subjects of philosophy. In comparison with such categories as motion, quantity, causality and substance all empirical things (including minds) were 'whirlpools in the ocean'. One of the smaller whirlpools of Alexander's 'ocean of reality' was his concept of emergent evolution. It was very briefly stated:

> As in the course of Time new complexity of motions comes into existence, a new quality emerges, that is, a new complex possesses as a matter of observed empirical fact a new or emergent quality (such as consciousness). . .The quality and the constellation to which it belongs are at once new and expressive without residue in terms of the processes proper to the level from which they emerge, just as mind is a new quality distinct from life, with its own peculiar methods of behaviour for the reason already made clear that the complex collocation which has mind, though itself vital, is determined by the order of its vital complexity, and is therefore not *merely* vital but *also* vital.[29]

To take a modern example, heredity is not *merely* molecular but *also* molecular. As J.S. Haldane put it: 'We must not mistake measurements of the balance of matter and energy entering and leaving the body, for information as to the manner in which this stream passes through the living tissues'.[30]

Is there, as Alexander suggested, such a 'compulsion of empirical fact' as to convince us of the necessity of a theory of emergences to account for evolution? Alexander provided no details of such evidence; he took it as axiomatic that below a certain level of complexity organisms had no consciousness. Others have taken it as axiomatic that consciousness is additive: no more than a function of the number of neural connections and possibly existing in a 'low-energy' form in inorganic matter. Nevertheless, there are a number of recognised evolutionary events which better fit the emergent mode than the additive mode. The most striking of these are the symbiotic interactions between different organisms which end as integrated individuals, such as the interaction between the mito-chondrial and host prokaryotes on the route to eukaryote evolution, and viral-host DNA interactions that can alter the genomes of both. These are all-or-nothing events which do not proceed by minute steps but happen all at once; i.e. there is a strong prima facie case for the distinction. What still has to be argued is the general relevance of emergence theory as opposed to its specific relevance to a few known or postulated extraordinary events.

Alexander had some difficulty in defining the new emergent quality of life: 'organisation' was insufficient, since atoms and crystals were organised. Self-regulation, the property of plasticity, the power of self-reproduction were the different ways in which the distinctive quality of life exhibited itself, 'but no advantage is gained by substituting the details comprehended under life for the simple quality of life itself'.[3] This is a perennial problem; statements such as 'life is a self-regulating and reproducing organic complex which has negative entropy' have an alluring reductiveness, but in Alexander's view would have been no better than a descriptive definition consisting of a list of physiological functions.

Alexander argued that both organism and machine were similar in that they both acted within the limitations of their structure. The difference lay in the comparative rigidity of the machine and the plasticity of the organisms, but a machine might be invented that had organismic plasticity. The real antithesis was between the vital and the material, and the vital needed explanation in terms of emergence, an explanation not forthcoming from simplistic materialism.[32] This is still within the bounds of 'naturalism', having the sense that things happen by natural causes rather than by the supernatural; but E.G. Spaulding, in *The New Rationalism* (1918) noted that naturalism had been adopted as the form of positivism applied to the natural sciences, and argued that despite its popularity among scientists it was vague and neglectful of certain problems. 'Indeed, very frequently, some of the most important philosophical problems either are ruled out of court with a high hand or are not recognized at all because of sheer ignorance.'[33] This naturalism was defended by R.W. Sellars's *Evolutionary Naturalism* (1922). An international commission for philosophical nomenclature, to assign priority of meaning and provide lists of 'trivial' equivalents would be useful.

This is not to say that emergentism tended to be realistic or naturalistic. Alexander's emergentism, as applied to organism, occupied only a short chapter sandwiched between slices of the metaphysics of space-time and the metaphysics of Deity. For Alexander, time was the *soul* of matter. God could be defined in two ways. The religious definition: that which is worshipped and satisfies religious need, he conceded could be no more than a mental construct. The metaphysical definition was 'that which possesses deity [but] which need not be the object of religious sentiment'.[34] Deity was next in the series of emergents above mind, and subject to the same law as other empirical qualities. Nevertheless, the levels of existence below deity were manifestations of God, who existed as space-time even before deity emerged. Alexander's God therefore resembled Bergson's primordial consciousness. Alexander was obviously influenced by C.L. Morgan, both in the realistic treatment of mind and consciousness, which Morgan began to write about in the 1890s, and in the concept of emergence itself. In turn he prepared the way for the broader and more biological treatment given emergentism by Morgan in his Gifford lectures.

B.D. Brettschneider asks the most cogent question of Alexander's emergence doctrine.

> If the newly emerged quality is a 'higher complex', what is the nature of the rearrangement process? Does this rearrangement involve a selection of elements from the lower orders of finite existence? Or is there an inexorable advance that sweeps along with it all the patterns of organization and qualities of preceding levels of existence? Alexander prefers the second alternative. As new qualities emerge, the old pattern of lower levels of complexity are somehow or other integrated into the new.[35]

Emergence to new levels of complexity and integration must involve several simultaneous changes and how the rearrangement occurs is the most important problem. Brettschneider's questions provoke another thought: emergences do not wait to happen until everything at a given level of existence has been perfectly adapted inside and out. Imperfections may be dragged up with the emergence, protesting in the manner of the infected vermiform appendix, and other anatomical malfunctions. The strength of the emergence may permit the continued existence of these imperfections just as the vigour of a phenotype may cover the adaptive inferiority of several alleles in the genotype. As Arthur Koestler and P. MacLean have argued, the emergence of the mammalian neocortex with its cognitive functions had more than enough intrinsic value to offset the disadvantages of other psychophysiological functions which could at times clash and even override detached objective intelligent analysis.[36] Emergence appears to guarantee neither anatomical perfection nor immediate reproductive success and is therefore difficult to equate with neo-Darwinian fitness. Brettschneider's opinion of emergence is that it followed ontological patterns established by Hegel's dialectic:

> a qualitative unity of component theses and antitheses of lower orders of being and complexity. The synthesis of any dialectical set becomes a thesis of the next higher order. Every thesis, therefore, contains within its dialectical unity the structural components and qualities of preceding levels without residue.[37]

The shift from ontological systems to phylogenic systems was consolidated by C. Lloyd Morgan in his *Emergent Evolution* (1923).

Conway Lloyd Morgan (1852-1936) was a solicitor's son who entered the School of Mines intending to train as a mining engineer.[38] As a post-graduate student he came in contact with T.H. Huxley, who had earlier encouraged his interest in philosophy as illuminated by biology. From 1878 to 1883 he taught in South Africa, and on his return to England was appointed to the chair of zoology and geology at the University of Bristol. He was the first Fellow of the Royal Society to be elected for work in psychology, to which he contributed 'Lloyd Morgan's Canon',

> In no case is animal activity to be interpreted as the outcome of the exercise of a higher psychical faculty, if it can be fairly interpreted as the outcome of his exercise of one which stands lower in the psychological scale.[39]

To judge from the biographical literature Morgan is best remembered as a comparative psychologist.

According to his own account of a discussion with T.H. Huxley of Mivart's *Genesis of Species*, Lloyd Morgan favoured, even as a post-graduate student, the notion of a leap from animal to human consciousness, and he taxed Huxley with his ambivalence concerning the possibility of saltatory evolution.[40] He also noted in retrospect that the term *selective synthesis* used in his *An Introduction to Comparative Psychology* (1895) was equivalent to *emergence*.

> Selective synthesis is illustrated alike in inorganic, organic, and mental evolution. . .there is time and again an apparent breach of continuity. . .not a gap or hiatus in the ascending line of development, but a point of new departure.[41]

He was at that time ambivalent about the concept since he believed that evolutionary novelty must be congruous with the pre-existent internal and external environment. This stumbling block had been overcome by 1915 when he began to argue, with regard to the evolution of the nervous system, that

> There are two factors: (1) constitutive structure, (2) the conditions under which its functioning is called forth. If intrinsic structure and external conditions are both, in any two cases strictly similar, nothing new emerges. But if with like intrinsic structure the conditions are different, or vice versa, something new may emerge. And if genuinely emergent (as contrasted with G.H. Lewes's distinction) it may be unpredictable.[42]

These remarks, coupled with the caution that there was no need to invoke a 'separate interacting agency' such as entelechy, influenced Alexander's views on emergence. Morgan did not take the time to develop fully his concept of emergent evolution until after his retirement from the University of Bristol in 1919, delivering his Gifford Lectures on the subject in 1922.

He set out by confirming that Lewes and J.S. Mill had been the most important contributors to the concept of emergence, with Leibnitz as a possible precursor. He regarded Henri Bergson as a strong supporter of the concept. Morgan was careful to stress that additive characters, and their resultants co-existed with the constitutive characters, as emergents.

> Resultants give quantitative continuity which underlies new con-
> stitutive steps in emergence. And the emergent step, though it
> may seem more or less saltatory, is best regarded as a qualitative
> change of direction or critical turning-point, in the course of
> events. In that sense there is not the discontinuous break of a gap
> or hiatus. It may be said, then, that through resultants there is
> continuity in progress; through emergence there is progress in
> continuity.[43]

Morgan rejected *élan vital* and entelechy, but at the same time noted
that he was looking for 'an ultimate philosophical explanation,
supplementary to scientific interpretation', necessitated by the
acknowledgement of God.[44] The alternative, i.e. acceptance of the
facts as they are found 'in the frankly agnostic attitude proper to
science', was not good enough. While Morgan did not propose
divine intervention he held that emergent evolution was the
expression of divine activity and 'omnipresent and manifested in
every one of the multitudinous entities within the pyramid'.[45] While
I wish to look at emergence in the frankly agnostic attitude it would
be a distortion to omit this element of Morgan's thought.

The questions to be asked at any stage of emergence were

> What is the new kind of relatedness that supervenes? What are the
> new terms and relations? What intrinsic difference is there in the
> entity which reaches this higher level, and what difference is there
> in its extrinsic relatedness to other entities?[46]

The problem of what might bring about the new relatedness was not
even posed by Morgan, and this is the major flaw in his development
of the doctrine. Brettschneider classes Alexander as an idealist mas-
querading as a realist, and following Brettschneider's logic Morgan
would also have to be put in the same pigeon-hole. However their
emergentism has a realistic significance which should not be discounted because of an underlying idealism. The question of whether
or not life, mind and deity are inherent in the universe puts even the
frank agnostic on a sticky wicket. Positivism permits us to ignore it,
but that is no solution. All that we can say is that while a universe
may have the potential to produce life, mind or even higher
emergents, mind is our highest vantage point. Our universe has

produced mind, and realistically, the most important problem is how. Emergentism indicates that qualities of the higher emergents are not present in the lower, whereas gradualism implies a hylozoic continuity, such that mind must be present in fundamental particles. Morgan's preoccupation with the metaphysics of emergence meant that he did not try to trace particular phylogenetic steps of emergence, a regrettable omission since he had the qualifications as a biologist and geologist to have done so. There was little attempt to relate the emergence of cognition, memory and mind to general evolution theory, and what little there was was reflective rather than explicit.

Animal behaviour was taken in a cursory manner as an ontogenic model for the phylogenetic process. While there had to be genetic continuity of the neural apparatus, consciousness could be seen to emerge in ontogeny:

> The cardinal issue is this: Does a few-hour old chick, a new-born infant, or any other sample of primitive mind. . .start with some apprehension of source to which there is initial reference; or is reference to source quite a late product of reflective thought?[47]

Projicience was coined by Sherrington, to signify the consciousness that the sensory signals from distance receptors such as eyes and ears refer to objects which were 'out there'. In Morgan's view projicience was a process of very gradual development that began with the evolutionary emergence of mind or consciousness due to the correlation of the psychical system with the physical system of the brain. Until there was projicience there was no external world.

Alternatively, mind could be an epiphenomenon of a projicience possessed by the most primitive organisms, present at the outset.

> We are once more at a crucial parting of the ways. And I think at bottom it comes to this. One route leads to the view that mind is emergent in the course of evolutionary history. The other path leads to the view that mind is not emergent. It is not an evolutionary stage in the natural history of the psychical correlates of physical events. It enters the world endowed with an original capacity for apprehending that world, with its several categories, through the use of sense-organs and brains, evolved to that end in a manner which it is for biologists to disclose. This apprehension is part of the mind's inherent activity which, with the conduct it

subserved, affords instances of a kind of causality elsewhere not to be found in nature. The two views are. . .irreconcilable.[47]

Morgan's own position was a 'behaviouristic' realism, accompanied with the belief that cognitive projicience arose from both a physiological foundation and a behavioural one, the behavioural precursor being 'behaviour towards *this* or *that* thing'.[48] From this conceptual centre his thought developed in a different direction, that of 'organic selection', whereby behaviour is the precursor of genetic assimilation, to be discussed in Chapter 12. Morgan was not to know that a behaviouristic selectionism would eventually discard emergence of mind as an epiphenomenon of the gradual accumulation of neurones and interneuronal connections.

Despite this contention that God, as *activity*, was the *nisus* through which emergence occurred, Morgan believed that emergence could be brought under the rubric of natural causality; indeed, as the expression of an orderly and progressive development it was one of the laws of nature. Emergences were unpredictable but determinate, the unpredictability relating to our only practical knowledge of the foundations and process of emergence. He agreed with J.S. Mill that 'the cause is the sum total of the conditions, positive and negative, taken together'.[49] Causation could be *immanent*, having its ground within the system, or *transeunt*, coming from without the system. Selectionism later simplified matters by projecting immanent causation on to a transeunt agency, effectively creating a new entelechy, natural selection.

8 HOLISM AND EVOLUTION

> The creation of wholes, ever more highly complex organized wholes, and of wholeness generally as characteristic of existence, is an inherent character of the world. . .And the progressive development of the resulting wholes at all stages — from the most inchoate, imperfect inorganic wholes to the most highly developed and organised — is what we call evolution.
>
> J.C. Smuts 1926[1]

Jan Christian Smuts's *Holism and Evolution* (1926) was the culmination of a period of reflection that commenced in his under-graduate years with a study of the poet Walt Whitman, and an unpublished essay, 'An Inquiry into the Whole', written in 1910.[2] Smuts (1870-1950) was born in South Africa and completed most of his education there, graduating from Victoria College, Stellenbosch in 1896 with honours in science and literature.[3] At Cambridge he studied law, and after his graduation practised briefly as a barrister. Smuts returned to South Africa in 1895 but his law practice in Cape Town did not immediately prosper, and he entered politics as a supporter of Cecil Rhodes. After the Jameson Raid he left the Rhodes party and moved to Johannesburg, where he became State Attorney of the South African Republic in 1898. A year later the Boer War broke out, and Smuts served as a commando general. He participated in the settlement made with Britain in 1902, in the formulation of a new constitution for South Africa in 1906, and in the writing of the Act of Union of 1910, serving during that period as Minister of Education and Minister of Defence. At the outbreak of World War I he shared command with General Botha in the successful South West Africa campaign, and in 1916 he took full command of the British forces in East Africa. After subduing the German effort there he was called to Britain to become a member of the War Cabinet. In 1917 he made a speech in which he defined the 'British Commonwealth of Nations'. In 1918 he proposed the formation of the Air Ministry, and was chairman of the committee which drew up the appropriate Bill as a result of which the Royal Air Force came into being. Also in 1918 he published a protocol for the formation of the League of Nations, which had been proposed by

Woodrow Wilson. Wilson adopted Smuts's detailed proposals, and Smuts was elected to the Commission of the League of Nations. He also participated in the Treaty of Versailles, and in 1919, in discussions with De Valera and other rebel leaders concerning the future of Ireland, he arbitrated an armistice agreement. In 1919 Smuts became Prime Minister of South Africa, following the death of Botha. The violent Rand strike by the white Mine Workers Union led to a Nationalist-Labour coalition against Smuts, and his party was defeated in 1924. It was over the next two years in the political wilderness that he finally found the time to work on *Holism and Evolution*.

While he named Plato as the progenitor of his concept of holism, Smuts regarded Bergson's *Creative Evolution* as the best contemporary philosophical treatment of evolution theory. Like J.S Haldane he felt that in the act of giving birth to life or mind, matter showed itself in an entirely unsuspected character, so that biological concepts must intrude into the physical order. Nineteenth-century materialism was mistaken in trying to apply the rigour of physics to the indefinite vagueness of life, and as a result concepts were narrowed down to their most luminous points, and the rest of their implications treated as non-existent:

> the most outstanding feature in the first situation was isolated and abstracted and treated as the cause of the most outstanding feature of the next situation, which was called the effect. . .This logical precision immediately had the effect of making it impossible to understand how the one passed into the other in actual causation.[4]

This view of causality had been expressed by G.H. Lewes in *Problems of Life and Mind*, and it finally found a niche in ecology. 'Organophosphates kill insects' is an illustration of such simple-minded causal thinking, reinforced by the rigorous simplicity of economics. 'Whole situations' were ignored, resulting in the alienation of farmland and the death or debilitation of other organisms. Smuts blamed this kind of simplistic causality for what would now be called depersonalisation, or existential gloom: 'The world becomes to us a mere collection of *disjecta membra*, drained of all union or mutual relations, dead, barren, inactive. unintelligible.'[5] This left the door open to spirits and vital forces conjured up to replace the real continuities of which the mind had been deprived.

Smuts proposed to replace simplistic causality with a kind of biological theory of relativity, by which each whole organism would be perceived as existing in a spatio-temporal field of causality. Structure and function were inadequate to fully understand the organism.

> The organism much more than the physical body is an historic event, a focus of happening, a gateway through which the infinite stream of change flows ceaselessly. The sensible organism is only a point, a sort of transit station which stands for an infinite past of development for the history and experience of untold ancestors, and in a vague indefinite way for the future which will include an indefinite number of descendants. . .From that centre radiates a field of ever-decreasing intensity of structure or force, which represents what has endured of the past, and what is vaguely anticipated in the future. The organism and the field is one continuous structure which, beginning with an articulated sensible central area gradually shades off into indefiniteness.[6]

In this description of the organismic field Smuts drifts close to absolute holism: the concept of the universe as a unified transcendent whole that is expressed in its simplest form by the metaphysical system of taoism. It is anti-intellectual in the sense that contemplation of the acausality of the cosmos leaves no room for the abstraction of simple causes and effects and the elucidation of *whys*.

It is not clear how Smuts intended his perception of the causal field of an organism to be used by biologists, except as a warning against over-simplification. Like the new realists Smuts applauded the necessity for and successes of analysis, but he warned against taking analytic abstraction as the only reality. Whole-making was an inherent characteristic of reality, and evolution proceeded by the progressive development of more organised wholes. Although Smuts explicitly rejected vitalism and focused his attention on 'small natural centres of wholeness in Nature and in thought', rather than on absolute wholes, he ultimately adduced transcendental 'tendencies'. The progressive evolution of wholes began with the inorganic and went through five major emergences. Inorganic wholes, consisting of atomic particles synthesised into chemical compounds, emerged as living wholes in which there was active cooperation between the parts. From primitive life emerged a more distinctly self-regulated form with coordinated centres such as nervous systems; then conscious wholes; then social wholes; then 'ideal wholes, or holistic Ideals, or absolute Values, disengaged and set free from human personality, and operating as creative factors on their own account in the upbuilding of a spiritual world.'[7] This goes far beyond the simple realistic statement that the essence of the whole was a reciprocal influence; which, while it was formed of parts it influenced the parts and affected their relations and functions.

Because of his distaste for the gene theory Smuts used the term 'elementary units', suggesting that they were transmutable in a way analagous to the decay of radioactive elements, and that their mutation was thus by quantum jumps. De Vries's mutation theory and the neo-Mendelian view of genetics were among the foundations of Smuts's holistic evolution, and as his criticism of T.H. Morgan's generalisations on the gene theory appeared as a footnote it is unlikely that he was familiar with the development of genetics in the United States, despite the publication of a review by Morgan and others in 1915. Although Smuts was also unfamiliar with the development of C. Lloyd Morgan's emergentism, his insistence on the importance of the 'emergent new' was very similar. The important question was how the effect came to contain the new and to transcend its cause.[8] As an example of holistic novelty he cited J.S. Haldane's work on how the respiratory system can adjust to

> abnormal condition, to situations artificially brought about, which it has probably never had to face in all time. No 'experience' or hereditary 'memory' can guide it here; and yet it rises to the occasion every time within a wonderfully wide range of adaptability and plasticity.[9]

Similarly, the immune system is able to respond to circumstances that the organism has never encountered in its phylogenetic existence, and other explanations than historical selection of adaptively useful responses have to be adduced.

A relatively recent advance in cellular biology provides us with an illustration of holism which Smuts would have appreciated. Pondering the paradox of how the organism could concentrate and organise energy in a universe dominated by the Second Law of Thermodynamics, which demands that energy must become dissipated and entropy increased, he suggested that a biological equivalent of Maxwell's Demon could somehow sort the energy of the body.[10] The mitochondria perform this role by converting the energy from food and storage compounds into ATP, which provides the energy for exothermic biological reactions. The mitochondria as energy converters, as independent, holistic organelles which at one time in the evolution of life seem to have lived as independent organisms, would have fascinated Smuts. It is true that they are empirically regarded as microscopic machines by the biochemists who have elucidated much of the mystery of proton transfer, which is the physicochemical basis of ATP production. But this transfer could not occur in a simple physicochemical environment except as a rapid heat-effusion or explosion. The controlled, step-wise production of biological energy can only occur in the presence of a selectively controlled intramitochondrial environment, within a controlled cytosol, in the presence of a coordinated sequence of enzymes and co-enzymes whose synthesis is also regulated, and in the framework of a complex anatomy of mitochondrial membranes. The elucidation of mitochondrial mechanisms and the understanding of the overall function and 'field of causality' is an outstanding example of the marriage of mechanistic and holistic thinking that Haldane and Smuts advocated.

Smuts argued that Darwinism was inadequate to the coordination or correlation of anatomical and physiological responses to individual varations:

> And here it is where Natural Selection breaks down completely. The whole body is a system of co-ordinated structures and functions, and its origin and development can only be represented as a complex movement forward in time of a mass of associated variations which have resulted in the most marvellous co-adaptation of structures and co-ordinated functions.[11]

The error lay in the study of isolated characters as if they were mere mechanical components of the whole. The whole was not a mechanical aggregate indifferent to and without influence on its parts, but an active factor in controlling and shaping the functions of the parts: 'The individual and its parts are reciprocally means and end to one another in the moving dynamic equilibrium which is called life.'[12] By holistic interpretation the 'selfish genes' would be obliged to subordinate their egos to that of the whole organism. *Holistic selection* was different from natural selection, having elements of organic selection and genetic assimilation:

Sometimes. . .the organism has long before the appearance of the variation begun to move in its direction. The functioning of the organism has anticipated its future structure. It has for many generations devoted a part of itself to a particular use; the part has in consequence undergone modification. . .When finally in the course of time this modification is superseded by and merged into an organic variation, it is in direct harmony with the needs and the practice of the organism as a whole.[13]

Anticipating L.L. Whyte's 'internal selection', Smuts observed,

The Holistic Selection which acts within each organism in respect of its parts *inter se* is essentially different from the Natural Selection which operates between different organisms which is more appropriately called the struggle for existence. Holistic Selection is much more subtle in its activity; it puts the inner resources of the organism behind the promising variation, however weak and feeble it may be in comparison with other characters. . .In the organism the battle is not always to the strong, nor is the struggle an unregulated scrimmage in which the most virile survive. The whole is all the time on the scene as an active friendly arbiter and regulator, and its favours go to those variations which are along the road of its own development, efficiency and perfection.[13]

Smuts was a little more specific about the whole as a regulator, noting that it repressed certain characters while allowing fuller expression of others; the foetalisation of humans was an example. He did not, however, follow the argument through to deal with changes in the order of complexity and coordination — the emergences which he mentioned earlier; nor to ask the question: what manner of variation is it that raises the whole to a new level of organisation?

Smuts viewed the cosmos as an ordered whole which could allow the individuation of lesser wholes in the form of organisms. Mind was a 'rebel whole' which emerged from mechanical precursors and began to re-order its own existence and future, transcending the organic limitations of heredity and inheriting its acquired wisdom through education and tradition. Mind was the most important constituent of personality, which Smuts regarded as the highest evolutionary emergent; and personality was a creative novelty in each individual, which psychology failed to take properly into account. Smuts proposed a holistic *science of personology* which would encompass psychoanalysis, clinical psychological analysis and endocrinology with appropriate doses of epistemology, anthropology and theology. This neo-*Novum Organum* had a fussy Baconian comprehensiveness and egocentricity, revealed when Smuts proposed that the lives of carefully chosen great men should provide important data. His son's biography observed that Smuts had used Whitman and Goethe as the models from which he planned personology, and that he was also an admirer of Christ as a 'very remarkably gifted young man'.[14] In other regards, however, *Holism and Evolution* was the antithesis of the *Novum Organum*. Smuts constantly advocated the comprehension of wholes, with analysis as a means to that end. Bacon proposed that for many investigators analysis should be an end in itself. This Smuts regarded as a degenerate condition, a depersonalisation and enslavement of personality. There is little doubt whose organon prevails, though holism has made some progress.

There was a strong moralistic current in *Holism and Evolution* which was reminiscent of William Blake's learning by excess, and Faustian character-testing intermingled with a quest for purity and freedom. The chapter on 'Functions and Ideals' reads like an intellectual version of Baden-Powell's *Scouting for Boys* (1908), which is not surprising considering the socio-political ethos and the training ground of the Boer War, which Baden-Powell and Smuts had shared. This quotation should have a familiar ring to the present generation preoccupied with personal development or self-actualisation:

It would be a mistake to look upon the ideal of personal holistic self-realisation as merely egoistic. No doubt in some cases the subjective selfish features may predominate; but earnest men will always find that to gain their life they must lose it, that not in self but in the whole (including the self) lies the upward road to the sunlit summits. We mostly move in the channels worn by social usage or convention and are influenced by personal, social impuses such as ambition, patriotism, love or money or power. But Holism is deeper than any of these. The inner call of Holism is to none of these things in themselves and for their own sake, but to its own victory in the personal life; to unity; freedom and free plastic power for the Personality.[15]

Organic evolution was only a mile post in this journey. However, the messianic flavour of holism as a quasi-religious force was distasteful to Smuts's materialistic contemporaries, and detracted from its value as an inductive guide.

Holism and Evolution, which was written over eight months beginning in 1924, ran to three editions, with some revision in the second edition and a translation into German. Smuts never embarked on the hoped-for rewritten fourth edition as he remained active politically for the rest of his life. Smuts was a scientific generalist, with interest in physics, geology, botany and ecology. As President of the British Association for the Advancement of Science when it met in Cape Town in 1925 he enthusiastically endorsed the Wegener hypothesis of continental drift, documenting his case with notes on the distribution of the diamond fields in South Africa and India. He also proposed that the temperate flora of the South West Cape Province was not due to a migration from Europe, but evolved there when it was joined to the land mass which later separated off to become the Antarctic continent.[16] This theory was later developed by Leon Croizat as the theory of panbiogeography, which in dealing with the dispersal tracks of related plants and animals blissfully disregards the modern maps of the globe.[17] Smuts's interest in anthropology led him to a simplistic view of human evolution in Africa; bushmen and Bantu he regarded as lower grades of human being. These ideas were translated into a paternalistic segregation proposal, benevolent in the sense that it was designed to allow independent and natural evolution of the biology and culture of the Bantu. The segregation was also to ensure 'public health, racial purity and public good order'.[18] As a practical scientist Smuts's chief interest was botany, especially plant taxonomy and biotic community structure, which put his philosophical holism to practical use. 'What am I without ecology and holism, and ecology and holism are virtually spiritually synonymous?'[19]

In 1931 Smuts presided over the Centenary Meeting of the British Association in London. In his main address, 'The Scientific World Picture of Today', he promulgated holism as an overview which brought physics and biology together, comparing quantum mechanics with the emergents of evolution.[20] Although he had not read Lloyd Morgan's *Emergent Evolution* before *Holism and Evolution* was complete, he added a note in manuscript to affirm general agreement with Lloyd Morgan. Holism also entered explicitly into Smuts's political philosophy.[21] He had seen in the punitive Treaty of Versailles an imbalance which would later affect political wholesomeness, and in 1934 he identified this effect in the rise of fascism.[22] The League of Nations and the British Commonwealth he saw as the outcome of social evolution to larger, better integrated wholes. After World War II, during which he directed the South African war effort with the rank of field marshal, he threw his energy into the founding of the United Nations, writing the preamble to the Charter which was passed at the San Francisco Conference in 1944.

When *Emergent Evolution* and *Holism and Evolution* were published Darwinism had been in eclipse for three decades, and both its gradualistic aspect and the sufficiency of its mechanism had frequently been criticised. However, although emergentism and evolutionary holism suggested an epistemological alternative to Darwinism, they offered no specific mechanisms, nor even a realistic focus for induction. Their transcendental implications were distasteful to most empirical biologists; the mechanistic alternatives provided by mutationism and neo-Lamarckism attracted most attention. A realistic holism had been effectively applied to physiology by Claude Bernard, and later Sherrington and J.S. Haldane. Its central concept was called *homeostasis*, propounded by Walter B. Cannon with the support of Sir Joseph Barcroft and F.H. Starling, who first aphorised it as 'the wisdom of the body'.[23] These experimental physiologists, like Bernard and Haldane, had medical backgrounds, and tended to see comparative physiology as the study of those aspects of non-human physiology that might have some bearing on the human condition, not as a vehicle for phylogenetic comparisons. This attitude is understandable, although it occasionally gets in the way of a proper comprehension of the evolution of function. The vague concept of physiological evolution as a form of self-adaptation, as Henslow had called it, was the only generalisation available, apart from Smuts's holistic internal selection.[24] There was no serious effort to resolve this problem, although a number of biologists including H.F. Osborn, C.M. Child and E.S. Russell recognised its existence. Cannon noted that the gradual evolutionary acquisition of homeostasis

is interestingly paralleled in the development of the individual. Indeed, a suggestive addition to the group of facts which support the ideas. . .that ontogeny recapitulates phylogeny, is found in the absence or deficiency of homeostatic regulation in babies during a considerable period after birth.[25]

This resulted in temperature and blood sugar fluctuations and regulation of these parameters was 'only gradually developed, perhaps as a consequence of exercise and training'.[26] This was an over-simplification, since the functional anatomy of the homeostatic system is genetically determined and is all ready in the new-born child to be habituated, or switched on. The remark does, however, return our attention to how the different stages in homeostasis might have emerged.

Cannon agreed with Barcroft that the freedom of the homeostatic organism was essentially a freedom for 'the activity of the higher levels of the nervous system and the muscles which they govern. By means of the cerebral cortex we have all our intelligent relations to the world about us.'[27] 'The full development and ample expression of the living organism' would be impossible in the absence of such freedom. It is not worth trying to understand how this freedom evolved instead of treating it summarily as an epiphenomenon of natural selection?

Ritter and Bailey (1928) believed that emergent evolution and the organismal conception were the same thing looked at from different directions, emergent evolution addressing the origin and development of living beings and the organismal conception focusing on their morphology and physiology. They particularly recognised C.L. Morgan's role in the development of emergentism in relation to psychology, but regarded behaviourism as fundamentally elementalistic, and gestalt theory as 'an aborted organismal theory, the abortion appearing to be due to the lack of an adequate biological foundation, especially in the realm of metabolism.'[28]

A brief survey of emergence doctrine was written by William Morton Wheeler (1865-1937), an entomologist who became interested in social insects while Head of the Department of Zoology at the University of Texas.[29] Subsequently he became Dean of the Faculty at the Bussey Institute of Harvard. While his earlier inclinations were towards the mechanistic ideas of Jacques Loeb, he disliked the Weismann gene theory and was sympathetic to Bergson's concept of creative evolution, including its Lamarckist implications. Wheeler also agreed with Anton Dohrn's assertion that behaviour was an important influence on subsequent evolution, and his chief criticism of the developing disciplines of experimental embryology and genetics was that these did not take behaviour and environmental circumstances into account. Emergence was 'a *novelty of behaviour*, arising from the specific interaction or organization of a number of elements whether inorganic, organic or mental which thereby constitute a whole, as distinguished from their mere sum, or "resultant"'.[30] Emergence could also involve simplification, as in the cases of neotenous and parasitic organisms. Hence, the transcendental implications of Bergson's, Alexander's, and Morgan's emergentism were unacceptable: 'To the observer who contemplates the profuse and unabated emergence of idiots, morons, lunatics, criminals and parasites in our midst, Alexander's prospect of the emergence of deity is about as imminent as the Greek Kalends.'[31] Nor did the grand categories of life, consciousness and mind seem to him the products of single emergencies:

> Emergence must be more ambulatory, or at any rate less saltatory. If the naturalist is to accept both genetic continuity and novelty in evolution, the viable novelty at each emergence must be very small indeed. This is attested both by the extraordinary slowness of phylogeny and the very subtle transition in even the most rapid ontogenies. Even metamorphosis in organisms is only superficially saltatory. Novelties such as life and mind, concerned in a wholesale fashion, are of such magnitude that we can regard them only as representing the final accumulative stages of a very long series of minimal emergences.[32]

From this statement it would almost seem as if Wheeler regarded gradualism in evolution as axiomatic, but he believed that the concept was particularly valuable in the assessment of associations such as parasitism, symbiosis, predatism and biocoenoses, as well as insect societies and human societies. Although vital forces and other transcendental aspects of the emergent were supposed to be directive and foresightful, 'the whole multimillenial course of evolution with its innumerable impasses and culs-de-sac, its abject and tragic failures, would seem rather to be one vast monument to their colossal and hesitating inadequacy, blindness and stupidity'.[33] This assessment of the biological past set the tone for Wheeler's pessimistic view of the future of human society.

In his essay 'Present Tendencies in Biological Theory' (1929) Wheeler noted that biological studies occupied two epistemological areas: the ideographic (historicism) and the nomothetic (naturalism). The former was exemplified by the historian whose research was narrative and qualitative, focusing on the irrational behaviour of individuals, intuitive, and closed to experimentation. The nomothetic, characterised by the physicist, was empirical, quantitative, exact and positive. Evolution theory was 'obviously. . .a creation of the ideographers' though borrowed by the nomothetists.[34]

> It is only where a true philosopher like Hans Driesch enters the biological pasture that the ideographic sheep and the nomothetic goats are made to realize the full iniquity of their contingencies. He informs the sheep that they are nothing but hairy mechanists, and that he can make them all lie down and ruminate together if they will only permit him to bring in some of his queer creatures, the entelechies and psychoids, from the metaphysical barnyard to act as go-betweens.[35]

On the other hand, 'while the nomothetes among the biologists were prostrating themselves before Mechanism, some of the more bolshevistic physicists very stealthily carried it off and dropped it into the sea'.[36] He also noted that even physicists retained a longing for mechanicism which had all the tenacity of original sin. Wheeler felt that a realistic theory of holism, which he equated with emergentism and organicism, might provide a synthesis between ideographic and nomothetic interpretations of life. Because emergence was a creative process, in the sense of producing novelty, the causes that biologists, psychologists and sociologists were concerned with differed from those of physicists and chemists, though the physicochemical laws were not annulled in organismic emergents. While the psychical shock or feeling or surprise experienced in the face of the emergent justified the ideographic approach Wheeler warned that 'regarding things as wholes, no matter how much aesthetic satisfaction or mental repose one may derive from their contemplation, is not scientific explanation', and the nomothetic analytical approach was also necessary.[37] Wheeler was one of the earliest biologists to emerge from the simplistic vitalism v. mechanicism controversy with a realistic sense of the need for analysis and a holistic appreciation of synthesis. It is tempting to take his writings as the epitome of enlightened biological opinion of the time. However, turning to R.S. Lull's *Organic Evolution* (1929) as an indicator of the times, as recommended by G.G. Simpson, one finds no references to holism, emergentism, organicism, entelechy, homeostasis, creative evolution, or any of the authors of these concepts.[38] There are extensive references to orthogenesis and Lamarckism, as well as the expression of doubt that natural selection was the whole story. Lull was a better indicator than Wheeler of the prevailing opinions and interests of professional biologists and students. Only theoretical considerations pertaining to mechanisms were relevant. As Wheeler remarked, holistic concepts were *not scientific explanation*, and could be disregarded entirely. Darwinian, Lamarckian and Eimerian (orthogenetic) concepts were scientific explanation, or at least seemed so at the time. Once something has been filed, pigeon-holed or philosophically broken up the categories of non-scientific explanation and scientific explanation, there is the danger that the category most easily grasped is the one dealt with first, sometimes to the exclusion of the more elusive category. Bernard was warning generations of French medical students of the same problem half a century before: you must take it apart and put it together again; taking it apart is easy; the re-synthesis is the problem. Francis Bacon, three hundred years before Bernard, had legitimised this specialised neglect of intellectual responsibility by proposing that different areas of scientific investigations were to be conducted by different individuals. There were to be problem-finders, collectors of data, inductive philosophers, experimenters, seers-of-the-light and so on.[39] Science was a task that could be carried out by inferior intellects. When Wheeler said that holism was not scientific explanation, he neglected to stress that biology without synthesis or the understanding of the specific properties of wholes was incomplete, and the positivistic analytical approach has since become more and more productive of a jejune reductionism in biology. Wheeler was a great propagandist for the nascent science of ecology, a discipline which cannot proceed without holistic understanding. But when I put it to modern students of ecology that their subject is philosophically different from other branches of biology they become uncomfortable, and are eager to assert that in time the complexities of the biosphere will be as reducible to mathematical abstraction as the bond angles of DNA.

Herbert Spencer Jennings (1868-1947) regarded emergentism as a haven from the storm of selectionism. On the basis of pure-line breeding experiments on *Paramecium* Jennings had concluded that while natural selection was capable of sorting out successful pure lines in a heterogeneous population, it had nothing more to do once the pure lines were established.[40] Size and form variations he thought he had proved to be environmental effects. Mutation thus had to be the source of variation in pure lines, either large mutations or small ones, but he thought that in general mutation was a meagre source of variation.[41] Under criticism from the biometricians Jennings conducted further experiments with *Difflugia*, and began to put more weight on additional variations supplied by genetic recombination.[42] By 1917 Jennings had pragmatically committed himself to Darwinism, concluding that the best interpretation of the experimental evidence was that minute heritable variations, so minute as to represent practically continuous gradations, were well demonstrated.[43] Jennings still believed that large steps were possible; orthogenesis was still possible; though not proved. However, he had a hankering for broader, epistemological interpretations: 'the different ways of conceiving the evolutionary process have diverse bearings upon one's attitude to the world; upon the temperament and outlook of the student of science; upon the course that science takes'.[44] Emergence doctrine, by rejecting simplistic mechanisms, would have an ameliorating, moderating, mitigating and uplifting influence.

> It holds that the conception of the universe as nothing but a set of one or a few kinds of particles moving according to a few immutable laws exemplified at any time and anywhere that particles occur, is pitiful in its inadequacy. The notion of computing the entire further course of evolution from the situation at a given moment it considers one of those raw and naively incompetent ideas to which at early and unsophisticated periods of culture man is prone.[45]

Jennings's definition of emergence conformed broadly with that of Alexander and Lloyd Morgan, and he set out to answer the questions 'What *of* it? What difference does it make?' First, he reassured his audience that according to emergentism observation and experiment were the primary and final methods of science. Instead of making a few observations and then sitting back to apply them theoretically in accordance with the laws of physics and philosophy, observation and experiment were needed to winkle out the secrets and surprises of emergent novelties. The emergence doctrine was 'The Declaration of Independence for biological science', independence, that is, from the tyranny of reduction to physics and chemistry. Since man could be regarded as an emergent type there was no *a priori* ground for sneering at the notion that man in some respects acted differently from other animals. Biologists were thus relieved of the responsibility

> of speaking oracularly on the problems of human life. . .of explaining to man what is wrong with him and what he must do to correct the evil situation that he gets into. . .if emergent evolution is correct doctrine, the proper study of man is man.

But the student had also to be

> an experimenter. . .an economist, a politician, a historian, as well as a physiologist. . .If it is indeed in social organization that we find emergent evolution most manifestly at work; if it *is* here that that which is new in principle most frequently and conspicuously appears, then we shall be cautious in accepting the advice of even the king of the termites on our own social problems; we shall use discretion and take his advice at most as suggestion toward experimentation. For any organism of society separated from others by steps in emergent evolution, the only possible method for progress is by trial and error.[46]

If, instead of being submerged by neo-Darwinism, emergentism had survived only to act as a restraint on the excesses of behaviourism and sociobiology, it would have served a useful purpose. Curiously enough, E.O. Wilson, observing that while the 'burning subject' of emergentism was 'temporarily eclipsed by the triumphant reductionism of molecular biology' it has re-emerged as quantitative holism, and is included as one of the elementary concepts on which he bases his *Sociobiology* (1975).[47] This has not tempered his frequent consultation with the king of the termites.

Jennings agreed with Ritter that emergentism emphasised also the uniqueness of the individual, not just that of types. This inspired in him an optimism in the face of mechanistic fatalism which had eradicated ideas, ideals, purposes and beliefs.

> Mingle this perfect doctrine of mechanism as has been done, with equal parts of natural selection, and you get a potion, a cocktail, with a kick that is warranted to knock out ethics and civilization. Warfare and destruction have been the means of advance; the laws of nature are immutable, this then must continue. . .gentleness, pity, humility, and the rest of the 'slave virtues' are mere weaknesses deserving of destruction. . .The tree that bears all these handsome fruits has its roots in the determinism of events, as conceived by mechanism.[48]

Arthur O. Lovejoy (1873-1962), one of the major figures in the area of epistemology known as the history of ideas, presented a less romantic assessment of emergentism in 1927.[49] He was concerned that 'emergence' was used with various meanings, and that since it indicated important philosophical issues it required clear and methodical formulation. He wrote,

> emergence may be taken loosely to signify any augmentative or transmutative event, any process in which there appear effects that, in some one or more of several ways yet to be specified, fail to conform to the maxim that 'there cannot be in the consequent anything more than, or different in nature from that which is in the antecendent'.[50]

Lovejoy argued that some objections to emergentism arose from the Scholastic principle; *e nihilo nihil fit*: (nothing can be made out nothing); 'the feeling that there would be something queer about a universe in which substantive increments popped into existence.'[51] The scholastic principle, not emergentism was at fault. Lovejoy noted that emergence doctrine could be applied non-deterministically in the way that Driesch's 'undetermined evolution' worked, i.e. sometimes it might happen under a given set of conditions, and sometimes not in the same set of conditions; but he preferred the alternative that it was determinate, with a rational explicability, as Jennings had suggested. Emergentism was in part inspired by the same motives as the idealist reaction to Darwinistic mechanism, especially as propounded by Spencer.[52]

> The emergence doctrine is a revolt of temporalistic and usually realistic philosophers against the same features of the older evolutionism. The union of the conception of evolution with the conception of reality as a complex of which all the parts are theoretically deducible from a very small number of relatively simple laws of the redistribution of a quantitatively invariable sum of matter and energy — this union the 'emergent evolutionist' now declares to be a *mésalliance*, of which the progeny are hybrid monsters incapable of survival.[53]

But the doctrine of emergent evolution was manifestly important if true; and there were, in Lovejoy's opinion, good reasons for believing it true.

In an attempt to formalise emergence doctrine Lovejoy wrote that emergent evolution could be said to have occurred if

> upon comparison of the present phase. . .of earth history. . . with any prior phase, there can be shown to be present. . .any one or more of the five following features lacking [in the earlier phase]:
>
> 1. Instances of some general type of event admittedly common to both phases. . .of which instances the mode of occurrence could not be described in terms of, nor predicted from the laws which would have been sufficient [for the earlier phase].
> 2. New qualities and especially classes of qualities. . .attachable as adjectives to entities already present [in the earlier phase].
> 3. Particular existents not possessing all the essential attributes of those [in the earlier phase] and having distinctive types of attributes of their own.
> 4. Some types of event — irreducibly different in kind.
> 5. A greater quantity or number of instances not explicable by transfer from outside the system, of entity common to both phases.[54]

This scheme suggests the possibility of identifying a graded series of biological changes in terms of their impact upon evolutionary progress. It rounds off an epistemological gyre that returns us to the search for the unknown factors of variation. Once the most influential changes have been identified, the characteristics that give them their evolutionary impact can be examined in detail. Instead of casting about for significant alterations in 'selection pressure' that 'cause' evolution we are brought a significant step closer to the outstanding problems of evolutionism by Lovejoy's approach. This will provide a theme for my concluding chapter. Some final words are due to C.L. Morgan, since his *The Emergence of Novelty* appeared in 1933, later enough for him to take acount of the advances in psychology, philosophy, genetics and population biology. In the area of psychology he rejected Marxist behaviourism, but acknowledged William James's belief that there was no need to regard consciousness as a transcendental quality, since it was adequate to deal with it in terms of functional relatedness. Morgan stressed that the solipsistic model of emergentism was the most immediately understandable.

There are genetically emergent steps in the recurrent advance of human mentality. Prior to birth the infant is probably sentient only. From birth to the age of some 30-36 months he is perceptive also. Thenceforward he is increasingly reflective. But it nowise follows that with advance in reflection, advance in perception (or indeed in sentience) ceases. That leaves it open to us to seek evidence of continuant and concurrent development within each level in the course of individual life. And on looking within, each of us can ask the question: Does this or that come with the cachet of surprise which marks the advent of something new and unprecedented in my experience?[55]

Morgan was convinced that emergences must have a genetic basis. While he conceded that geneticists had to focus on small superficial changes that might occur in any generation he cautioned that 'for evolutionary *advance* one must reckon with the occasional occurrence of original novelty, however frequent such occurrences may have been during the vast span of continuant organization in the evolution of living organisms.'[56]

The Haldanes, father and son, had both been critical of emergentism; J.S. had accused Alexander of producing the real world, 'very much as a conjurer produces rabbits from a hat. The rabbits are real enough and not shams; but in reality they were here from the beginning'.[57] In other words, matter, life and consciousness were inherent in the universe from the beginning and emergences were artificial surprises. J.B.S. Haldane had remarked that the more extreme forms of emergence doctrine were hostile to true scientific progress.[58] Morgan inferred that J.B.S. Haldane could be construed as a closet emergentist since he admitted that mutations

> often exhibit quite novel behaviour in a new relational setting; that the behaviour of any one gene has multiple effects and may be expected to affect the organism as a whole, even if its most striking effect is on some particular organ or function.[59]

He concluded his comments on the significance of genetic emergences with the question:

> Where then in the biological affairs of life, should original novelty be sought; in the relations of the organism to its external

environment, in the intrinsic processes within each several organism; in its constituent cells as interdependent units; or yet lower in the chromosomes and their genes as the determinants of 'unit characters'? Should we not seek it wherever it may perchance be found throughout the whole gamut of vital organization?[59]

Commenting on F.R. Tennant's *Philosophy of the Sciences* (1932), which equated epigenesis with emergence, Morgan remarked that, 'many of those who fully accept epigenesis as a "scientific fact" will tell you that emergent novelty is no better than a pestilent whimsy.'[60] Tennant used epigenesis in a metaphorical sense as a form of history that developed in the same way as an organism did, not determined strictly by the unfolding of the preformed but by interaction between the germ and the environment, or as Tennant more succinctly put it, 'the growth of something into something else, in which the nature of the "something else" is partly determined by the "something"'.[61] But Morgan's emergentism, couched in terms of epigenesis, focused on an epigenetic novelty that was something beyond normal epigenetic development, just as intellectual development included not only the 'aggregation of the stuff of knowledge', but also the tacit discovery of novelty that was hailed with 'glad surprise'.[62] Who, asked Morgan, could look back on the origins of life, the evolution of life, and personal experience, and 'proclaim with Tyndall that he could descry in the precedent fire-mist the promise and potency of all that subsequently happened, nay more, of all that will happen in the future?'[63]

9 THE SHADOW PARADIGM

> Those who rate Lamarck no higher than did Huxley in his contemptuous phrase 'buccinator tantum', will scarcely deny that the sound of the trumpet had carried far, or that the note was clear.
>
> William Bateson, 1909[1]

The nature of progress in the biological sciences and the question of its conformity with Thomas Kuhn's 1962 theory of scientific revolutions is an issue not confronted up to this point on the grounds that one controversy at a time is quite enough. Kuhn's critics have demanded that his theory perform like a law of physics, with all its predictions fulfilled with precision, and some scholars have looked to biology for its falsification.[2] One author finds twenty-one different usages of 'paradigm' by Kuhn, who admits to two; the first signifies 'the entire constellation of beliefs, values, techniques and so on shared by the members of a given [scientific] community'; the second 'denotes one sort of element in that constellation, the concrete puzzle-solutions which, employed as models or examples, can replace explicit rules as a basis for the solution of the remaining puzzles of normal science'.[3] The paradigm can be the new model that opens a new era of research, or it can be that model taken together with its scientific, social and psychological effects.

The burden of Kuhn's theory is that science is not the solemn, well-coordinated, gradual accumulation of knowledge that Francis Bacon prescribed, but is instead revolutionary, discontinuous, punctuated by crises, and is altogether a human activity. While few biologists would dispute this, the textbook history of evolutionism presents the Baconian view of steady and stately enlightenment, together with a fierce resistance by Darwinists to radically different ideas. Most biologists are *normal* scientists, the *puzzle-solvers* who get on with the job of working out the implications of the new major or minor paradigms.

> No part of the aim of normal science is to call forth new sorts of phenomena; indeed those that will not fit the box are often not seen at all. Nor do the scientists normally aim to invent new

theories, and they are often intolerant of those invented by others.[4]

Indeed, there are so many niches of normal biology that most of us can go about our business without any reference whatsoever to theoretical crises, and respond only when shadow paradigms such as Lamarckism and special creationism, whose ghosts have never been truly laid, begin to haunt the living. While some biologists might resent being classified as puzzle-solvers, Kuhn assures us that most problems undertaken by the best of scientists are of the normal variety. Normal science does, however, tend to be the art of the soluble, especially when the quest for research funding is uppermost in the mind; and with the concentration on soluble problems, and especially useful soluble problems, some inhibition of scientific progress is inevitable. I support Mayr's 1972 argument that Darwinism has been the only universal paradigm of biology. The shadow paradigm, Lamarckism, has presented a strong challenge to Darwinism, and a number of useful Lamarckist concepts, though damned by this association, deserve a counter-inductive airing. Some definitions of terms are in order here: *Lamarckism* is the doctrine proposed by Lamarck, a doctrine followed strictly by only a few who came after him. *Neo-Lamarckism* is an eclectic doctrine, which by A.S. Packard's comprehensive definition,

> gathers up and makes use of the factors both of the St Hilaire schools, as containing the more fundamental causes of variation and adds those of geographical isolation or segregation (Wagner and Gulick), the effects of currents of air and of water, of fixed or sedentary as opposed to active modes of life, the results of strains and impacts (Ryder, Cope and Osborn), the principle of change of function as inducing the formation of new structures (Dohrn), the effects of parasitism, commensalism, and of symbiosis — in short, the biological environment; together with geological extinction, natural and sexual selection and Hybridity.[5]

Its practitioners called themselves neo-Lamarckians. The idea that habits, modes of life and environment were the influences that shaped organs preceded Lamarck's theory of *transformism*, which in its early conception simply gave Aristotle's *scala natura* an evolutionary dimension. The theory presented by the *Philosophie Zoologique* (1809) had two separable elements — a *gradation* or progression dictated by the laws of nature, and the adaptive modification of structure and function in response to the environment. Lamarck was a geological uniformitarian, a gradualist:

> For Nature time is never a difficulty; she always has it at her disposal, and it is for her a power without bounds, with which she makes the greatest things like the least. . .For all the evolution of the earth and of living beings, Nature needs but three elements — space, time and matter.[6]

He denied saltations, including catastrophes in living nature, which was one of the reasons for Cuvier's low opinion of him. His laws of evolution were:

First Law: 'Life by its own forces tends continually to increase the volume of every body, as well as to increase the size of all the parts of the body up to a limit which it brings about.'
Second Law: 'The introduction of a new organ or part results from a new need, which continues to be felt, and from the new movement which this need initiates and causes to continue.'
Third Law: 'The development of organs and their force or power of action are always in direct reaction to the employment of these organs.'

This law was further developed as,

'In every animal which has not passed the term of its development, the more frequent and sustained employment of each organ little by little strengthens this organ, develops it, increases it in size, and gives it a power proportioned to the length of its employment; whereas the constant lack of use of the same organ insensibly weakens it, deteriorates it, progressively diminishes its power, and ends by causing it to disappear.'

Fourth Law: 'All that has been acquired or altered in the organization of individuals during their life is preserved by generation, and transmitted to new individuals which proceed from those which have undergone these changes.'[7]

It was the fourth law that was the main focus of late-nineteenth-century discussion. The inheritance of acquired characteristics, or *somatogenesis* as Weismann and his colleagues called it, is still used simplistically to characterise Lamarckism, and although it did not originate with Lamarck it is the essential feature of his theory, since none of the other laws would apply directly to evolution if the fourth were not true. Lamarck qualified the fourth law thus: 'Circumstances influence the form and organization of animals. . .But I must not be taken literally, for environment can effect no direct changes whatever upon the form and organization of animals.'[8] Extrinsic environmental factors did not directly cause biological changes. Intrinsic responses by the organism to the environment were the causes of change. He equivocated on this subject, saying that plants and lower animals could be directly affected by the environment. Since the neo-Lamarckians wanted to include the direct effects of the environment on higher organisms, they ignored these distinctions.

In his discussion of the second and third laws, Lamarck used the term 'besoin' or 'want' for the stimulus of change;

> Great changes in circumstances bring about changes in the wants of animals. Changes in their wants necessarily bring about parallel changes in their actions. If the new wants become constant or very lasting, the animals form new *habits*. . .If new circumstances, becoming permanent in a race of animals have given them new habits, there will result the preferred use of such a part and, in certain cases, the total lack of use of such a part as has become useless.[9]

The French 'besoin' has the same spectrum of meaning as 'want' in the English of Victorian translators; 'need' would be less equivocal in modern English. It signifies what an intelligent animal or human might perceive as its necessities, as well as unperceived but necessary requirements for existence. When Lamarck talked about 'needs being produced by circumstances' he did not necessarily imply perception of the need by the organism. The use of the active verb 'to want', as in 'the same bird, wanting to fish without wetting its feathers', implied more strongly a voluntary or conscious wish, and at the avian or mammalian level voluntary action was part of the scheme, but Lamarck, in discussing the meaning of the *sentiment intérieur,* said that it could be effective 'without any of those resolutions that we call *acts of will* being necessary.'[10] Modern interpreters of Lamarckism, such as Boesinger (1974), Burkhardt (1977) and Mayr (1982) treat Lamarck's use of 'besoin' as entirely metaphorical, and regard his concepts of adaptation and gradation as thoroughly mechanistic, despite his occasional reference to the first cause, or the sublime author.

Cuvier abandoned his policy of retaining 'an absolute silence' over systems which were not based on absolute facts when he wrote the eulogy for Lamarck. Previously he had merely joked about Lamarck in lectures and dismissed his books as 'nouvelles folies', but the 'eulogy' dumbfounded its listeners with an attack on the use of 'want' and 'wanting' by Lamarck.[11] Cuvier interpolated 'desire' as well as 'want', reading an explicit literal consciousness of the problem by the organism that did not appear in the original, and this was emphasised in the English translation of 1836. Cuvier was more influential than Lamarck, both in France and in Britain, and respected for his usually detached, scientific appraisal of the facts and accurate observation of detail, and orthodox opinion was, like Cuvier's, non-evolutionary, prevailing until Darwinian evolutionism had become established.

One can understand Darwin's own possessive jealousy of evolutionism, and his reluctance to mention Lamarck until Lyell criticised his statement in the manuscript of *Origin of Species* that, 'the most eminent of naturalists have rejected the view of the mutability of species'. Lyell wrote to Darwin, 'You do not mean to ignore G. St Hilaire and Lamarck'.[12] Darwin made the excuse that he had been referring to living naturalists, corrected the manuscript accordingly, and in the third edition of *Origin of Species* he included the 'justly celebrated naturalist' Lamarck, in the list of evolutionists who had preceded him.[13] In a later letter to Darwin, Lyell rubbed it in by remarking that it would be proper to refer to evolution theory as Lamarck's views 'improved by Darwin's'.[14] Huxley originally wrote,

> The Lamarckian hypothesis has long since been justly condemned, and it is the established practice for every tyro to raise his heel against the carcass of the dead lion. But it is rarely either wise or instructive to treat even the errors of a really great man with mere ridicule. . .[15]

However, he later ridiculed Lamarck with 'buccinator tantum': all bleat and no wool; and observed, 'half of what Lamarck had said was obsolete and the other half erroneous or defective'.[16] Although he helped to perpetuate Cuvier's misrepresentation of Lamarck and the suggestion of plagiarism of Erasmus Darwin by Lamarck, his comparisons of Lamarck and Cuvier were more objective:

> Here we see the men, [Lamarck and the nature philosophers] over whose minds the coming events of the world of biology cast their shadows, doing their best to spoil their case by stating it; while the man who represented sound scientific method is doing his best to stay the inevitable progress of thought and bolster up antiquated traditions.[17]

Lamarck had anticipated such critics:

> Those who would conclude that in the study of nature we must always limit ourselves to amassing facts resemble an architect who would advise always cutting stones, preparing mortar, wood, iron-work etc., and who would never dare to employ these materials to construct an edifice.[18]

Huxley may have believed his continued deprecation of Lamarck was justified as a defence of Darwinism in the face of the neo-Lamarckist revival, just as some modern critics of selectionism continue to assault Darwin. In the heat of the early debates some unrestrained invective was to be expected, but it lingers on in the present with criticisms of Lamarck's character and personal habits that suggest an epistemological force of mythic proportions.

Opinions as to who was the first neo-Lamarckian differ. A.S. Packard, himself an enthusiastic neo-Lamarckian, as well as F.W. Hutton, an anti-neo-Lamarckian, considered Herbert Spencer to be the first to work along neo-Lamarckian lines.[19] Spencer believed in the direct effect of the environment, 'direct equilibration', and the indirect effect of natural selection,[20] and his confrontation with Weismann over the adequacy of natural selection made him appear a natural ally of the neo-Lamarckians. The American biologists Cope and Hyatt both expressed neo-Lamarckian sentiments in publications in 1866 before either of them had studied Lamarck in any depth.

Samuel Butler (1835-1902), the grandson of Darwin's headmaster at Shrewsbury School, made an interesting contribution to the dispute between the neo-Lamarckians and Darwinists.[12] After a classical schooling he became a sheep farmer in New Zealand (1860-4), which gave him the setting for his best known work

Erewhon (1872), enough capital to return to England, and the leisure to read *Origin of Species*, which impressed him sufficiently to make a pilgrimage to Darwin's retreat at Down. Prior to the publication of his first biological book, *Life and Habit* in 1877, painting was an important occupation for him, but he gave this up to concentrate on a series of evolutionary works. When *Evolution Old and New* appeared in 1879, Butler fell out with Darwin. The academic critics were hostile, though in later years Francis Darwin, Marcus Hartog, William Bateson and E.S. Russell were sympathetic.

Life and Habit was intended for the general reader. It began with the observation that quite complex human activities started out as conscious, voluntary activities that with practice could be executed unconsciously: driving a car would be a modern example. He then considered those automatic actions like breathing and digestion, which required no practice, and wondered if these 'hereditary instincts' represented activities that at some time in the evolutionary past had been voluntary, had become unconscious from practice, and then somehow genetically fixed. Arranging human activities on a spectrum going from conscious voluntary activities, through activities which though involuntary are acquired after birth, such as breathing and eating, to unconscious activities, he concluded, 'There is something too much like method in this for it to be taken as the result of mere chance.'[22] Even the least conscious of involuntary actions, such as blood circulation, might represent ancient conscious behaviour that became genetically fixed. The germ cells provided the continuity of unconscious memory from the original primordia, and individual personalities were

> the *consensus* and full flowing stream of countless sensations and impulses on the part of our tributary souls of 'selves' who probably know no more than we exist, and that they exist as part of us, than a microscopic water flea knows the results of spectrum analysis.[23]

Embryonic development was recapitulative because the embryo was a slave to its unconscious cellular memories. Death was really 'loss of memory'. A grain of corn eaten by a hen was put in an unfamiliar position, disoriented, its memory unable to aid it:

> The first minute or so after being eaten, it may think that it has just been sown, and begins to prepare for sprouting, but in a few

seconds, it discovers the environment to be unfamiliar; it therefore gets frightened, loses its head, is carried into the gizzard, and comminuted among the gizzard stones. . .Once assimilated, the grain ceases to remember any more as a grain, but becomes initiated into all that happens to, and has happened to, fowls for countless ages. There it will attack all other grains, whenever it sees them; there is no such persecutor of grain, as another grain when it has once fairly identified itself with a hen.[24]

One cynic suggested that by Butler's theory kleptomania would be the evolutionary outcome of picking pockets.

Since the memory of an individual rapidly lost small details with time, the unconscious memory remembered only repetitious details. Some combinations of germ cells might have conflicting memories, a possibility that Butler exploited to explain hybrid sterility. Concerning a hybridisation experiment mentioned by Darwin, involving crosses between chickens and pheasants, Butler noted that only twelve chicks survived from five hundred eggs:

no wonder the poor creatures died, distracted as they were by the internal tumult of conflicting memories. But they must have suffered greatly; and the Society of the Prevention of Cruelty to Animals may perhaps think it worth while to keep an eye on the embryos of hybrids and first crosses. Five hundred creatures puzzled to death is not a pleasant subject for contemplation.[25]

The evolution of instincts fitted naturally into Butler's memory model. After generations of being taught, complex behaviour would become fixed in cellular memory. Darwin had remarked that 'habitual action does sometimes become inherited', and pangenesis was a possible mechanism of unconscious memory.[26]

While *Life and Habit* had originally been intended as a popular complement to the Darwinian theory, Butler read Mivart's *Genesis of Species* when the manuscript was almost complete, and on referring to the most recent edition of *Origin of Species* found that Darwin had changed his mind about the possibility of the inheritance of habit. This shook his faith in the master, and it was only at this point that he realised that his own views were thoroughly Lamarckist, and that Mivart's criticisms of Darwinism could all be answered by Lamarckism.[27]

Evolution Old and New (1879) attempted to revive interest in the work of Erasmus Darwin, Buffon and Lamarck, who had received too little recognition for their pioneering work. *Origin of Species* had been made to appear as a 'literary Melchisedek, without father and without mother in the works of other people', and Butler's loss of respect for Darwin could no longer be disguised.[28] When *Evolution Old and New* was published, Darwin, to avoid accusations of slavishly reacting to Butler, wrote to inform him that he had already commissioned the translation of an earlier article by Krause on Erasmus Darwin. The translation appeared as the second part of Darwin's *Life of Erasmus Darwin* (1879). Butler, suspecting that he had been plagiarised, then proceeded to teach himself German so that he could read the original Krause essay. Apparently Krause had

> helped himself — not too much, but enough; made what other additions to and omissions from his article he thought would best meet *Evolution Old and New* and then fell to condemning that book in a finale that was meant to be crushing.[29]

The only direct mention of Butler in the Darwin biography was a prefatory note to the effect that the original Krause article predated *Evolution Old and New,* which Butler took as an inference that he was plagiarising Krause, instead of the other way around. Confronted with Butler's suspicions, Darwin replied that the original Krause article had indeed been enlarged before it had been translated into English, and that if the book should go to a second printing he would append a note to the effect.[30] Nothing short of a public apology or a printed erratum for the first edition of the *Life of Erasmus Darwin* would satisfy Butler, but his accusations, published in *The Athenaeum,* brought no response from the Darwinists.[31] Butler's own account of the controversy was related in *Unconscious Memory* (1880), a work which was chiefly a comparison of Butler's theory with a similar one developed by the German physiologist Hering in 1870. Marcus Hartog later remarked that Hering's formal scientific language was more likely to be listened to than the upstart's satire.[32]

Luck or Cunning (1886), Butler's final evolutionary book, contrasted the random luck of selection theory with the cunning activity of the organism. A polemic against 'the mindless theory of Charles-Darwin natural selection', it also challenged G.J. Romanes, who had come round to a similar theory of unconscious memory a few years after remarking it 'simply absurd, to suppose that it could possibly be fraught with any benefit to science'.[33] Since Romanes had also dropped natural selection as the primary cause of evolution, Butler accused him of trying to run with the hares and hounds at the same time. As for A.R. Wallace's assertion that Lamarckism had been refuted, it had no more substance than Darwin's claim that natural selection caused evolution; both were like the Bellman's 'Rule of Three' in 'The Hunting of the Snark':[34]

> Just the place for a Snark! I have said it twice;
> That alone should encourage the crew.
> Just the place for a Snark! I have said it thrice:
> What I tell you three times is true.[35]

Martin Gardner, commenting on the 'Rule of Three', points out that the cyberneticist Norbert Weiner who had written that a computer's effectiveness may be checked by asking it the same question several times, or asking the question of several other computers, had speculated that the human brain might have a similar checking mechanism, and noted the similarity with the Bellman's rule of three.[36] J.B.S. Haldane also confessed to Bellmanship:

> I give annual. . .lecture courses. I introduce an idea with such words as 'A possible explanation of these facts is. . .' Next year this becomes 'the most probable explanation. . .', and after I have said it three times it becomes 'the explanation'. What is worse, when I write a text book I use this last phrase. I fear that a good many scientific theories originate in this way.[37]

Butler's anti-intellectual attacks on scientists and other academics were bound to provoke a hostile response. Ray Lankester wrote:

> That such an attempt to advocate the discredited speculations of Lamarck should be made is an illustration of a curious weakness of humanity. Not infrequently after a long contested cause has triumphed and all have yielded allegiance thereto, you will find, when few generations have passed that men have clean forgotten what and who it was that made that cause triumphant, and ignorantly will set up for honour the name of a traitor or an imposter or attribute to a great man as a merit deeds and thoughts which he spent a lifetime in opposing.[38]

Lankester, writing soon after the death of Darwin, may have still been influenced by the Victorian aura of Praise for Great Men, as well as by fears for the adequacy of natural selection.

Butler's influence was more literary than scientific. His friend George Bernard Shaw expressed the frustration with Darwinism that still affects the teleological soul:

> The Darwinian process may be described as a chapter of accidents. As such it seems simple, because you do not at first realize all that it involves. But when its whole significance dawns on you, your heart sinks in a heap of sand within you. There is a hideous fatalism about it, a ghastly and damnable reduction of beauty and intelligence, of strength and purpose, of honour and aspiration to such casually picturesque changes as an avalanche may make in a mountain landscape, or a railway accident in a human figure.[39]

A casual non-biological reader of Butler's evolutionary book must wonder if Butler was a man ahead of his time, and a victim of authoritarianism. Is there anything to the unconscious memory theory in terms of modern biology? It would be stretching metaphor to claim that our DNA molecules are 'memories' of our parents, since they are rather mixed copies with occasional fortuitous novelties, but our knowledge of how memory works is not much better than it was in Butler's day. Memory models are an interesting illustration of how mechanical models are defined by contemporary technology. Twenty years ago the cybernetic model of a reverberating neural circuit, triggered to fire a conscious memory by suitable stimuli, was popular. This was replaced or rivalled with the memory molecule (magnetic tape recorder) model, with RNA and polypeptides acting as the memory tapes, and which was closer to Butlers's and Hering's theories. A synthesis of these two earlier ideas is found in a recent molecular model that suggests that a brain-cell membrane enzyme, controlling potassium flux, and hence the electrical stimulation of the cell, may be temporarily activated by the calcium, or permanently activated by a proteolytic alteration.[40] These two activation states correspond to short-term and long-term memory. Rehearsing increases the number of activated cells. In elderly rats the degree of activation is reduced. The spread of the memory through the brain is also taken into account by another more general memory model, which the holographic *engram* model (laser technology), which suggests how memories can be retained even when the apparently appropriate portions of the brain have been damaged. The term 'engram' was coined by Haeckel's student Richard Semon (1859-1919), who believed that the germ plasm reacted to the conditions of the body, and that even transitory influences might leave an engram impressed upon the body.[41] Some of the total store of imprints were acquired and some inherited, and those acquired could become inherited, such as the callosities that appeared in embryos before they ever came in contact with abrasion.[42] This would now be regarded by as a quasi-Lamarckist genetic assimilation, which will be discussed in Chapter 10. Semon's mneme theory also adumbrated the concept of the conditional reflex developed by Pavlov.

The American neo-Lamarckians listed by E.D. Cope in *Primary Factors of Organic Evolution,* included himself, A.S. Packard, A. Hyatt, F.A. Ryder, W.H. Dall, and W.B. Scott.[43] He also claimed H.F. Osborn, who although he had Lamarckist leanings at first, finally rejected both Lamarckism and Darwinism.[44] Cope also claimed Spencer, Cunningham, Henslow, Giard, Perrier, Semper, Eimer and Nägeli; but his first short list is a more accurate representation of those neo-Lamarckians who would have welcomed the title and whose ideas were mutually consistent.

Edward Drinker Cope (1840-97) conducted independent studies on the lower vertebrates and palaeontology of North America in his early career.[45] Several years of field studies with the US Geological Survey were terminated because of a running dispute with O.C. Marsh, who became chief vertebrate palaeontologist with the Survey in 1879. Following the loss of his private income Cope became a travelling lecturer, and eventually was appointed to a position in the department of geology at the University of Pennsylvania in 1889. His first major work on evolution was *The Origin of the Fittest* (1887), a compendium of essays and lectures which stressed the role of organisms in evolution. His epigenetic laws of acceleration and retardation and the law of repetitive addition have been discussed in Chapter 4. Cope's *doctrine of kinetogenesis* was Lamarck's third law of use and disuse. The response of an organ to use was exemplified by the expansion of a muscle with exercise, caused by a release of growth force and accompanied by an increase in the nutritive blood supply. Disuse brought a cessation of growth and degeneration. Experiments on the encouragement or suppression of the use by tadpoles of their gills had been shown to affect the onset of metamorphosis.[46] A vital *growth force* both repressed and activated of potentials, and was in turn directed by *grade influence,* a correlate of evolutionary progress; which acted through the nervous system and ceased to be exerted at the maturation of the organism, except for the case of periodic changes in sexual ornaments, for example in birds and deer.[47] The use of a part by an animal could be compulsory or optional, the former if there was an environmental change requiring a change in behaviour. Behavioural options would depend on intelligence and creativity.

> Thus intelligent choice, taking advantage of the successive evolution of physical conditions, may be regarded as the *originator of the fittest,* while natural selection is the tribunal to which all the results of accelerated growth are submitted. This preserves or destroys them, and determines the new points of departure on which accelerated growth shall build.[48]

In this way, the capacity for complex behaviour and intelligent choice in the higher animals took on the important role of channelling the imminent growth force through the nervous system.

In 'Consciousness in Evolution' Cope used memory as an analogue of the hereditary accumulation of biological experiences, in the sense that the individual's memory is vast but unconscious until stimulated. Adaptive specialisation was a fixation of automatic or unconscious activities, resulting in reduced flexibility of response to change. Thus it was in the unspecialised ancestral types that the greatest conscious flexibility resided, expressed through archaesthetism (Lamarck's second law). Cope's conception of the evolution of involuntary function, delivered as a lecture in 1874 and published in 1875, was quite similar to Butler's, although it did not stress the memory metaphor so strongly. Hering predated both and was apparently unknown to them when they developed their ideas.[49]

To the *Law of Use and Disuse* or *kinetogenesis*, Cope appended the corollary *physiogenesis,* the production of change by physicochemical causes. In other essays Cope discussed the way in which the mechanical stresses on particular parts of the animal body would affect developing organs, i.e. function determining form, prior to their somatogentic induction. The latter were among his most plausible and less speculative ideas. *The Primary Factors of Organic Evolution* (1896) accepted Darwin's pangenesis as the mechanism of the inheritance of acquired characteristics (Lamarck's fourth law). Cope credited Ryder with kinetogenetic ideas pertaining to reductions of digits in the feet of mammals, and the evolution of dentition in mammals. The malacologist Dall interpreted hinge production in bivalve shells as a product of mechanical influences during development.[50] He also noted that ontogenic variation could not possibly be equal in every direction, otherwise the final structure would lack integrity. Hyatt's studies of the Cephalopoda had expanded the physiogenesis concept:

> The action of physical changes takes effect upon an irritable, plastic organism which necessarily responds to external stimulants by an internal reaction or effort. This action from within upon the parts of organisms modifies their hereditary forms by the production of new growths or changes, which are therefore adapted or suitable to the conditions of the habitat, and are therefore physiologically and organically equivalent to the physical agents and forces from which they directly or indirectly originated. In so far as they are different, they probably produce the differentials which distinguished series and groups from each other.[51]

This has a logical consistency and remains significant in terms of physiological evolution, even if question-begging about the inheritance of acquired characteristics is taken into account.

F.W. Hutton, in *Darwinism and Lamarckism Old and New* (1899), wrote that Cope's description of 'Primary Factors of Organic Evolution,' in his book of that name, as 'constructive and not destructive' appeared to mean 'that he has brought forward all the facts he knows that assist to build up his theory' and has left out all those which tend to destroy it'.[52] There was no *a priori* reason for supposing that physiogenesis produced useful variations, since the forces involved were beyond the control of the organism, and the future of such random variations therefore depended on natural selection.[53] He agreed that kinetogenesis was not fortuitous, but argued that it could not produce novelty, only emphasise or reduce a pre-existing organ. There was plenty of evidence to show that variation might take place without any change in the environment, and therefore that there had to be an alternative cause to physio-genesis. For example, variants of fossil trilobites must have lived in a stable marine-benthic environment. Why also was there such a variety of lung-like organs throughout the animal kingdom if air was the physical cause of each? Hutton countered kinetogenesis with examples of asymmetrical use of paired organs like elephant tusks that did not result in compensating asymmetrical growth, only more wear and tear on the preferred tusk. For every example of supposed evolution by direct mechanical cause Hutton could find an example where like mechanical causes did not have such an effect. The neo-Lamarckians had failed to prove a single example of heritable kinetogenesis, although the embryonic callosities cited by Semon were awkward for Darwinism. However, neo-Darwinism, or Weismannism, also consisted of more speculation than fact. Kinetogenesis was possible in the case of eye migration in flatfish, and disuse inheritance could not be excluded.

T.H. Morgan, writing in 1903, considered that the best positive evidence for the inheritance of acquired characteristics had been provided by Brown-Séquard's experiments on guinea-pigs.[54] Brown-Séquard's had observed that epilepsy and organ mulfunctions due to injury of the nervous system were transmitted to the offspring. Romanes had been able to reproduce some of the effects, but not the heritable limb-mutilation results that Brown-Séquard claimed.[55] This was, however, uncomfortable evidence, since the facts could have indicated a transmitted disease, and in any case it would be a great disadvantage for offspring to inherit the injuries of the parents.[56] Kidney and liver diseases in pregnant laboratory animals were known to be transmitted to embryos *in utero* if the diseased organs were interfered with surgically. Heritable diseases could of course have given the same spurious appearance of inheritance of acquired characteristics.

Another ideologue of neo-Lamarckism was a student of Louis Agassiz, Alpheus Spring Packard (1839-1905), who was involved with the US Geological Survey at the same time as Cope. His major work was in entomology, but he also studied arthropod evolution and cave fauna commencing in 1870. In association with A. Hyatt, E.S. Morse and F.W. Putman, Packard founded the *American Naturalist,* and was its general editor for twenty years.[57] In *Lamarck, The Founder of Evolution* (1901) he asserted that Lamarck was the first to state fully and authoritatively the causes of the primary factors of evolution, namely variations. However, he was deliberately vague about Lamarck's views concerning the direct influence of the environment on the germ plasm. Packard's study of cave insects concluded,

the characters separating the genera and species of animals are

those inherited from adults, modified by their physical surrounding and adaptation to changing conditions in life, inducing
certain alterations in parts which have been transmitted with
more or less rapidity, and become finally fixed and habitual.[58]

The peculiar physical conditions of caves, such as reduced light and
scarcity of food, were the direct causes of the characteristics of
cavernicolous animals. Correlated with the physically induced
changes were the ontogenic adaptability of the animals; their
isolation, which prevented the swamping of their new characteristic,
the use and disuse of particular organs, and heredity, which would
ensure the continuance of the new characteristics as long as the
physical conditions remained the same. These conclusions indicated
a strong Spencerian influence, but little that was distinctively
Lamarckian; the new school would have been more aptly titled neo-
Geoffroyian in this regard, as well as in the acknowledgement of the
secondary role of natural selection. However, like Cope, Packard
preferred Lamarck as the figurehead, and in 1885 coined the
expression *neo-Lamarckianism*, the shortened form of which
became common usage.[59] Unlike Cope, but like Delboeuf, Giard
and later L.S. Berg, Packard recognised that the direct effect of the
environment could influence enough organisms in a population in
the same direction to be able to dispense with the need for a tribunal
of natural selection.[60]

E.B. Poulton, an unrelenting Darwinist, was the scourge of neo-
Lamarckism. After an introductory slap at Lamarck's putative
plagiarism of Erasmus Darwin, he attacked Herbert Spencer as the
chief neo-Lamarckist offender, and builder of a hypothetical edifice
on the insecure foundation of the unproved assumption of the
inheritance of acquired characteristics.[61] Lamarckism was alluring
because it was the kind of process that a human designer would have
incorporated into nature, given the opportunity. However, Poulton
rejected the Duke of Argyll's criticism that natural selection had a
similar appeal, being a loose analogy with 'natural' expressing
familiarity of objects and processes, and 'selection' expressing a
familiar part of human behaviour: natural selection could not be
accused of seductive appeal since so many people had
misunderstood it.[62] E.B. Poulton relayed a satirical verse from
Courthope's *The Paradise of Birds* (1870) concerning the duck-
billed platypus:

For he saw in the distance the strife for existence,
That must his grandchildren betide,
And resolved as he could for their ultimate good,
A remedy sure to provide.
He [sic] laid as a test four eggs at a time.
On the first he sat still and kept using his bill,
That the head in his chicks might prevail:
Ere he hatched the next young, head downwards he slung,
From the branches, to lengthen his tail.
Conceive how he watched till his chickens were hatched,
With what joy he observed that each brood,
Were unlike at the start, had their dwellings apart,
And distinct adaptations for food.
From the bill, in brief words, were developed the Birds,
Unless our tame pigeons and ducks lie;
From the tail and hind legs in the second-laid eggs,
The Apes and — Professor Huxley.[63]

The particular error exposed by Courthope was that the Lamarckist process can only see what is in front of it, and so lead to specialisation, without making any provision for future changes. Similarly, selection can also only see what is immediately useful or detrimental, and ignores what is neutral or only potentially useful. It is the direct causal relationship between the environment, the organism and its germ plasm that is the very limitation of neo-Lamarckism, as well as its allure, a point that was rarely made. However, the choice of fortuitous adaptive variation is open to the same criticism.

Noting that the lively behaviour of an autotomised lizard tail or lobster claw was triggered by the loss of the appendage, Poulton asked how somatogenic induction could be affected by an organ which was no longer part of its anatomy. There was also the problem of how somatogenic induction knew how to stop when the most effective level of response had been reached. Poulton correctly pointed out that many of the so-called instincts discussed by neo-Lamarckists were actually intelligent behaviour; on the other hand he was prepared to claim that 'Natural Selection acts on the nervous system of the caterpillar, and thus compels it to make the cocoon in a certain way'.[64]

The final phase of neo-Lamarckism prior to the resurgence of neo-Darwinism was dominated by Paul Kammerer. A strict chronological treatment ought to discuss first of all two other Teutonic Lamarckists, Nägeli and Eimer. However, both of these proposed corollaries to the Darwinist and Lamarckist laws which will be considered separately in Chapter 10. Kammerer (1880-1926) studied music at the Vienna Academy and then switched to zoology at the university.[65] He had already published studies on reptiles and amphibians when he was appointed to the Institute for Experimental Biology in 1902. His first series of experiments dealt with the effects of unusual climatic conditions on *Salamandra,* and demonstrated marked behavioural and developmental effects. Offspring from the affected salamanders showed varied degrees of difference from the original type.[66] Kammerer was a Weismannist until he conducted these experiments and interpreted their results as an indication of the inheritance of acquired characteristics.[67] Other experiments induced colour modifications in salamanders and over several generations these changes became more pronounced. Salamanders raised and bred over several generations in an environment with a dark background lost their yellow spots and those exposed to a light background expanded their yellow colouration. Professor Przibram, Kammerer's director, confirmed that the modifications required the salamander's ability to see the background. These results might now be regarded as a special case of extranuclear inheritance by the accumulation of pigments or pigment organisers in the appropriate proportions in the eggs.

The experiment which caused Kammerer's major problem later in his career concerned the midwife toad, *Alytes,* a terrestrial animal which breeds on land, the male carrying the egg strings round its hind legs. Kammerer induced the toads to mate in water, and after several generations under these conditions the males developed nuptial pads, horny growths at the bases of their digits which enabled them to hold on to the females. This result was published first in 1909 and elaborated in a 1919 paper.[68] Kammerer admitted in a 1923 lecture at Cambridge that the nuptial pad trait could have been atavistic, i.e. could have appeared by environmental inducement of a trait which the animals had never lost from the genotype, but which did not appear phenotypically under normal circumstances.[69] A similar phenomenon was exhibited in the eyeless, cave-dwelling newt *Proteus.* Kammerer found that *Proteus* reared in red light did not develop the black pigment that normally arrested eye development, and so his specimens developed normal eyes. He made no claims that a saltatory novelty or mutation had been discovered. Although these experiments made Kammerer's research a *cause célèbre,* he regarded as his crucial Lamarckist experiment the one concerning *Ciona intestinalis,* a sea squirt, whose siphons regenerated when cut off, and become elongated with repeated amputation. He bred from these and claimed that the elongated siphons were inherited by the offspring.[70] There was some debate over the reproducibility of the siphon elongation, due to Kammerer's critics' failure to follow Kammerer's techniques. However, the regenerative elongation of the *Ciona* siphons had first been made by Mingazzini in 1891 at Naples, and was confirmed by Jacques Loeb, then a colleague at the Institute for Experimental Biology in Vienna.[71] No subsequent confirmation of the inheritance of siphon-elongation has ever been obtained.

The story of the controversies aroused by Kammerer's research, and especially William Bateson's role in stirring up the trouble, has been so well recounted by Koestler that I will provide only the barest outline here, to illustrate the complex paradigmatic crises that characterised this period. In 1910 when Bateson, who was then working on his book *Problems of Genetics* (published in 1913), wrote to ask Kammerer for the loan of a preserved specimen, Kammerer replied that he only had live specimens and suggested that Bateson look to other aspects of his work for confirmation of the inheritance of acquired characteristics. In the same year Bateson visited Kammerer in Vienna, and privately noted a 'suspicion of humbug' about the man.[72] By the time *Problems of Genetics* was published Bateson's suspicions still remained, and he made the ambivalent statement concerning Kammerer's work that

> Many of the results that are described, it is scarcely necessary to say, will strike most readers as very improbable; but coming from a man of Dr Kammerer's wide experience, and accepted as they are by Dr Przibram, under whose auspices the work was done in the Biologische Vesuchsanstalt at Vienna, the published accounts are worthy of the most respectful attention.[73]

During World War I the live specimens at the Vienna Institute were neglected and died off, since the assistants had been conscripted, and Kammerer was co-opted to the military censorship department. Kammerer's response, in 1919, to earlier criticisms was the plausible one that he had been caught in the dilemma familiar to all biologists who have worked with rare live specimens. Does one preserve some in good condition to have something to show later, or does one save the largest possible number so as to obtain repetitive or statistically significant experimental results? Sometimes biologists find themselves in the position where one properly preserved specimen, if it were only available, would resolve a crucial question, and lacking that specimen, the only alternatives are to leave it unanswered or go to a great deal of expense and time to answer it. In this light Kammerer's reluctance to supply specimens on demand was understandable. Professor E.W. MacBride took up Kammerer's case in a letter to *Nature*, suggesting that the best way to criticise Kammerer was to repeat his experiments.[74] Bateson's reply insinuated that the salamander experiments were fraudulent, and that even if the *Alytes* males had developed nuptial pads, they need not be interpreted as proof of the inheritance of acquired characteristics, a point that Kammerer had freely conceded to him years before.[75] Bateson was sent slides of the sectioned nuptial pads by Przibram in 1920, but was not prepared to accept them as bona fide evidence, although Przibram had given assurances that he had been present when the slides were prepared.[76]

In 1923 Kammerer visited England, bringing a collection of preserved experimental material, including the last *Altyes* male, and he gave lectures at Cambridge and London. Bateson did not attend the Cambridge meeting, but went finally to the Linnaean Society meeting in London. According to MacBride's letter to *Nature,* he 'accepted his published results as genuine, claiming, however — as he had the full right to do — to differ from the deductions which Dr Kammerer drew from them'.[77] Bateson could not, however, rest easy in this position of neutrality, and entered into a hostile exchange of letters, demanding again to see the specimen which had been available to him at the Linnaean Society.[78] In Koestler's opinion Kammerer's reputation was further damaged by sensational press coverage of his lecture visits to England and the United States. Whatever his motivation, G.K. Noble, who examined the remaining *Alytes* specimen in 1926 in Vienna, found that it had been faked with an injection of Indian ink. He exposed the fraud in *Nature*, which simultaneously published a letter from Przibram suggesting that the original nuptial pads might have been sloughed off due to age and frequent handling. However, it was clearly shown that the pigment was ink and not natural pigment. Kammerer shot himself six weeks later, an action interpreted as an admission of guilt by most biologists. Koestler, who went into the psychological and political implications of the case as well as the biological ones, suspected that a well-meaning supporter of Kammerer at the Vienna Institute might have faked the specimen after the loss of the nuptial pads for the reasons suggested by Przibram.

Whatever satisfaction Bateson derived from the controversy, he had the negative effect of directing attention from the general biological implications of Kammerer's work to focus on the not very relevant aspect of the *Alytes* nuptial pads. Richard Goldschmidt, incensed by the Russian film *Salamandra,* which romanticised and fictionalised Kammerer, relayed suspicions that Kammerer had a fraudulent streak.[79] Looking at Kammerer's work from the vantage point of the present, we must credit him with remarkable experimental skills. Some of his results remain interesting, particularly those concerning colour change in *Salamandra,* and the induction of physiological and behavioural changes in salamanders by manipulating their environments. However, none of the experiments can be regarded as proof of the inheritance of acquired characteristics.

Kammerer's chief supporter in Britain was E.W. MacBride, Professor of Zoology at the Imperial College of Science and Technology at the University of London. *An Introduction to the Study of Heredity* (1924), which Alister Hardy dismissed as 'perhaps the most biased of all books', gives us some idea of the residual strength of English neo-Lamarckism.[80] As well as Kammerer's 'proof' of the inheritence of acquired characteristics, MacBride cited Pavlov's experiments with trained white mice which seemed to pass on their learned behaviour to their offspring.[81] He cited also the early studies by Johanssen and Jennings which had been critical of neo-Darwinism, without mentioning their later, revised opinions. He was critical of T.H. Morgan's 'hypothesis' concerning sex linkage, and argued that mutations of the type that Morgan found in *Drosophila* were nothing but pathological deviations. To support this idea he drew upon Tornier's theory that the various 'mutant' forms of goldfish had been caused by the poor conditions under which they were kept: the 'vigour' of the embryo was weakened by hypoxia during the first few days of its life.[82] The form of the embryo during the development could be altered by the amount of water absorbed by different parts of the embryo, resulting in deformed adults. Colour changes could be affected if the embryo, deprived of certain resources, had to draw on pigment as an energy source. While he admitted that these changes obeyed Mendelian rules in future generations, MacBride agreed with Tornier that it was the degree of germ-weakening that was inherited, not the structural consequence of that weakening. In other experiments Tornier treated the eggs of axolotls with anoxic sugar solutions, which had the effect of distorting the embryos. Though Tornier was unable to breed from the anomalous axolotls MacBride concluded:

> If then the hypothesis of Tornier be accepted — and we think that the cumulative evidence in its favour is irresistible — it throws a brilliant light on the nature and origin of mutations. As they are all pathological in character, the result of weakened germ energy, they can have played no part in the process of evolution, for in competition with the type they would be inevitably weeded out by natural selection, and to explain evolution we are thus driven back on the inheritance of acquired habits and of the structures that result from these. When we consider how Morgan reared his flies, confined in glass tubes with bits of rotten banana, we can form a pretty confident guess as to the causes of the production of the wretched pathological mutants with which he has worked.[83]

According to MacBride's eugenic ideas,

> attempts to favour the slum population [consisting of Celts, Latins, blacks, etc.] by encouraging their habit of reckless repro-duction in throwing the support of their children on the state, places a heavier burden on the shoulders of the Nordic race, who form the bulk of the taxpayers. The prospect is such as to make a patriotic Englishman [sic] shudder. . .the idea that education and environment, acting through one or two generations, can cancel the work of thousands of generations is singularly futile.[84]

Curiously enough, the same biological principles espoused by MacBride were used to derive the opposite conclusions by certain Marxist dialecticians, and with more drastic consequences than MacBride ever foresaw.

MacBride's and Tornier's belief that physicochemical conditions encountered during development could effect heritable changes came under the category of *chemical Lamarckism,* proposed by J.T. Cunningham's *Hormones and Heredity* (1921). It assumed that adaptive characters were somatogenic and that only gametogenic characters, which were non-adaptive, conformed to Mendelian principles. The somatogenic adaptations could, however, be impressed on the germinal material by hormones induced in the somatic cells by environmental stimuli. It was known that hormonal changes, such as occurred in the reproductive cycle, could be so elicited, and Cunningham had justifiably argued that the explanation of their origin and function was at the time beyond the explanatory scope of the infant gene theory. Although the advances of genetics were leaving Lamarckism little room for manoeuvre, except in this area of chemical Lamarckism, Seba Eldridge's *The Organization of Life* (1925) responded that the reproductive hormone activity depended on sex, which was a hereditary character; hence Cunningham's logic was unsound; nor was there any proof that any hormones were able to induce heritable change in the germ plasm, tempting though the thought might be. Eldridge was more sympathetic to a hypothesis of Guyer and Smith which had developed from a series of immunological experiments conducted in the early 1920s. Of special interest as an experiment in which fowl serum antibodies for rabbit lens-protein were injected into pregnant rabbits, resulting in eye abnormalities in the young, which were inherited through eight generations.[85] This was possibly caused by placental transmission of the antigenic factors. Eldridge doubted that such a serological mechanism, if valid, could be responsible for larger morphological changes; instead, a holistic approach that would integrate the chemical hypotheses with explanations of organismic plasticity and adaptibility was required to advance Lamarckism. He stressed the importance of mutual adaptation of internal physiological systems, but also believed that adaptability expressed a traditive accumulation of adaptive experiences of particular environments. Eldridge was most critical of 'atomistic' theories such as the form of gene theory proposed by Guyer, which had related the chromosomes to the production of enzymes and consequently to the control of cellular activity:

> We are not told what it is that determines the production of the right sorts and amounts of the enzymes, not to mention the proteins and other specific substrates of the cell; nor are we told what arranges the chromosomes in such a way that they can feed out the enzymes at the proper rate.[86]

This limitation to reductionist gene theory remains relevant to the present time. With realistic holism Eldridge had little more patience, although he appreciated that it recognised significant organisational relationships. His personal inclination was towards a Bergsonian immanent, primordial, vital factor.

Eldridge's discussion of chemical Lamarckism is particularly interesting since it is this materialistic hypothesis that has doggedly survived to the present, and immunological observations continue to dominate serious consideration of the concept. Eldridge's holistic emphasis on the importance of development processes and the co-adaptation of internal physiological systems was unusual, because the focus of Lamarckism was on narrow relationships between organisms and particular environments. The processes that most Lamarckists studied were those which led to specialisation, and they rarely took account of Lamarck's *gradation,* nor the progress of evolution expressed as improvements in homeostasis. Neo-Lamarckism, which flourished for about forty years from the 1880s into the 1920s, had a growing space afforded by the decline of Darwinism, the immaturity of genetics, and a greater tolerance for speculative science than is found at the present time. It was also a doctrine that was diminished through the inconclusive or falsifying nature of its experimental results. Many corollary ideas associated with Lamarckism are worthy of salvage; the most significant of these include the general emphasis on the role of the organism, especially in its behavioural responses to the environment, and Cope's epigenetic ideas. These do not require the prerequisite of inheritance of acquired characteristics to be relevant to evolution theory. The concept of physiogenesis, i.e. direct effects of the environment on organisms, that both Cope and Eldridge regarded as a crucial factor in evolution, will later be considered in relation to physiological evolution, though divested of its implications of inheritance of acquired characteristics.

10 LAWS OF GROWTH

> Organisms develop in definite directions without the least regard
> for utility, through purely physiological causes, as the result of
> organic growth.
>
> T. Eimer 1898[1]

> We may waive our applications of these facts to theories, but let
> us not turn our backs to the facts themselves.
>
> H.F. Osborn 1895[2]

Darwin introduced the 'laws of growth' to account for some
phenomena that he could not explain in terms of natural selection.

> A trailing bamboo in the Malay Archipelago climbs the loftiest
> trees by the aid of exquisitely constructed hooks clustered around
> the ends of the branches, and this contrivance, no doubt, is of the
> highest service to the plant; but as we see nearly similar hooks on
> many trees which are not climbers, the hooks on the bamboo may
> have arisen from unknown laws of growth, and have been sub-
> sequently taken advantage of by the plant undergoing further
> modification and becoming a climber.[3]

This Darwinian expression will serve as a general title for hypotheses
that have suggested that certain traits have come into being and have
been modified or extended without their having clear and immediate
utilitarian advantage. Most biologists would call such processes
orthogenetic, and although the term has been given the implication
of mysticism by proponents and detractors alike, its materialistic
implications deserve some attention. Darwin himself made no effort
to explain what he meant by laws of growth, and did not seem to
understand that he was exposing his flank to unorthodox en-
croachments.

German evolutionists were among the first to explore this tributary
of evolutionism. In 1867 W. Waagen used the term *mutation* to refer
to gradual variations that seemed to occur within particular phyletic
lines, proceeding in the same direction for several geological
epochs.[4] For example, the lobate lines which traversed the shell

whorls of ammonites began in a simple undulating form, and with time became gradually more complicated and ultimately foliaceous. Waagen's associate Neumayer inferred that the gradual nature of these serial changes would allow them to fit within a Darwinian framework. Carl von Nägeli (1817-91), a Hegelian idealist and supporter of Oken's nature philosophy, had become converted to materialism by 1884 when his book *A Mechanico-physiological Theory of Evolution* was published.[5] Concerned with Darwin's lack of consideration of the organisation of the whole organism, and seeking some inherent principle that would provide a more fundamental explanation or organisation than natural selection, he proposed the *Vervollkommungsprincip*, or *perfecting principle*, a kind of organic law of inertia, whereby the developmental process, once set in motion, gathered momentum to move in a particular direction, resulting in more complex or more 'perfect' structure. *Completeness of organisation* was a function of time, so the most complex series of organisms such as the mammals must have the longest evolutionary history, and the simplest organisms such as the infusoria must have come into existence most recently. At each stage in evolutionary progress there would be adaptations by means of indirect and perhaps direct responses to the environment, and the perfecting principle could also provide adaptational drive, leaving the elimination of certain evolutionary lines by competition as the only role for natural selection.

That some varieties of plants could exist in a number of different environments without any apparent adaptive modifications to those environments, and that several varieties of plants could co-exist in the same environment yet retain their distinct differences, suggested to Nägeli an internal non-adaptive factor, responsible for the basic forms. He was aware, as were his critics, that some plants would differ from their progenitors if planted in a different environment. Nägeli thus attributed to the perfecting principle all grade evolution and some adaptive evolution, and therefore was closer to Lamarck in his thinking than any of the neo-Lamarckians, who had discarded gradation. Nägeli, while emphasising progress to completeness, admitted the direct influence of the environment and somatogenic induction as minor causes of adaptation. The formation of a thick coat of hair in a mammal was due to the direct effect of a cold climate; offensive and defensive organs like horns and tusks were produced by the physical stimulation of the appropriate parts of the body; the organisation of tissues in the stems of plants was due to mechanical stress. Adaptation and the perfecting principle worked together to produce the final product; no determinate action of the organism was required.

Nägeli proposed that protoplasm was made up of minute, invisible *Mizellen*, arranged in chains, constituting two kinds of protoplasm: idioplasm, which was like a microscopic picture of the macroscopic individual, and stereoplasm, which was differentiated protoplasm that possessed only some of the properties of the idio-plasm. According to Radl, this provided one of the foundations of Weismann's germ plasm theory, and also influenced De Vries, but a theory of modifiable particles circulating in the blood gives Maupertius a century of priority for this concept, and as Zirkle (1946) points out, the general principle is found in Hippocrates.[6]

T.H. Morgan commented that Nägeli had tacked on the effect of the environment to explain adaptation, which the perfecting principle could not accommodate, and both the perfecting principle and the *ad hoc* effect of environment were arbitrary and speculative, and fundamentally an ill-considered reaction to Darwinism:

> Despite Nägeli's protest that his principles are purely physical and that there is nothing mystical in his point of view, it must be admitted that his conception, as a whole, is so vague and difficult in its application that it probably deserved the neglect which it generally receives.

However, Morgan conceded that there must be something 'over and beyond the influence of the external world' that caused organisms to change, though it was going too far to suppose an internal driving force taking evolution in a particular direction.[7]

H.F. Osborn felt that Nägeli's perfecting principle contained the true implication that 'the trend of variation and hence of evolution is predestined by the constitution of the organism'. Such trends were illustrated by the fossil record, and 'definite tendencies in variation spring from certain very remote ancestral causes'.[8] He contrasted this gradualistic concept of autonomous variation with the popular saltatory theories of the time, but allowed for the action of natural selection at a time when the developing trait could be recognised as useful or detrimental.[9]

In *Organic Evolution as the Result of the Inheritance of Acquired Characters* (1890), Theodor Eimer (1843-98) presented a compendium of neo-Lamarckian thought to oppose the views of Weismann.

He believed that phylogenetic branching depended on the direct influence of external conditions; on the functional activity of organisms in relation to the external world; on the struggle for existence; on saltations; on gradual adaptation in response to long-continued exposure to the same environment, and on sexual mixing. The propositions that acquired characteristics could be inherited was the easiest of all his hypotheses to prove. For example, the children of very old parents had remarkably old-looking facial expressions, undoubtedly inherited from their senile parents. His critics were quick to denounce such proofs and to condemn his gullibility in believing stories about the inheritance of scars and mutilations.

Eimer's laws of organic growth consisted of a collection of analogies: just as a mechanical shock could produce dimorphic crystals, so a similar shock to the organism might activate a reconstitution of the parts to form a new whole, or new species. [10] The new conditions of organisation might be just as useful to the organism as if they had been due to natural selection:

> The claims for utility are accidentally satisfied by the results of evolution from internal causes [but]. . .From internal causes characters which are indifferent for the success of the organism, and even harmful characters may arise. . . [11]

This law of internal 'constitutional causes' was far from being a perfecting principle. All evolutionary changes which occurred onto-genically and phylogenically were to be interpreted in terms of changes in growth, and speciation was the result of changes in the conditions of growth. Species were suspended points in a progression which proceeded 'as though following a plan drawn out beforehand'. [12] Different species of the same genus could therefore be at different stages in the progress of the genus. Unlike Nägeli's scheme which had to assume that simple organisms had evolved more recently, Eimer's held that simple organisms had been *suspended* longer. Eimer argued that his laws of growth were inherent in proto-plasm that was alterable, responsive to stimuli, and had 'special growth tendencies' such as irritability, contractility and motility, which could be passed on to specialised organ systems as metazoan evolution progressed. Eimer's laws were holistic in a very general sense:

> by the inheritance of acquired characteristics there is established in every developing and every adult organism a relation of the

particles to one another, which finds its expression in their striving to form themselves into the whole, and to maintain or re-estabish the co-ordinated whole.[13]

On the basis of his studies of butterfly wing spots and felid coat markings Eimer concluded,

That the progressive evolution of a character in a definite direction. . .exhibits perfectly regular stages. . .That where new characters appear, the males. . .acquire them first and that the males transmit these new characters to the species. . .That the appearance of new characters always takes place at definite parts of the body, usually the posterior end [the 'law of undulation'].[14]

Eimer developed his most important law of growth, *orthogenesis,* which he discussed in an 1895 address, and published as a pamphlet in English translation in 1898, as more counter-propaganda to Weismannism. Orthogenesis, a term coined by Wilhelm Haake in *Gestalten und Vererbung* in 1893, meant 'definitely directed evolution', and was intended as a materialistic principle.[14] Weismann's view that 'directed evolution' was a result of natural selection contradicted the Darwinian proposal that selection made use of variations which occurred randomly and in every direction. Eimer pointed to sculpturing and colouration of mollusc shells, that might never be observed in nature by other animals, could have no conceivable utility. Since the fossil record indicated limited lines of development, instead of having the random multiplicity that Darwinism inferred, so orthogenesis must have been a more powerful agent than natural selection. Utility could, of course, coincide with orthogenesis, and natural selection would then be effective. The cause of orthogenesis was not an internal autonomous force, but the external environment. 'In my view development can take place in only a few directions because the constitution, the material composition of the body, necessarily determines such directions and prevents indiscriminate modification.'[15] Only through the agency of outward influences could the limited epigenetic alternatives be expressed.

Although Eimer emphasised the distinction between the perfecting principle and orthogenesis, biologists have confused the two, attributing autonomous causation and progressionism to orthogenesis. Both Nägeli and Eimer protested that their laws were materialistic, but they were largely based on analogy rather than analysis. This does not mean to say that they were talking nonsense about nothing. They appear to have had an intuitive grasp of a real phenomenon that they simply were unable to explain clearly, but which indicated some aspects of evolution that required explanation.

The most basic problem is that the fossil record shows apparent evidence of directed evolution; in certain phyletic lines particular morphological features changed progressively in a particular direction, and the record shows no alternative evolutionary routes for these structures. Nor does the line of evolution suddenly change direction and develop along an alternative line: it may stop, and have living representatives; it may become extinct, and not appear in the recent geological strata. Some oysters have not changed in their shell morphology for hundreds of millions of years: a related line of evolution of *Gryphaea* oysters apparently underwent orthogenesis and became extinct. Nägeli could have argued that a momentum carried the characteristic to extinction: by Eimer's interpretation the environment forced a deleterious orthogenetic drift. Despite the semantic and logical miasmata that Nägeli and Eimer created, or had imposed on them by their critics, the fossil evidence remained to confront biologists, and as H.F. Osborn recognised, this was why palaeontologists unconsciously dissented from Darwinism.

> Our palaeontological series are unique in being phyletic series. They exhibit no evidence of fortuity on the main lines of evolution. New structures arise by infinitesimal beginnings at definite points. In their first stages they have no 'utilitarian' or 'survival' value. . .The main trend of evolution is direct and definite throughout, according to certain unknown laws and not according to fortuity.[16]

Orthogenesis was subsumed under a more general law of growth by L.S. Berg (1876-1950), an ichthyologist and limnologist, who held the chair of geography at the University of Petrograd/Leningrad.[17] *Nomogenesis, or Evolution Determined by Law,* was originally published in 1922 and the first English edition appeared in 1926. Nomogenesis meant, 'development in accordance with definite laws',[18] which were inherent in the organism. Evolutionary events were not random or fortuitous, but something in organisms produced them in a form appropriate to the circumstances.

'Purposive structure and action are thus a fundamental property of the living being.'[19] The conscious choice of the appropriate response to circumstances was a refinement of an unconscious ability inherent in the simplest organisms. Berg saw no vitalistic implications in this; at the cellular level 'purposive' qualities implied no more than irritability, contractility, self-nourishment, and reproduction. Since the directive qualities of ontogenic modifiability were inherent, 'primordian fitness' was inherent, and selection theory redundant.[20] Like many critics of Darwinism, Berg pointed to the limited number of morphological types found both in living organisms and in the fossil record, in contrast to the potential infinity of variability suggested by Darwinism. In Berg's holistic view an organism was a stable system, and most deviations from a limited number of lines of development would be detrimental or lethal. Autonomic orthogenesis was apt to disrupt the harmony between the organism and the environment, but D'Arcy Thompson's allometric transformation had revealed the activity of some kind of inherent laws.

Berg noted that the elongation of the fins in flying fish, adaptive in the sense that it allowed them to escape their predators successfully, might be regarded as pathological in other fish. Cases of extreme elongation in the fins of roach, perch and other fresh-water fishes were analogous to acromegaly in man, brought about by pituitary malfunction, with no utilitarian benefits.[21] By contrast, a number of fish with normal fins were capable of the leaping behaviour that led to flight in flying fish. He concluded,

> in the modelling of the form of the fins, the act of flight. . .had played no part. And if the fish has made use of its pectoral fins for a definite purpose, that is a matter of instinct, *i.e. skill*, psychical adaptation.[22]

Berg argued against the concept of orthoselection which had been noted by T.H. Huxley and elaborated by Plate.[23] Undeviating, directed evolution could not have gone on for millions of years in ammonites and dinosaurs, finally leading to extinction, by the agency of natural selection. The phenomenon of convergence also required a more adequate explanation than selection theory provided. How could whales and fishes have assumed the same form by chance variations and adaptation? Darwin had written,

> as two men, working on the same invention, sometimes indepen-

dently of one another hit on the same invention, so it appears that natural selection, working for the good of each being, and taking advantage of analogous variations sometimes modifies nearly in the same way two organs in two living beings, which have inherited but little in common from their common progenitor.

Berg judged this to be tantamount to argument from design.[24] He minimised the role of natural selection in producing generalised change, such as in size, colour, and shell sculpture in particular geographical areas, inferring that the environment itself 'acts in a determined direction simultaneously affecting all the organisms subject to it', a concept close to Cope's physiogenesis. Mimicry, in which the mimicked organism had no advantage to offer its imitator, Berg also regarded as nomogenic. For example, black and yellow striping appears in a number of hymenopteran groups more primitive than the Vespidae (wasps), as well as many other unrelated insects, and Berg agreed with Handlirsch that insects had a predisposition to assume a wasp-like colouration, and with Punnett, that there must be a limit to the colour combinations possible in relation to insect heredity, and therefore the same patterns might appear in unrelated families.[25] As well as the more easily recognisable convergences it was likely that many supposedly monophyletic groups were convergently polyphyletic. Since nomogenesis could be saltatory, gradualism was not essential, as it had been in Eimer's orthogenesis.

The disappointment of *Nomogenesis* is Berg's failure to carry the causal analysis of nomogenesis beyond a descriptive definition and an exhaustive presentation of putative examples of its effects. He intuitively stressed evidence from ontogenesis that might be related to phylogenesis, but he was not a recapitulationist. He argued that autonomous causes were materialistically 'connected with sterechemical properties of the albumens of the protoplasm of a given organism, which induce the latter to develop in a determined direction'.[26] He did not attribute evolutionary novelty to autonomous causes but to choronomic (geographical) causes. His comparison of Darwin's *tychogenesis* (evolution by chance) with his own *nomogenesis* is reproduced here.[27]

I (tychogenesis)	II (nomogenesis)
1. All organisms have developed from one or a few primary forms, i.e. in a monophyletic or oligophyletic manner.	Organisms have developed from tens of thousands of primary forms, i.e. polyphyletically.
2. Subsequent evolution was divergent.	Subsequent evolution was chiefly convergent (partly divergent).
3. Based on chance variations.	Based upon laws.
4. to which single and solitary individual are subject.	affecting a vast number of individuals throughout an extensive territory.
5. by means of slow scarcely perceptive, continuous variations.	by leaps, paroxysms, mutations.
6. Hereditary variations are numerous and they develop in all directions.	Hereditary variations are restricted in number and they develop in determined direction.
7. The struggle for existence and natural selection are progressive agencies.	The struggle for existence and natural selection are not progressive agencies, but being on the contrary conservative, maintain the standard.
8. Species arising through divergence are connected by transitions.	Species arising through mutation are sharply distinguished from one another.
9. Evolution implies the formation of new characters.	Evolution is in a great measure an unfolding of pre-existing rudiments.
10. The extinction of organisms is due to external causes, the struggle for existence and the survival of the fittest.	The extinction of organisms is due to inner (autonomic) and external (choronomic) causes.

Theodosius Dobzhansky excused Berg's disaffection from natural selection on the grounds that he was writing at a time when mutation was not understood, and World War I and the Russian Revolution had cut off Russian scientists from current research in America and Western Europe. While the criticism that Berg greatly overemphasised polyphyletic and convergent evolution was valid, Dobzhansky's conclusion that 'The assumption of inherent purposiveness remains, however, the Achilles' heel of nomogenesis, as it is of any theory of evolution not based on natural selection' not only disregarded the question of purposiveness, but also the problem of clarifying those aspects of biology and evolution for which natural selection is insufficient if not redundant.[28] Despite Berg's failure to analyse nomogenic principles, the search is still valid.

Berg's compatriot, N.I. Vavilov (1887-1943), proposed a Law of Homologous Series in 1920 to explain the regularity of variation found in related species that was beyond the explanatory scope of natural selection of random and fortuitous mutations.[29] Berg had pointed out the same regularity of variation in distantly related taxa as well as species of the same genus as a product of nomogenesis and bore upon parallel and convergent evolution. As Vavilov stated, the phenomenon had been mentioned earlier by Darwin, Mivart, and others. Lankester, in 1874, had called this phenomenon *homoplasy*, and Osborn elaborated it in 1902 into his Law of Latent or Potential Homology. The mammalian tooth was affected by homoplasy, and after the divergence of the ancestral mammals the addition of new cusps to the basic tritubercular condition followed a similar plan in twenty-three orders of placentals and seven orders of marsupials.

> These homoplastic cusps do not arise from selection out of fortuitous variation because they develop directly and are not picked from a number of alternatives. Neither does it appear that the mechanical inheritance theory, if granted, would produce such a remarkable uniformity of result. We are forced to the conclusion that in the original tritubercular constitution of the teeth there is some principle which unifies the subsequent variation and evolution up to a certain point.[30]

S.J. Gould writes, 'Vavilov, overenthused with this principle, became intoxicated with the notion that his law might represent a principle of ordering that would render biology as exact and experimental as the "hard" sciences of physics and chemistry'.[31] Vavilov wished to construct a biological periodic table of known species and their characteristics that would allow predictions to be made concerning the existence of undiscovered species. He revised his opinions in 1936, saying that he had given an insufficient role to natural selection and that he had since discovered that some of the variations that he had discussed earlier were analogues rather than true homologues. However, Gould has also admitted 'that in his imperfect way Vavilov had glimpsed something important'.[32] Many potential variations cannot be permitted by natural selection if they interfere with developmental pathways, and thus the limited number of permissible variations and their phenotypic correlates in one species are likely to be permissible in related species. Gould reports an experiment in which mice were selected for large body-size alone. An increase of 20 per cent in body-size was accompanied by a 'non-adaptive' 40 per cent increase in brain-size. Thus, in nature, selection for the advantageous body-size in various species would produce a parallel series of brain-size increases. Gould, like Eimer, indicates the existence of such parallel series in shell sculpturing and other anatomical features that cannot be easily explained in terms of adaptation.

The introduction to the original English translation of *Nomogenesis* was written by D'Arcy Wentworth Thompson, who had used 'laws of growth' in much the way Berg used nomogenesis, though with a more mechanical bias.[33] While he rejected many of Berg's conclusions, he regarded *Nomogenesis* as a healthy stimulus for discussion. Thompson's *On Growth and Form* (1917) was sceptical of the adequacy of natural selection, and he looked forward to the time when mechanical principles might explain biological phenomena.

> Of how it is that the soul informs the body, physical science teaches me nothing; and that living matter influences and is influenced by the mind is a mystery without a clue. Consciousness is not explained to my comprehension by all the nerve-paths and neurones of the physiologist; not do I ask of physics how goodness shines in one man's face and evil betrays itself in another. But of the construction and growth and working of the body, as of all else that is of the earth earthy, physical science is, in my humble opinion, our only teacher and guide.[34]

Thompson's mathematical derivations of allometric change from one form to another in related animals, and his coordinate transformation diagrams showing these relationships are well known. His recognition that some evolutionary changes were expressions of relative changes in growth rates is most cogent:

> In short it is obvious that the *form* of an organism is determined by the rate of *growth* in various directions; hence rate of growth deserves to be studied as a necessary preliminary to the theoretical

study of form and organic form itself is found, mathematically speaking, to be a *function of time*.[35]

Thompson claimed no originality for this idea, and recorded that it appeared in the writing of Francis Bacon and the proportional diagrams' of Dürer. It was also implicit in the preformationism of Haller (1766) which suggested that pre-existing organs could be proportionately changed by differential growth rates during their *evolutio*.[36] The nineteenth-century embryologists His and Roux had also noted that embryonic folds were caused by unequal growth rates, and also by the action of physical forces. On the other hand Darwin had cast doubt on the supposition that 'a difference in structure between the embryo and the adult [was]. . .in some necessary manner contingent on growth.'[37] The agents of differential growth discussed by Thompson included temperature and osmotic influences, in combination with internal pressures, elasticity, resistance and surface tension; 'autocatalysis' or enzyme action was also of fundamental significance; Loeb had observed that the rate of nuclein synthesis in early cell division increased in proportion to the amount of nuclear protoplasm already present, which Loeb attributed to enzyme action. Thompson suggested that in some simple organisms growth curves were found to fit enzyme activity curves, and growth could be stimulated or inhibited by simple elements or ions. He paid particular attention to the role of sex hormones in producing sexual dimorphism, the pituitary hormones in producing dwarfism and giantism, and the thyroid gland in controlling metamorphosis. These factors were not only developmental and ontogenic but also evolutionary, explaining the origins of the monstrous dinosaurs and tertiary mammals, while natural selection could only explain, and then only partially, their extinction. While Thompson argued that the eludication of physicochemical growth factors closed off the last refuge of neo-vitalism, he was drawn to Bergson's evolutionism, though not his repudiation of mathematical analysis.

The world of things living, like the world of things inanimate, grows of itself, and pursues its ceaseless course of creative evolution. It has room, wide but not unbounded, for variety of living form and structure, as these tend towards their seemingly endless but yet strictly limited possibilities of permutation and degree. . .Environment and circumstances do not always make a prison, wherein perforce the organism must either live or die; for the ways of life may be changed and many a refuge found, before the sentence of unfitness is pronounced and the penalty of extermination paid.[38]

Thompson's conclusions concerning the growth of organisms can be paraphrased as follows: except in minute organisms form is a function of growth or a consequence of differentiated growth rates, which form gradients from one part of the organism to another. In each type of organism the relative growth rates of different parts maintains a characteristic constant ratio, but the relative growth rates are alterable. Rate of growth is highest in early development, is affected by physical conditions, and is accelerated or retarded at physiological climacterics, e.g. birth, puberty, metamorphosis; it may even be negative. Regeneration is characterised by a low growth acceleration.[39]

Although he recognised the practical value of reduction, Thompson was in no doubt that the holistic view was necessary for adequate synthesis, and was critical of the Darwinists' view that natural selection could act on any part of an organism independently of the rest of the organism. The coordinated parts could be seen to fit the function of the whole, and the whole could be seen as the outcome of the system of physical forces which originated and affected the development of the whole. His co-ordinate derivations of morphological relationships led him to conclude that discontinuous variation in gross morphology was natural, and that sudden changes were bound to have occasionally given rise to new types. Because of the relatively small number of feasible mathematical expressions of change of form, 'such mutations, occurring on a comparatively few definite lines, or plain alternatives, of physico-mathematical possibility, are likely to repeat themselves'.[40] Thus, historically, parallel and convergent evolutionary events had been repeatedly construed in terms of laws of growth.

Thompson's bias towards an over-simplified mechanistic epigenesis and against a phylogenetic explanation was an error that modern biologists can understand since mutation had not then been elucidated, and gene theory was giving difficulty even to committed geneticists. For his biological mathematics Thompson is hailed as a pioneer of theoretical biology, but few of the cut-and-dried relics that are found in the modern literature recall the creative holism that characterises *On Growth and Form*. However, it is difficult to understand how even a neo-Darwinist can look at the familiar diagram of the relationship between *Diodon* and *Orthagoriscus* and still insist that evolution can be interpreted only in terms of point mutations of structural genes, and their differential reproduction.[41]

Figure 10.1: The Morphological Relationship between *Diodon* and *Orthagoriscus*

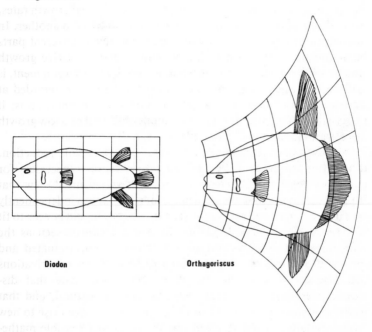

Diodon   Orthagoriscus

Source: After D'Arcy Thompson 1917, courtesy Cambridge University Press. Thompson demonstrated by use of coordinate grids that some allometric growth changes followed mathematically consistent patterns.

The evolutionary theory of J.C. Willis had two elements, the *differentiation theory*, and the *age and area theory*. These were more concerned with plant and animal distribution than with growth, but they also encompassed an autonomous driving force, or momentum, that produced evolutionary change without reference to utility and therefore without reference to natural selection. *Age and Area* (1922) stated, 'age forms a measure of dispersal when one is dealing with allied and similar forms'.[42] Willis attributed a similar thought to Lyell, who had written,

As a general rule, however, species common to many distant provinces or those now found to inhabit very distant parts of the globe, are to be regarded as the most ancient. Numerically speaking, they may not perhaps be largely represented, but their wide diffusion shows that they have had a long time to spread themselves, and have been able to survive many important revolutions in physical geography.[43]

By studying populations of plants which had taken residence in the hollow tops of pollarded willows, micro-environments which had been opened to invasion at known times, and by other similar studies, Willis concluded that while dispersibility by birds, wind, explosive seed pods, etc. was an aid to distribution under certain circumstances, as an aid in the invasion of unoccupied ground it was rarely significant, and the most cosmopolitan genera lacked such dispersion adaptations. Any plant, regardless of dispersibility, would cover an entire land mass in a short period of geological time, though this process could be greatly delayed by a number of physical and biotic obstacles. The age and area theory was also stimulated by Willis's discovery that the endemic plants of Ceylon were usually confined to small areas within it. The orthodox Darwinist view that such species must be highly adapted to those areas was questionable: it could be that the area covered by the plants was proportionate to the age of the species, and that the area of distribution of a plant could be predicted if the age of its species were known. Such predictions for the flora of New Zealand appeared to fit the hypothesis.

The age and area theory seemed to mesh perfectly with a different hypothesis that had been independently formulated by Willis and H.B. Guppy. Guppy in 1906 had called for rejection of the axiom that varieties were incipient species, species incipient genera, genera incipient families and so on up, and that it was possible to project from varieties to the entire scheme of evolution. Guppy, like Willis, saw variety formation as the last twitch of an evolutionary process, or a fraying out of the specific twigs of the greater systematic branches. This 'theory of differentiation of generalised types' had a formal resemblance to Richard Owen's compromise between creationism and evolutionism, which had proposed that species could be derived by natural selection from created archetypes.[44] Guppy's theory had, however, a purely evolutionary foundation. Willis, who credited Geoffroy St Hilaire with the origination of the differentiation theory, concluded that the 'generalised types' must have come into being by large mutations. This then was how the higher taxons like families came into being. They would, of course, have simultaneously been species, by definition, but would have the distinction of having a higher order of phylogenetic potential than normal species. The new types could then undergo cladogenesis by a sort of 'mechanical' process, including further mutation, producing families, orders, genera and finally the terminal species and varieties. These mechanical processes or 'permanent forces', as he called them at one point, qualify as autonomous laws of growth.

In *The Course of Evolution* (1940) Willis insisted that the mutation theory had 'come to stay'; but he also focused on the adaptability of newly emerged types

> the parent of the genus still possesses great adaptability, or suit-ability to a considerable range of conditions. This will enable it to move far with less difficulty than usual, and as at the same time its structural revolution, which has probably little or no relation to adaptation, will be going on, it will give rise to more and more species. These will probably inherit their parents' general suit-ability to conditions, but it is quite probable that it may all the time be getting less (perhaps at each mutation), so that each new species may be liable to become more localised than its pre-decessor[45]

Thus, he presented an evolutionary process analogous to a clock winding down with less energy after each tick.

The evidence of Willis's endemic flora studies can be explained in a variety of ways — localised hybridisations, genetic drift, chromosomal mutation, relict types — they do not all have to be products of an age and area differentiation process. However, because there are alternative explanations it does not follow that the age and area theory in its simplest expression is wrong. If one type had existed longer than another, and the environment has been stable and the adaptive qualities of the two types similar, then indeed the first to appear would likely have a larger distribution. Add to this the tactical advantage the founding population of the environment: 'getting there first with the most', as Nathan Bedford Forrest expressed it, the second arrival having less success in establishing itself may appear only as a highly localised endemic.[46]

Leon Croizat took Willis to task for the naïveté of the age and area hypothesis because it ignored the broader aspects of distribution and the possibility that some endemic species are either relicts, or are products of adaptation to local conditions in the strict Darwinist sense. Indeed, this aspect of 'Willissism' could, as Croizat demanded, be defenestrated into lasting oblivion, but the remainder, including Willis's wealth of botanical observation, opposition to selection theory and differentiation theory, should be recorded and cherished.[47] Even neo-Darwinists, when they argue that it is possible to estimate rates of speciation and extinction are tacitly accepting an aspect of differentiation theory. Calling Darwin's laws of growth *orthogeny* and *oriented evolution*, Croizat wrote,

> Call it *orthogeny*, call it *a genetic pool with restricted mutations*, call it even *orthoselection*, in the 'ortho-' sense the matter is forever the same. . .a stream of far-reaching inheritance which running in a certain rather than in a certain other direction. . .frays at the margins along its course into particular 'adaptations'. These 'adaptations', whether or not clearly amenable to 'natural selection' cannot of course be understood properly and fully unless against the whole course of the stream of which they mark but byproducts.[48]

To illustrate the nature of the stream he noted that,

> upon a primordial cell have been superimposed infinite genetic characteristics, along particular lines of directional development. But the preconditions of these lines are the intrinsic requirements of structure, and, I might add, function. The underlying mechanism of orthogeny is a harmonious allometric rate change, as D'Arcy Thompson had proposed. The *Gryphaea* was produced in this way until its form fitted an available soft substrate, whereupon it was 'selected' for that environment. Selection, while brutal and sharp in its action, was the secondary deed.[49]

One of the most formidable opponents of orthogenesis, 'a hypothesis held in common by most vitalists and finalists' is G.G. Simpson, whose particular target is the putative orthogenesis in horse evolution.[50] The evolution of the horse did not, in Simpson's opinion, follow a direct line, but frequently diverged into side branches which ultimately became extinct. Even the direction of size increase was contravened from time to time. The canine teeth of the sabre-tooths varied considerably from one group to another, but

> varied about a fairly constant average size, which is exactly what
> would be expected if the size were *adaptive* at all times and there
> were no secular trend in adaptive advantage, but only local and
> temporary differences in its details[51]

The orthogenic line of the sabre-tooths was thus a 'scientific legend'.
Simpson also considered the various lines of *Gryphaea* oyster
evolution, which involved spiral shell thickening to the point where
the upper valve was apparently unable to open, to be a trend that was
adaptive to the soft silt of the oyster's environment, the argument
being that the shell sits in the mud with the heavy valve providing
anchorage and holding the living portion out of the turbidity: 'Death
by self-immurement came only to extreme variant individuals in
their old age'.[52] Extinction also came, though the kind of environment to which they were supposed to have been adapted has always
been available. Simpson nevertheless believes that support for the
'straight line' evolution of orthogenesis is merely 'evidence that
some scientists' minds tend to move in straight lines, not that
evolution does'.[53]

Simpson has been challenged on the question of horse evolution
by P.-P. Grassé, who accuses him of, 'only taking into account the
facts that fit with the theory, strictly excluding those which indicate
its weaknesses'.[54] Grassé uses Simpson's interpretation of horse
evolution to support his own hypothesis of serial parallel homology
in mammalian evolution. Agreeing that horse evolution was bush-
like, rather than developing in a single orthogenetic series, Grassé
points out that independent lines of hippomorphs in the Old World
and in the New World showed a number of parallel, progressive
evolutionary features such as tooth height, general size increase,
allometric facial increase, limb length, disappearance of leg bones,
predominance of third toe, etc. Moreover, the anatomy of the hoof
and its ultrastructure are features that require a great measure of
coordinated novelties for their final 'adaptation' to the gait of the
horse. Grassé challenges the assertion that identical environments
were the agents of the evolution of these characteristics. He also
demonstrates the parallel series of new variations in the evolution of
the various lines of theriodont reptile evolution. One of these
ultimately gave rise to the Mammalia, but the other showed
independent parallel evolution of many of the same progressive

features. Grassé concludes that the natural selection of random mutations played no part in this kind of *creative evolution*.[54] He believes that there are two under-estimated, but important factors of evolution, first that evolution is limited to a small number of limited possibilities, second that 'The organisms in which the new plans were embodied have partly lost their evolutionary capabilities, these will not recur in their descendants'.[55] The first point has been noted by a number of authors with an interest in progressive evolution. The second suggests a regressive element in evolution. Taking the example of the insects, Grassé notes that although they have not declined as a whole, various orders have been lost and no new orders have appeared in the last 70 million years. As a corollary to this he observes that the mutations of *Drosophila* do not alter the specific structural pattern, and implies that such mutations have nothing to do with creative evolution. He writes,

> The 'manoeuvering space' of evolution has never stopped decreasing. The genesis of the phyla stopped in the Ordovician; of the classes in the Jurassic; of the orders, in the Palaeocene-Eocene. After the Eocene, the evolutionary 'sap' still flowed through a few orders, since mammals and birds continued to specialize in various directions and invaded all the terrestrial and marine biotopes previously occupied by reptiles. The extent of evolutionary novelties gradually changed. They no longer affected the structural plan but only involved details. The only form which evolution took was speciation. . .Evolution after its last enormous effort to form the mammalian orders and man, seems to be out of breath and drowsing off. . .Aren't the small variations which are being recorded everywhere the tail end, the last oscillations of the evolutionary movement? Aren't our plants, our animals lacking some mechanisms which were present in the early flora and fauna?.[56]

Another formidable opponent of orthogenesis and autonomous forces in evolution has been Bernard Rensch, whose major exposition, *Evolution Above the Species Level* (1959), argues that those factors involved in 'infraspecific evolution', i.e. the production of races and varieties, are the same as are involved in evolution above the species level (transpecific evolution). These factors are mutation (of genes, chromosomes, genomes and plastids); fluctuations of population size; natural selection; isolation;

hybridisation, and their interactions. This leaves little room for disagreement, but demands a much broader scope for the interpretation of evolution than conventional neo-Darwinism.

The first example of orthogenesis discussed by Rensch is Cope's Rule which refers to progressive increase in body-size. He concedes that such lines of evolution are found in most of the major phyla, and led to an array of giant snails, oysters, cephalopods and other invertebrates; kangaroos, sloths, armadillos, rodents, beaver, elk, etc. However, if increase in body-size is useful under some circumstances, then some line of size increase 'caused by natural selection' is bound to appear under these circumstances. Larger animals are more vigorous, more resistant, retain body-heat better, and are swifter of foot. Their allometry, i.e. proportions of sizes of organs, is different from smaller related members, so that organs of defence and attack may be larger, and physiological advantages may accrue. Therefore, utility can be postulated at each stage of the way and orthoselection is a satisfactory causal explanation. This is a plausible alternative. As selectionists are usually quick to point out, plausibility is not a conclusive criterion; however, once a causal explanation that fits selectionism has been found, they look no further.

Increased body-size also produced curious correlated allometric effects, which viewed in isolation would appear to be meaningless orthogenesis, but which make sense if considered as epiphenomena of a useful size increase. This had been found to be a useful sophism by H.W. Conn in 1900:

> By the skilful use of this principle of correlation many of the difficulties arising in connection with the problem of utility may be made to disappear. If a character which is manifestly of no utility can be found in an animal and we are unable to explain it on the ground of utility, it is only necessary to say that it is correlated with some other character which *is* useful and the difficulty disappears.[57]

Rensch considered the antlers of the Irish elk to be be such a case, and Julian Huxley came independently to a similar conclusion. S.J. Gould measured the proportions of antlers to body size, as represented by skull-size, in seventy-nine fossil specimens, and found that with unit increase in body-size there was an allometric ($\times 2.5$) increase in antler size.[58] Gould assumes that the antlers had a direct sexually selective value, the set of the antlers indicating a display function rather than a combat function. His explanation of the species extinction — the onset of a new ice age! He concludes, 'Darwinian evolution decrees that no animal shall actively develop a harmful structure, but it offers no guarantee that useful structures will continue to be adaptive in changed circumstances'.[59] One of the curiosities of selectionism is that it allows mutually contradictory hypotheses to flourish, so long as they wear the badge of natural selection. This 'joyous debate' illustrates how open selectionism is to intellectual challenge, but when Darwinism decrees. . .the fun stops.

However, Gould has more recently shifted ground, and now acknowledges the benefit of thinking in terms of parallel series of homologues explained in the more holistic terms of epigenetic evolution:

A complete theory of evolution must acknowledge a balance between 'external' forces of environment imposing selection for local adaptation and 'internal' forces representing constraints of inheritance and development. Vavilov placed too much emphasis on internal constraints and downgraded the power of selection. But Western Darwinians have erred equally in practically ignoring (while acknowledging in theory) the limits placed upon selection by structure and development — what Vavilov and the older biologists would have called 'laws of form'. We need, in short, a real dialectic between the external and internal factors of evolution.[60]

The various Laws of Growth are summarised in Table 10.1.

There is a possible materialistic mechanism of evolution that might operate beyond the scrutiny of natural selection and bear upon the various expressions of the laws of growth. To paraphrase D'Arcy Thompson and R. Goldschmidt, coordinated reaction velocities in early development determine the form of the mature organism. If there is a change in the relative rates, and that change can be equilibrated so that its effects are not lethal, then a modified structure is produced. Its complement of structural genes — the ones responsible for the production of the array of proteins that are the basis of the structure and functions of the organism — need be no different in the new organism. If such an epigenetic change occurred

Table 10.1: The Laws of Growth

Author	Name of Law	Cause	Consequences
Darwin 1859	Laws of Growth	Unspecified	Development of non-utilitarian characters
Kölliker 1865	Theory of Heterogeneous Generation	Radical developmental changes	Saltatory evolution
Cope 1868	Law of Retardation and Accleration	Vital growth force	Allometric changes, life-cycle alterations
Cope 1871	Law of Repetitive Addition	Vital growth force	Segmentation and differentiation
Nägeli 1884	Perfecting Principle	Autonomous momentum	Orthogenesis and grade evolution
Cope 1896	Cope's Rule	Vital growth force	Orthogenetic increase in body size
Osborn 1895	Law of Latent or Potential Homology (= Lankester's homoplasy)	Autonomous variation	Othogenesis, parallel evolution
Eimer 1895	Orthogenesis	Direct effect of environment	Direct-line evolution
D'Arcy Thompson 1917	Coordinated transformation	Hormonal induction of differential growth-rates	Morphological and parallel evolution
Vavilov 1920	Law of Homologous Series	Materialistic ordering principle	Parallel and convergent evolution
Guppy 1906 } Willis 1922 }	Theory of Differentiation of Generalised Types	Materialistic, autonomous driving force	Cladogenesis
Berg 1922	Nomogenesis	Purposive, autonomous laws	Saltatory evolution, parallel evolution

again in the same direction then the modifications would be orthogenically exaggerated. The important question is, how could an epigenetic change keep occurring in the same direction? If a developmental gene can be shown to mutate at a faster rate than others, the possible mutation might be constrained in a single direction by the limited epigenetic possibilities. On the other hand, it is conceivable that a given gene might *duplicate* at a faster rate, over many generations, than others involved in the development of the organism, resulting in doubling of the rate of reaction controlled by the duplicated gene, and consequently a relative rate change. A reduplication of the same gene would produce an exaggeration of its effects in the same direction as before, assuming that the original duplication did not have lethal or seriously detrimental effects. This kind of event falls squarely under the definition of materialistic orthogenesis, and the same duplicative tendency of the same gene in related groups could produce parallel homologous series. Sooner or later in the course of this kind of evolution the results might coincide with internal or environmental exigency, and then be seen to be adaptive by an eye trained to ignore anything else. One of the best known cases of advantageous gene duplication at the developmental level is discussed by S. Ohno.[61] In the nucleolar organiser of the frog *Xenopus laevis* there are four hundred duplicated copies of the pair of genes responsible for ribosomal RNA reproduction. Furthermore, during oögenesis the nuclear organiser region of the chromosome is duplicated many times, so that eventually the egg nucleus contains about one thousand copies of the organiser, each with four hundred and fifty duplicates of the gene. The advantage of this is that reactions requiring ribosomal RNA, i.e. protein production in general, can be greatly accelerated at an early stage of development. If such acceleration is seen to be advantageous then the selectionist interpretation is that natural selection has operated at every step of the way, selecting the properties for nucleolar gene duplication. But the propensity for duplication has to exist before it can be selected; and there are possibly many cases where such a propensity effects changes with no immediate adaptive value. The real problem here is how to stop the process before it leads to hypermorphosis and extinction. To say, for example, that modern oysters lost the propensity for duplication leading to shell thickening and coiling, while *Gryphaea* oysters did not, explains nothing. But if the selectionist interpretation is adhered to we must pretend that hypermorphosis is spurious and extinctions occur through climatic change. Ohno remarks that 'the fact that a segment of the chromosome can engage in repeated DNA relocation and disseminate its free copies, while the rest of the chromosomes are not involved in DNA replication, has far reaching implications'.[62] The implication of a materialistic mechanism for orthogenesis is not one that he addresses, but it requires consideration. However, as Croizat lamented,. . .'these tendencies will never be studied with the earnestness they amply deserve by students propagandistically induced to lump them in the end with 'vital urge' and 'cosmic goals' implicity at least discredited as anti-scientific'.[63]

11 THE STRUGGLE FOR EXISTENCE OF SELECTIONISM

Darwinian Natural Selection as the final arbiter of control is sound and clear of objections. But Darwinism as the all-sufficient or even causo-mechanical factor as the sufficient explanation of descent is discredited and cast down.

Vernon Kellog 1907[1]

By 1876 Charles Darwin had declared *nunc dimittis* on his evolutionary writing, and T.H. Huxley's attention shifted to other issues.[2] Darwin died in 1882, and the phoenix of neo-Darwinism rose in August Weismann's declaration of the all-sufficiency of natural selection. His 1883 essay, 'Heredity', denied Lamarckian inheritance on the grounds that continuity of the germ plasm and its isolation from somatic effects made it impossible for somatogenic induction to occur. Only in the structure and behaviour of the chromosomal apparatus could exist the mechanism for the transmission of genetic information from parent to offspring and its distribution from cell to cell within the developing organism. This was further elaborated in his book *The Theory of the Germ Plasm* (1893). The chromosomes of the germ cells were supposed to be composed of 'ids', each one representing characters of ancestral organisms. The ids themselves were composed of *determinants,* which were made up of molecular *biophores.* In the production of the reproductive cells the number of ids was halved (as the number of chromosomes was halved) and the fertilisation of the egg brought the id and chromosome complement back to the original number. *Amphimixis,* the shuffling of the ids and chromosomes, was the mechanism that produced variation. Weismann proposed to sweep away the revisionism of pangenesis and neo-Lamarckism, and out-Darwin Darwin by proclaiming the all-sufficiency of natural selection, with the origins of variation explained in terms of chomosome behaviour.[3] This challenge earned Weismann the title of neo-Darwinist, which Samuel Butler had sarcastically used of Charles Darwin himself, claiming to prefer 'the original Darwinism of Dr Erasmus Darwin'.[4] Apart from the suggestion that the biophores were divided up among the different cell lines,

Weismann's theory was remarkably insightful and close to the truth, since chromosomes carry molecular genetic information which has an evolutionary history going back to the origins of life, and sexual reproduction is a source of a good deal of variation, even in the absence of mutation. But the germ plasm theory ran into opposition from almost every quarter, from Mivart, Romanes, the neo-Lamarckians, Driesch and especially from Herbert Spencer. Indeed, Weismann's epistemological significance becomes very important when one considers that much of the research and speculation discussed in the previous chapters was partly in reaction to his mechanistic theory of heredity. Much of the rest was influenced by Mivart's initial exposure of the inadequacies of selection theory in *The Genesis of Species*. These two biologists are important keys for unlocking the epistemological puzzles of the late Darwinian period. The Spencer-Weismann controversy unfolded as an exchange of provocative essays in the *Contemporary Review* in 1892 and 1893, the years of the publication in English of Weismann's essay on heredity and his book on the germ plasm theory. Weismann's chief problem in having his ideas accepted was that he was arguing against the authority of Darwin's pangenesis and Darwin's de-emphasis of natural selection and trying, virtually alone, to shift the paradigm of evolutionism back to pure selectionism. A more complete proof of the kind provided by molecular biology could have easily have been dismissed as a special case with no general relevance, just as the Darwinists had dismissed the quasi-Mendelian observations that were known to them. Evidence is a weak siege-weapon for storming a belief-system. Until well into the 1920s most evolutionists wanted to believe that the environment could cause adaptation directly, though reluctantly admitting the lack of evidence. Hans Driesch was already conducting the experiments that disproved the disintegration of the ids, and in the absence of any knowledge of the hierarchical organisation and coded control of protein synthesis a molecular theory seemed fantastic. There were too many traits in the makeup of an organism for each one to be individually represented in germ cells even by particles as small as molecules. Thus, this first wave of neo-Darwinism broke and receded.

The new wave was long in gathering, but its scattered origins went back earlier than Weismann. Essentially selectionistic, its thrust was the quantitative treatment of heredity and evolution in populations and species. In order to circumvent the stumbling block of blending inheritance, several authors independently developed concepts of the separation of sub-populations that would allow the maintenance of diversity. Moritz Wagner took priority with an 1868 paper on 'The Law of the Migration of Organisms', based largely on entomological data, which maintained that 'the constant tendency of individuals to wander from the station of their species is absolutely necessary for the formation of races and species'.[5] This is now called *allopatric speciation*. J.T. Gulick's 1872 paper on 'Diversity of Evolution under One Set of External Circumstances' expressed a similar idea. 'Separation generation' was a necessary condition for divergent evolution; but not for the transformation of all the survivors of a species in one way. Separation did not necessarily imply physical barriers nor geographical isolation, and differences of external conditions were not necessary for diversity; separation and variation were all that were necessary. 'Seperation' meant anything, in the species of the environment, that divided the species into two or more sections that did not freely interbreed, 'whether the different sections remain in the original home or enter new and dissimilar environment'.[6] Separation of sections 'in the original home' now has the apt term of *sympatric speciation*.

G.J. Romanes formulated, in 1886, a hypothesis of sympatric speciation as 'Physiological Selection: An Additional Suggestion on the Origin of Species':

> If a section of a species is prevented from intercrossing with the rest of its species we might expect that new varieties (for the most part of the trivial and unuseful kind) should arise within that section, and that in time these varieties should pass into new species. . .Some individuals living on the same geographical area as the rest of their species, have varied in their reproductive systems so that they are perfectly fertile *inter se* while absolutely sterile with all other members of their species. . .the barrier, instead of being geographical, is physiological.[7]

Gulick had earlier developed the concept of *segregate fecundity* to imply the same thing.[8] In addition to an exhaustive exploration of the possible varieties of segregation Gulick provided a mathematical analysis of the consequences of different degrees and combinations of segregation, anticipating the quantitative approach of the second wave of neo-Darwinism.

The 'obstacle' of blending inheritance also had the heuristic property of stimulating the quantitive approach to evolution. The biometricians, who sought to find solutions to problems of inheritance and diversity in correlations that might have been missed by the casual observer, were in the vanguard of neo-Darwinism's second advance. Francis Galton, seeing the need to use mathematical analysis to surmount subjective obtuseness in the interpretation of nature, was an early contributor to the development of biometrics and to quantitative neo-Darwinism. His statistical analysis of the efficacy of prayer is an amusing if flawed example of his art.[9] This period of evolutionism is succinctly described by William Provine in his *History of the Origins of Theoretical Population Genetics* (1971). Karl Pearson (1857-1918) was a mathematician and physicist who was inspired by Galton's idea that statistical correlation could bring the elusive causation of biology under control. Pearson was also influenced by his associate, the biologist W.F.R. Weldon (1860-1906), whose early research turned from embryology to the correlation between evolutionary changes in adults and larval forms. Galton's *Natural Inheritance* (1889) suggested an analytical tool for the study of such correlations, and Weldon ultimately became the prophet of quantitative neo-Darwinism:

> It cannot be too strongly urged that the problem of animal evolution is essentially a statistical problem, that before we can properly estimate the changes at present going on in a race or a species we must know accurately: a. the percentage of animals which exhibit a given amount of abnormality with regard to a particular character; b. the degree of abnormality of other organs which accompanies a given abnormality of one; c. the difference between the death rate percent in animals of different degrees of abnormality with respect to any organ; d. the abnormality of offspring in terms of the abnormality of parents and *vice-versa.* These are all questions of arithmetic, and when we know the numerical answers to these questions for a number of species we shall know the deviation and the rate of change in these species at the present day — a knowledge which is the only legitimate basis for speculation as to their past history and future fate.[10]

Weldon and Pearson began to work together in 1893, founding a Royal Society Committee for Conducting Statistical Inquiries into the Measurable Characteristics of Plants and Animals. Co-members of the committee were Francis Darwin, E.B. Poulton, R. Medola and A. Macalister, with Francis Galton as chairman. Their early attempts to formalise selection theory bogged down, partly because of Weldon's inclination to bite off more than he could chew, and partly because of disagreement about whether evolution was continuous or discontinuous. William Bateson took arms against the biometricians, since his book *Materials for the Study of Variation* (1894) was reviewed by Weldon, who recommended that Bateson study biometry.[11] The subsequent public and private feud between Bateson and Weldon and his supporters was a paradigm of scholarly disagreement: detachment led to debate, which broke down into mutual insult. The history of evolutionism is punctuated with such exchanges, which illuminate only the fact that many biologists, then and now, rate rhetorical victory over scientific proof. One of the outcomes of the feud was the enlargement of the Statistical Inquiries Committee, its renaming as the Evolution Committee, and its dominance by the new member William Bateson, due to the resignations of the biometricians and the shift in emphasis of the committee's efforts to support Bateson's breeding experiments. Bateson thus had a base from which to launch neo-Mendelism in 1900. The first inclination of the biometricians was to regard Mendel's uncongenial results as a special case.[12] In Provine's judgement the feud between Bateson and the biometricians postponed the development of population genetics by about fifteen years, although the heat went out of it with the death of Weldon in 1906.[13] He does not put all the blame on one side, remarking that if the biometricians had been appreciative of Bateson's 1904 throw-away line: 'when the unit of segregation is small, something mistakeably like continuous evolution must surely exist', the progress of the subject might still have been accelerated.[14] The more traditional Darwinists were regarded by the biometricians as weak sisters:

> The Committee. . .contains far too many of the old biological type, and is far too unconscious of the fact that the solutions to these problems are in the first place statistical, and in the second place statistical, and in the third place biological.[15]

While many Darwinists had bobbed in the wake of Darwin's vacillations, E.B. Poulton was made of sterner stuff, and his influential support of Darwinism in the darkness before the dawn was clouded with none of the doubts and revisionism of the early Darwinists. Here is the Selectionists' Creed, abstracted from his essays.

The Darwin-Wallace theory of Natural selection has three factors: *Individual Variation:* all individuals are different and some of the differences are inherent. *Heredity:* ensures that inherent variations are passed on to the next generation. *Struggle for Existence:* more individuals are born than can survive; these must compete and strive to survive and reproduce.[16]

Selection theory does not explain the causes of variation; but 'So long as as individual variation is present, so long as it is hereditary, it does not signify how it is produced. . .So long as it is there it is available, and Natural Selection can make use of it.'[17] (Poulton may not have been the first to claim that natural selection was the umpire who said 'They ain't nothin' till I calls 'em', but the message was loud and clear.)

The Creed admitted that natural selection could not account for the appearance of incipient stages of new organs. The appearance of new organs was rare,

Organs. . .are formed by the modification of pre-existing organs. . .a single origin accounts for several successive organs, or at any rate several functions instead of one. . .Unguided variation can never be the 'origin of the fittest'. Even if large variations lead to fresh positions of stability Natural Selection is still invoked in order to arbitrate between different mutations, as well as between these and the parent species.[18]

Poulton's criticisms of the saltationists, neo-Mendelians and mutationists bordered on the vituperative. William Bateson was the focus of this attack, for writings 'injurious to Biological Science, and a hindrance in the attempt to solve the problem of Evolution'; for dogmatism concerning work with which he was evidently imperfectly acquainted; assumptions made on the slenderest evidence; the appropriation under the name of Mendel of results owed to Weismann; the exaggerated importance of Bateson's work on Variation, and Mendel's work; the contemptuous depreciation of other lines of investigation inspired by Darwin and Wallace. All of these resulted in a widespread belief among the ill-informed that the teaching of the founders of modern biology had been abandoned.[19] Moreover, Poulton added that he had consulted 'a number of the leading zoologists and botanists in this country', who had agreed with his opinions without a single exception. 'Whoever studies the distinctions of geographical varieties closely and extensively will smile at the conception of the origin of species *per saltum*'.[20] The creed thus closed with the axiom that only those with close and extensive personal experience of intraspecific varieties or population biology were in a position to make any judgements on evolution theory. As a general polemic Poulton's attack on Bateson has the familiar mixture of accusations of ignorance, unfair attribution of priority, dogmatism, irrelevance, monomania and sacrilege, backed with appeal to authoritative consensus. All are still to be found in responses of modern selectionism to its critics, but Poulton's essays are distinguished by a refreshing vigour and lucidity, together with originality of invective.

Poulton was in the right when he cautioned that great care must be taken in assessing discontinuity, giving the example of the colour of butterfly wings, 'It is in every way probable that the chemical changes by which a pigment is transferred are excessively minute, although the impression on our senses is so great'.[21] He was justifiably suspicious of De Vries's explanation of the exceptionality of *Oenothera,* namely that the period of mutability in a species was brief and rare, and inclined to agree with G.A. Boulenger that *O. lamarckiana* was a viable hybrid capable of disjunction into the parental types with the appearance of saltation.[22] Poulton's main objection to large mutations like Darwin's was that they could not be adaptive, whereas step-by-step selection of minute stages would be adaptive, an argument developed later by R.A. Fisher. 'Sports' such as the ancon sheep, that was perceived to be immediately useful by breeders because its short legs allowed it to be easily confined, were irrelevant. If we opened our eyes we would be 'driven to realize the uselessness for evolution of many a result which the breeder can attain.'[23] He prophesied that Darwinists, together with biogeographers would prove that 'the relations of an organism to its environment are so accurately and elaborately adjusted that any advance by large variations is only possible as a very rare occurrence'.[24]

Poulton demonstrated some understanding of the recently rediscovered Mendelian genetics, but because Bateson had turned them into the neo-Mendelian doctrine, or as Poulton termed it, the Batesonian doctrine, he played down their importance, noting that they offered an 'arithmetical precision we did not possess before.[24] The usefulness of Mendel's work was limited to understanding the prevention of blending; some characters might be non-Mendelian, and even those that were might prove 'in the long run to be less fixed than they are in the early series of generations'.[25]

He did not have long to wait before his confidence in the continuity of variation was vindicated. He had overlooked some ammunition already in print. William Castle, who had at first been a mutationist, had bred a race of polydactylous guinea pigs that he observed 'was not *created* by selection, though it was *improved* by that means'.[26] However, by 1911 Castle's work was beginning to show that by selection an apparently continuously variable coat pattern could be produced.[27] Mendel had himself suggested that if a trait was determined by two or more independent genetic factors an appearance of continuity of variation would be the result. H. Nilsson-Ehle provided proof of Mendel's supposition in 1908-09 with the results of his multiple-factor inheritance-breeding experiments with oats. He concluded that 'many mutations, above all in exotic plants, are only new groupings of already present factors and really represent nothing new, especially in such cases where they throw back';[28] This argument could apply not only to De Vries's *Oenothera* results, but also to some of the controversial 'sports' known to animal breeders.

T.H. Morgan (1866-1945), having completed his doctorate at Johns Hopkins University in 1890 went to Europe, where he was associated with Anton Dohrn, founder of the Naples Marine Station, and collaborated with Hans Driesch.[29] In *Evolution and Adaptation* (1903) Morgan expressed sympathy with Mivart's criticism of Darwin, and agreed with Mivart and Driesch concerning Weismann's mechanistic neo-Darwinism:

> Weismann had piled up one hypothesis on another as though he could save the integrity of the theory of a natural selection by adding new speculative matter to it. The most unfortunate feature is that the new speculation is skillfully removed from the field of verification, and invisible germs whose sole functions are those which Weismann's imagination bestows on them, are brought forward as though they could supply the deficiencies of Darwin's theory.[30]

He was particularly incensed by Weismann's view that death was a product of natural selection: 'This insidious form that the selection theory has taken in the hands of its would-be advocates only serves to show to what extremes its disciples are willing to push it'.[31]

One of the interesting features of this period of the history of genetics is the spectacle of biologists eating their words, being forced to make radical changes in their own ideas because of their own experimental results. This is greatly to their credit, and it puts human faces on the textbook effigies. One of the early adjustments Morgan had to make was his view of the mutation theory. He had been favourably impressed with the examples of discontinuity of variation provided by Bateson, and was greatly taken with the way that the mutation theory met the major difficulties of Darwinism. This was before *Drosophila,* although the fruit fly had already been used as a subject for genetical research by William Castle, and earned the epithet of 'miserable little saprophyte' from W.M. Wheeler.[32] In 1909 the 'fly room' opened for business at Columbia University, with the undergraduates C.B. Bridges and A.H. Sturtevant and post-graduate H.J. Muller in attendance the following year.[33] By 1912 the *Drosophila* results had shown that there were small, stable, heritable variations of the kind needed as the raw material for natural selection. *A Critique of the Theory of Evolution* (1916) revised many of Morgan's earlier ideas, and introduced the term *mutation* in a revised form, observing that an indifferent or injurious mutation could be eliminated from a population, but a beneficial mutation would appear in more and more individuals. In 1925 a revised edition, *Evolution and Genetics,* became the first textbook of classical genetics, with chapters on Mendelian genetics, mitosis, meiosis, chromosomes, linkage groups, sex linkage, crossing over — all the material that a modern biology student is still required to know before proceeding to molecular and population genetics.

In his historical introduction Morgan described 'the four great cosmogonies, or the four modern epics of evolution'.[34] These included the three discredited hypotheses of direct change brought about by the environment; self-induced change through the responses of the organism, plus the inheritance of acquired characteristics, and the perfecting principle. Only Darwin's theme of the survival of those individuals whose chance variations fitted them for their environment remained supportable since the new evidence from genetics demonstrated slight mutations that might be beneficial. While worries about blending inheritance were over, there was still the problem of small beneficial mutations being swamped, but, on current evidence,

> If the offspring of individuals did continue to share as wide a range of variability about the new average as did the original population then it would follow that selection could slide successive generations along in the direction of selection.[35]

He noted that Darwin himself had been unusually careful on this subject.

> It is rather by implication than by actual reference that one can ascribe this meaning to his views. . .We should not forget that just this sort of process was supposed to take place in the inheritance of use and disuse. What is gained in one generation forms the basis for further gains in the next generation.[36]

This marks a shift from typological thinking to demological thinking, that is, instead of the individual evolving the population evolved. This is one of the distinctions made between Darwinists and neo-Darwinists.[37] Morgan noted that a good deal of variation in individuals occurred by recombination, and was complicated by 'modifying genes', i.e. where traits were affected by more than single factors. But since all genes obeyed Mendelian laws there would come a time when selection ceased to produce further effects and the genetic material would then be homozygous, and only changes in the genes themselves would produce new characteristics, 'transcending the extremes of the original population'.[38]

The recessive mutant allele could spread randomly in the population until it came into homozygous condition, and then spread by selection if it were 'better suited to the old environment; or, if better fitted to a new environment within reach it will then give rise to a new type leaving the original type in possession of the old station'.[39] Any new mutation of any kind was in the position of Goldschmidt's 'hopeful monster', hopeful both that there would be enough of its kind for its perpetuation, and hopeful of an adjacent environment that would suit it better than the original allele. Morgan admitted that physiological changes were more important than superficial morphological changes, but held that since the two often went together research on superficial traits was justified. He also showed how selection could give the appearance of directed evolution. 'Whenever a variation in a new direction becomes established the chance of a further advance in the same direction is increased. . .so long as the advance does not overstep the limits where further change is advantageous'.[40]

In *The Theory of the Gene* (1928) Morgan stated that the characters of the individual were referable to paired elements (genes) in the chromosomes, held together in a definite number of linkage groups; the members of each pair of genes separated during development of the germ cells so that each germ cell came to contain one set only. Although the members of different sets assorted independently, an orderly interchange, crossing-over, also occurred in corresponding linkage groups, and the frequency of cross-over demonstrated the linear order of the elements in each linkage group, and the relative positions of the elements on the chromosomes.[41] Morgan tried to estimate the size of a gene on the data from crossing-over frequencies. The result seemed too large for any molecule, even though Morgan had at last been inspired by 'the fascinating assumption that the gene is constant because it represents any organic chemical entity'.[42] Muller's complementary work on the effects of X-rays showed not only how these greatly increased gene mutations, but also offered another way of measuring gene size. His estimations were of the same order of magnitude as Morgan's.[43]

While the roots of modern selection theory were slowly taking hold due to the results of selective breeding experiments and the application of Mendelian principles, Morgan thought that

> These theoretical considerations do no more than suggest certain possibilities concerning the theory of natural selection. Before we can judge as to its actual efficiency we must be able to state how much of a given advantage each change must add to give it a chance to become established in a population of a given number. Since only relatively few of the individuals produced in each generation become the parents of future generations numbers count heavily against any average individual establishing itself.

Although there were few data relevant to this difficult problem and theoretical analysis was in its infancy, Morgan observed that J.B.S. Haldane had developed, 'a partial analysis of the problem for a few Mendelian situations'.[44] The classical geneticists had finally come round to the biometrical approach proposed by Weldon more than thirty years before.

It is worth considering how much vigour remained in traditional Darwinism. The total disarray depicted by Jacques Barzun in *Darwin, Marx, Wagner* (1941) was symbolised by Vernon Kellog's remarks given in the leading quotation. Barzun wrote,

> Whatever the merits of the case of Lamarck, Erasmus Darwin, and Samuel Butler, as against Charles Darwin and his cohorts, little doubt could be left in the minds of thoughtful scientists that the orderliness in the facts of Heredity and Variation could not depend on pure Natural Selection, whose essence it is to make change produce adaptations, and whose operation can only begin after variations — however generated — have put in an appearance. Mechanism was doomed unless a new mechanism was discovered in the germ plasm.[45]

The biometricians had conducted a polemical campaign which had discredited some of them in the general view. Although they had a number of heirs who continued in the quantitative tradition, their ideas were yet to mature to the point where they would have a universal impact. Thus, the Darwinists and neo-Darwinists had the confidence to press on, but among the larger populace of biologists, generalists and casual onlookers there was doubt and uncertainty. Lull presented Darwinism as the central theme of evolutionism, but remained open to the possibility of neo-Lamarckian mechanisms and various laws of growth. The mechanism that neo-Darwinism needed was about to appear: while it was not the mechanism that Burzun claims is necessary for evolution theory, It was at least a mechanism.

T.H. Morgan's remarks on the inadequacies of quantative theoretical genetics set the scene for the entrance of the three wise men of theoretical population biology, J.B.S. Haldane, R.A. Fisher and Sewall Wright. There had been a good deal of back-stage preparation for this entrance. In 1908 the Hardy-Weinberg Law, proposed independently by its two authors, demonstrated that dominance did not effect the increase of the dominant allele in a population any more that recessiveness would cause the disappearance of an allele, thus showing that variability was preserved in a randomly reproducing large, population.[46] The emotional overtones of the word dominant had perhaps induced the conclusion that the allele would come to dominate the population numerically. The equilibrium of the population is theoretical, since it is upset by immigration, mutation and non-random mating. A publication by Weinberg predicting the distribution of a single and multiple locus with multiple alleles, as well as dealing with correlations in human populations was implicitly critical of Karl Pearson, and dismissed as a 'curiously ignorant account of the biometric treatment of heredity. . .it hardly seems needful to reply to criticisms of this nature'.[47] As a result Weinberg was not credited with priority (of six months) nor even co-authorship of the principle for some years. Provine's judgement of this is that it was due to 'ignorance' of Weinberg's work, but it is obvious from his account of the history of the biometricians that whoever was not for them was against them.[48]

It must be emphasised here that the later neo-Darwinists were dealing theoretically with natural selection in the sense of differential reproduction, with the differentiality being given a numerical value called a *selection intensity* or *selection coefficient,* its value based on observations on the distribution of alleles from one generation to the next. *Fitness* was a concrete expression indicating the degree of reproductive success of a given variant in a given population. We must accept that differential reproduction occurs (the alternative would be for all organisms to have the same number of offspring during their lifetimes); that the differential reproduction can be observed most clearly in terms of the frequency distribution of genetic traits; that the development of the mathematical treatment of allele distribution in populations is self-consistent and necessary for an understanding of the biology of populations; that there is no metalogical confusion at this level of discourse. This is not to say that we must accept the broader theoretical implications of selection theory as the explanation of evolution. While the development of the mathematical aspects of population biology was scientifically sound, the growth of neo-Darwinism as selectionism was characterised by an over-simplification that has become an obstacle to the investigation of the evolution of complex whole organisms.

In 1915 R.C. Punnett's *Mimicry in Butterflies* proposed that the evolution of mimicry must be discontinuous, on the grounds that Mivart and Bennett had argued. He believed that the slight incipient variation in the direction of the mimicked model would be worthless. Punnett's collaborator, the mathematician H.T.J. Norton, constructed a table which predicted the performance of a genetic factor subjected to a hypothetical series of *selection intensities.* From Norton's table the fate of a given allele at a given selection intensity, after a given number of generations, could be determined. The speed of the process surprised Punnett and vitiated his demand for discontinuity in the evolution of mimicry. This was the earliest quantitative demonstration that 'selection pressure' was a more potent agent of allele distribution than 'mutation pressure', and it stimulated the interest of J.B.S. Haldane and supplied support for T.H. Morgan's developing gene theory. The model assumes unrestricted gene flow, which is challenged by some modern evolutionists.

R.A. Fisher (1890-1962) was educated as a mathematician, becoming interested in genetics and evolution after he graduated from Cambridge. Because he was a Mendelist, Fisher's original co-operation with the biometricians soured, and he had some difficulty in publishing his first paper demonstrating correlations based on Mendelian principles.[49] The paper also showed how the effects of dominance could be distinguished from gene interactions and non-genetical environmental effects on the phenotype. Fisher subsequently developed ideas on the evolutionary consequences of Mendelian principles; concluded that for the survival of mutations the population had to be large; and that chance was less important than selective elimination of alleles. Fisher collaborated with E.B. Ford on the study of natural populations of moths in 1926, finding that large populations had greater variability as he had predicted; and the following year he proclaimed the synthesis of Mendelism and Darwinism, attacking Punnett's stubbornness in adhering to a discontinuity theory based on mimicry.[50] He demonstrated that apparent phenotypic saltations producing polymorphism could be explained in terms of an original phenotypic continuity, that by slow accumulation of modifications in two diverging lines, resulted in dimorphism in the population, i.e. two distinct forms of the species need not have appeared by saltation, but could have gradually diverged, according to Darwinist and Mendelian principles. A 1928 publication hypothesised how dominance could evolve, and *The Genetical Theory of Natural Selection* (1930) brought all his arguments together to provide the needed synthesis of Darwinism, Mendelism and the biometrical approach. Fisher likened the increase in fitness in a population to the performance of gases in accordance with the gas laws and the second law of thermodynamics, implying that the larger the sample the greater would be the probability of the statistical predictions coming true, again assuming rapid and unrestricted gene flow.

Darwinian blending inheritance had had the corollary requirement of a great deal of new variation in every generation to prevent swamping and return to the original centre of stability, as Galton had called it. Fisher pointed out that this explained the great stress put on the importance of variation by earlier authors. It does not, however, exclude the possibility that a large mutation could make natural selection redundant, that it might be especially significant in a small population, or that some variations towards improved homeostatic function would occasion positive differential reproduction regardless of circumstances. It must also be remembered that the saltationists in general were dealing not with pure supposition but with the evidence of 'sports' and the discontinuity of the fossil record. 'Sports' might be spurious saltations evolved by slow continuous selection combined with a masking dominance, evolved according to Fisher's hypothesis, leaping out only very occasionally as homozygotes. But this was hypothesis, not proof, and should have been considered in contrast to the hypothesis that large phenotypic changes might be reproduced by simple point mutation in genes controlling development.

Sewall Wright took gene interactions more seriously than did R.A. Fisher, and in his early work dealt with polygenic coat-colour traits. In 1917 Wright generalised that size differences in individuals were more dependent on factors that affected the whole body than factors that affected parts of the body.[51] By 1925 he had come to the conclusion that random drift in small populations was an important generator of new polygenic systems, and was therefore able to provide diversity of variations for natural selection. From the non-selectionist point of view such generation of new polygenic variants would be highly significant, and natural selection would be considered a secondary mechanism. Wright was particularly critical of Fisher's theory of the mass selection of single-factor variants: the fitness values assigned by Fisher were too rigid, since they ought to vary according to the gene interactions, a viewpoint that Mayr later called 'genetic relativity'.[52] Wright believed that a mosaic population with semi-isolation of small, inbreeding sub-populations would be more productive of rapid evolution.

> The consequence would seem to be a rapid differentiation of local strains in itself non-adaptive, but permitting selective increase or decrease of the numbers in different strains and thus leading to relatively rapid adaptive advance of the species as a whole.[53]

This view, which is unpopular with some orthodox population

biologists, appeared in 'Evolution in Mendelian Populations' in 1931.

J.B.S. Haldane (1892-1963) was the son of the anti-mechanistic holist J.S. Haldane. J. Maynard Smith has remarked that J.B.S. Haldane's original interest in biochemistry might have been a reaction to his father's 'peculiarly woolly ideas about the nature of living matter'.[54] Haldane's Oxford degree was in mathematics, classics and philosophy. His physiological experiments as his father's research assistant, often with themselves as subjects, and 'amateur' experiments with the genetics of guinea pigs with his sister, provided a practical background in biology. He had no formal education in natural science and, although 'he was not himself a good observer — and he was a terrifyingly bad experimenter. . .he read avidly and he listened to what people told him, and he had a knack of drawing conclusions which the observer himself had missed'.[55] During his World War I military service he kept up an interest in genetics by reading the journals, and in 1915 he submitted for publication a paper on linkage in mammals, based upon earlier studies of guinea pigs in collaboration with his sister, claiming the distinction that he had been the only officer of the Black Watch to have submitted a scientific paper from a forward position in the trenches.[56] His interest in linkage continued after the war, and Norton's table, which had aroused his interest in the mathematical consequences of selection, stimulated the studies referred to by T.H. Morgan, which culminated in *The Causes of Evolution* (1932).

Like Fisher and Morgan, Haldane was particularly interested in the effect of selection on single alleles, and regarded large populations as the most likely area of rapid evolution, but he recognised the importance of gene interaction and the grosser chromosomal alterations such as polyploidy. He was less convinced than Wright that large populations could be regarded as aggregates of smaller semi-independent populations. The *Causes of Evolution* opened with a defence of Darwinism both as evolutionism and as selectionism, and the dismissal of 'a vast mass of criticism. . .almost entirely devoid of scientific value'.[57] To illustrate the extent to which the rot had set in, Haldane remarked, 'The criticism of Darwinism has been so thoroughgoing that a few biologists and many laymen regard it as more or less exploded'.[58] Although he regarded mutation theory as a lost cause, Haldane placed strong emphasis upon the revised concept of mutation, and Muller's work on mutagenesis. His chapter on natural selection closed with the qualification,

A selector of sufficient knowledge and power might perhaps obtain from the genes at present available in the human species a race combining an average intellect equal to that of a Shakespeare with the stature of Carnera. But he could not produce a race of angels. For the moral character or for the wings he would have to await or produce suitable mutations.[59]

He put a certain emphasis on embryonic evolution, citing neoteny and human foetalisation in particular, and pondering the timing of developmental gene activity. Advantageous epigenetic changes could have subsequent disadvantageous effects in the adults, to the puzzlement of biologists studying adaptiveness on the basis of adult features alone. While natural selection was the main cause of change in a population it was unnecessary to revert completely to Darwinism,

In the first place we may have every reason to believe that new species may arise suddenly, sometimes by hybridisation, sometimes by other means. Such species do not arise as Darwin thought by natural selection. When they have arisen they must justify their existence before the tribunal of natural selection, but that is a very different matter.[60]

He was thinking of the hybrid vigour of certain alloteraploid plant hybrids, which bred true. It occurred to him that the vigour was intrinsic to the hybrid at the moment of fertilisation, but was sufficiently confused by the metaphor of natural selection as an external force that he did not realise that the explanatory value of the 'tribunal' was thereby redundant.

Haldane equivocated on evolutionary progress, implying that it was real but rare on the one hand, and that we tended to be anthropomorphic about it on the other.

Particularly hostile to true scientific progress are the extremer forms of the doctrine of emergence. According to these, a material system of a certain degree of complexity suddenly exhibits qualitatively new properties such as life or mind, which cannot be explained by those of the constituents of the system.[61]

While he saw an element of truth in this he objected to the anti-analytical emergentists and to those who adduced divine power, but he also thought that 'if we ever explain life and mind in terms of atoms I think we shall have to attribute to the atoms the same nature as that of minds of constituents of mind such as sensation'.[61] Also,

> If we are to have mind at all, it must probably conform to certain laws. There is no need to suppose that those laws, any more than those of biochemistry are products of natural selection. Selection no doubt accounts for certain details, but in all probablity not for the general character of mind.[62]

For Haldane science was an attempt to unify experience by reducing it to physicochemical laws.

Provine assesses the interrelationships between Fisher, Wright and Haldane thus:

> Haldane stated that the work of Wright 'resembles the work of Fisher more than that of Haldane'. Wright believed that the strong emphasis upon the deterministic effects of mass selection of single genes in the work of Fisher and Haldane distinguished their work sharply from his own, which emphasized the selection of interaction systems of genes. Fisher thought Wright and Haldane failed to appreciate the importance of very small selection pressures acting over long period of time in the evolution of natural populations. He often lumped Wright and Haldane together as critics of his views. Thus the relationship between the three appears to have been symmetrical.[63]

Wright was certainly the most holistic of the three, in the narrow sense of emphasising interactions rather than single gene distributions. He and Haldane were more open to the possibility of larger, saltatory events, which Fisher had argued were impossible to conceive of as adaptive. Haldane was the generalist who considered, if only briefly, many of the non-selectionist hypotheses and doctrines discussed in this book. Their extreme forms he rejected, but he left some room for the accommodation of Lamarckian inheritance, laws of growth, and discontinuous evolution. The three are presented in elementary texts as the three wise men who attended at the birth of neo-Darwinism. Their success in quantifying the behaviour of alleles in time and space, a task that seemed almost insurmountable to T.H. Morgan as late as 1925, was the catalyst. It signalled an end to the frustration caused by a paradigmatic theory that had been available for seventy years, but had never been quantified, so that biology had been left as another exercise in stamp collecting, in contrast to the exciting advances in quantum mechanics, nuclear physics and relativity enjoyed by contemporary physicists. These neo-Darwinists provided mathematical explanation; no-one else did. That they were explaining population biology rather than evolutionary progress was overlooked in the exhilaration of the chase.

Provine, in a later essay on 'The Role of Mathematical Population Geneticists in the Evolutionary Synthesis of the 1930s and 1940s' (1978) offers a further explanation of the influence for the three. The conventional wisdom, promoted chiefly by Theodosius Dobzhansky and P.M. Sheppard, was challenged by C.H. Waddington and Ernst Mayr, qualified by the acknowledgement by Waddington that, 'the outcome of the mathematical theory was, in the main, to inspire confidence in the efficiency of the process of natural selection and in the justice of applying this type of argument also the realm of continuous variation'.[64] Mayr also qualified his criticism with,

> It seems to be that the main importance of the mathematical theory was that it gave mathematical rigor to qualitative statements long previously made. It was important to realize and to demonstrate mathematically how slight a selective advantage could lead to the spread of a gene in a population.[65]

Waddington's objection to mathematical population genetics was that a productive mathematical treatment might

> reveal new types of relation and of process, and thus provide a more flexible theory, capable of explaining phenomena which were previously obscure. It is doubtful how far the mathematical theory of evolution can be said to have done this. Very few qualitatively new ideas have emerged from it.[66]

Similarly, I have criticised dogmatic reductionism because its adherents are satisfied with the statement of biological phenomena in physicochemical terms, even where such a reduction is not productive of new ideas about the unknowns of biology. Provine has demonstrated that while Fisher, Wright and Haldane defended themselves they were unable to give a substantial response to Waddington and Mayr. The conceptual difficulties with the selection of small variations were cleared up both theoretically and experimentally, to the extent that mathematical neo-Darwinism showed that the distribution of randomly chosen single-factor alleles follows those principles in space and time. The fact that four biologists had come independently to sufficiently similar conclusions provided the strong persuasiveness of numbers. The fourth quantitative population geneticist was the Russian Chetverikov, who had estimated the importance of small selection pressures in the 1920s. His publications were not well known in Britain and America, but he influenced T.H. Morgan's student Theodosius Dobzhansky, whose significance as a translator of the mathematical theories into a form comprehensible to the majority of biologists in *Genetics and the Origin of Species* (1937) was, Provine argues, the most important single factor in the success of mathematical neo-Darwinism.[67] Julian Huxley's *Evolution: The Modern Synthesis* was close behind in its general appeal, appearing in 1942, almost as soon as neo-Darwinism had begun to make an impression in biological consciousness.

Provine concludes,

> Perhaps the most pervasive way the models served to eliminate competing theories was by showing that many factors, especially the inheritance of acquired characters, were not logically essential for the evolutionary process. All of the ideas of directed evolution, and there were many, were dealt a severe blow.[68]

This did not however prove the all-sufficiency of natural selection; even as an induction by elimination it was weak. Special creationism argues with similar logical consistency that evolutionism is not necessary to explain the existence of living organisms. Provine judges that most evolutionists and scientists in general 'accept some variation of the argument from parsimony', and goes on to argue that selection theory was the best simple hypothesis for the mechanism of evolution, other hypotheses suffering in contrast.[69] However, if we must argue from parsimony the simplest hypothesis must at least refer to the phenomenon it purports to explain. I have some sympathy for the creationist who says that God created the universe and its contents in six days, observing that this is a more parsimonious hypothesis than natural selection's trial and error methods extending over four billion years; yet his argument has no more substance than the selectionist's in abusing the parsimony principle, begging the question, and plastering up his convictions with *ad hoc* additions. I hold no strong brief for mutationism nor

Lamarckism, but neo-Darwinism did not disprove these possibilities. Mayr sagely asked, 'Is it not perhaps a basic error of methodology to apply such a generalizing technique as mathematics to a field of unique events, as is organic evolution?'[70] His most crucial question pertaining to such unique events related to 'development homeostasis', or 'hormeorhesis' as Waddington called it.

> One might utilize the same phenomenon of developmental homeostasis to explain. . .the sudden, sometimes explosive breaking up of morphological types. . .such evolutionary explosions may take place after the parental type had been morphologically stable for periods of 20 to 40 million years. None of the explanations that are usually advanced, such as bursts of mutation or geological cataclysm is particularly convincing. It would seem more likely that some special event such as a sudden shift in ecological balance or the rare occurrence of a successful hybridization has led to the upsetting of the developmental homeostasis and has permitted new selective forces to act upon the newly available phenotypic variations. The history of domestic animals offers abundant proof for the emergence of rich stores of phenotypic variation as soon as the developmental and genetic homeostasis is broken down, owing to close inbreeding, extreme selection, or both. Why not apply this finding to the phenomenon of explosive evolution?[71]

Homeostatic improvements are the mark of the great emergences of animal evolution: insects; teleosts; dinosaurs; birds and mammals. They are characterised internally by a new level of physiological organisation, and externally by strong adaptive radiation; but the mechanisms of their emergence have largely been disregarded. It is only in the last ten years that orthodox evolutionists have begun to tentatively consider Mayr's question. But the strangulation of selectionism's umbilical cord has still to be removed.

12 THEORETICAL ORIGINS

Darwin has been called the Newton of biology, but it will be time enough to talk about the Newton of biology after our science has found its Galileo.

J.H. Woodger 1929[1]

Life might be defined as the art of getting away with it; and Theoretical Biology as the attempt painstakingly to explicate just how it is done.

C.H. Waddington 1969[2]

In the progress of twentieth-century evolutionary thought the decade 1925 to 1935 was a significant juncture. Neo-Lamarckism fell from grace, and neo-Darwinism was reborn in the form of theoretical population genetics. At the same time emergentism, holism, and the organismal conception raised holistic consciousness, but lacked the loaves and fishes to sustain substantial growth. The current of holistic thought that flowed on through the confluence is one that has attracted the convenient title of theoretical biology. It had sources in metaphysics, in the mechanistic and vitalistic expressions of classical embryology, and in the perennial exploration of the fundamental causes of evolution. Theoretical biology will be separated for ease of discussion into attempts to formalise biology, as proposed by Woodger and Bertalanffy, and into epigenetics, which arose more directly from the developmental studies.

The Janus of theoretical biology, contemplating the past and influencing the future at the nexus, was J.H. Woodger, who took what R.W. Sellars called an epistomeological route to evolutionary naturalism. Sellars had argued

organization is objectively significant and causally effective. Function and structure go together at every level. Function is but the active phase of structure. In no other way can we interpret evolution and bring mind and body together. From these considerations we are justified in regarding evolution as the active rise of new wholes with new properties. . .this rise of higher levels

222

must rest upon and but carry out the potentialities of the lower levels. It is the specificity and individuality of the organism which we must stress.'[3]

These opinions were shared by Woodger, who stressed formal logical and mathematical analysis of the internal relationships and organization of organisms. Like the new realists, he insisted upon clarity of definition of terms and cautioned against the dangers of abstraction and over-simplification; doctrinaire positivism had to be balanced with a proper consideration of metaphysics. *Biological Principles* (1929), begins with a discussion of *abstraction*, or simplification.

> We can say that an abstraction is used *intolerantly* if it is treated as though it were not an abstraction, and this occurs when what is abstracted *from* is regarded as non-existent, or unimportant, not only to the person who is abstracting from it but to everyone else! When the operation of pride in human relations is borne in mind. . .it is not difficult to to recognise a tendency in people to exaggerate the importance of the things in which they are interested, and to despise, denigrate, ignore or reject some things in which other people are interested and therefore do not abstract from.[4]

This referred to the gene theory in particular, and genetics in general. The wilder expressions of the modern selfish-gene concept illustrate his point, such as Dawkin's assertion in *The Selfish Gene* (1976) that the organism is nothing but a conglomerate of selfish genes, each one out for itself, but compromising for the moment in co-operation with other genic cohabitants of the organism, a means of transport which they have constructed to serve their selfish ends. This is an intolerant abstraction by Woodger's definition since it treats the conglomerate of selfish genes as the reality and the whole organism and its phenotype as unimportant. Woodger pointed out that the consequence of intolerant abstraction was to give some aspects of a problem undue importance, and to discourage thought and research on other aspects of the problem. Another fundamental philosophical issue was that biologists regarded metaphysics as a disreputable subject, without making it clear that they knew what metaphysics meant:

if you do not know clearly what metaphysics is you may easily fall
into it unawares, and if metaphysics is something it is desirable to
avoid it will be a misfortune to fall into it unawares. It seems very
frequently to be the case that what a person means by metaphysics
is the opinion on certain topics of other people who do not agree
with his. His own opinion he calls 'science'.[5]

Woodger used 'metaphysics' to mean the general synthesis of the
special sciences brought to bear upon the nature of reality, and
attributed the neglect of this synthetic holism both to Galileo's
favouring of the primary physicochemical qualities of nature over
the secondary sensual ones, and to Cartesian dualism.[6] Of the
secondary qualities, he wrote,

They have been the orphans of the world of scientific thought.
Nobody wants them but nobody can get away from them. The
belief that they are 'unreal' has been a cardinal point in all
materialistic metaphysics, because they constitute an obstacle to
all neat and tidy explanations of the world in terms of bits of stuff
pushing each other about. But this, at least until recently, has
been regarded as the type of scientific explanation *par excellence*
and its metaphysical success (as distinct from the methodological)
has been partly dependent upon the ease with which the secondary
qualities can be forgotten in natural science. This is one of the
ways in which natural science, if it is not on its guard, may
unwittingly slip into metaphysics. It is but a particular instance of
the general fact that when abstraction is made of anything the
latter does not thereby cease to exist but returns to claim a place in
the scheme later on when after successful analysis, we return to
the question of synthesis.[7]

The behaviourist's 'conscious automata' concepts were devices for
ignoring psychical qualities, and a result of confusing methodological models with metaphysical dogmas, and natural selection was a
product of the same kind of confusion.[8] Even if these models were
plausible and empirically successful this 'should not blind us to the
desirability of exploring others'. Phenomenalists such as Mach
pleaded only for relief from awkward questions so that they could
get on with the job of science, but this left critics such as himself with
the prospect of a guerilla war where the enemy could never be confronted. Instead of automatic rejection on positivistic grounds,

vague new ideas should be respected as the 'tender growing points of thought. . .and the sharp wind of criticism should not "visit them roughly". But if they are admitted into a theory with a serious claim while still in the woolly state, they only bring confusion with them'.[9] Woodger anticipated the application of Kuhnian epistemology to biology: once the paradigm was established it developed rapidly, and was generalised, with a great deal of enthusiasm, irrelevant metaphysical speculation, and intolerance of criticism. Inherent difficulties brushed aside in the first flush of enthusiasm slowly worked their way to recognition, though 'impeded by the glamour still glowing from the early dogmatic days of the theory.'[10] These arguments applied both to cell theory and selection theory. On the same subject he asked,

> Who but a very learned man would dream of conceiving the growth of a metazoan as a 'simple physico-chemical process'? And yet biologists no more ask for a 'proof' of this than fundamentalists ask for a proof of the infallibility of the Scriptures. Surely this is a shining example of what C.D. Broad calls a 'silly theory' in the sense of one 'which may be held at the time one is writing professionally, but which only an inmate of a lunatic asylum could think of carrying into daily life.' Now such silly theories are quite indispensible for scientific progress, but it is not only fatal to carry them into daily life, it is also fatal to take them too seriously in science.[11]

The worst damage was done when silly theories got into the hands of the 'camp followers' of original thinkers, with their subsequent petrification 'as tentative assumptions harden into dogmas which are never examined our thoughts become encrusted with layers of intellectual rubbish which require the labors of an intellectual Hercules for their proper purgation'.[12]

Woodger was attracted by epigenetic interpretations of the evolution of complexity, and he criticised contemporary embryologists for their failure to take the whole and its organisation seriously. Entelechy was 'an adventure in thought — the exploration of a possibility — and biologists ought to be grateful to Driesch for the patience and thoroughness with which he has carried out this explanation'.[13] Woodger convinced his fellow members of the Theoretical Biology club that in comparison with physics there was plenty of biological theory but no theoretical biology.[14] British biologists had fallen far behind German theoreticians, and there was a desperate need for both mathematical and epistemological methodology in biology.[15] His impact on his fellow theoreticians was one of intellectual stimulation. His own attempts at formalising organisation and cellular hierarchies of interrelations were initially obstructed with his belief that differentiation divisions occurred during development, and by his rejection of a molecular hypothesis for the gene, though he did qualify these attempts as being less concerned with clarifying 'particular muddles' than with urging 'the general desirability of devoting some little attention to the linguistic and logical aspects of biology'.[16]

According to Donna Haraway's *Crystals Fabrics and Fields* (1976), the historical period under discussion marked the end of a phase characterised by the simplistic mechanism v. vitalism controversy, represented by Roux and Driesch. The same period of biological history had also been characterised by the Weismannite v. Lamarckist evolutionary controversy and the gradualist v. saltationist controversy; and the holistic-emergentistic concept had associations with all the losing sides. In the area of developmental biology Haraway sees mechanicism and vitalism relieved by the more sophisticated combatants, reductionism and organicism. R. G. Harrison, pioneer of organicistic embryology rejected the emergentism of C.L. Morgan and J.C. Smuts as obfuscatory and obstructive, and requiring acts of special creation. Emergent novelty could also be done away with. Taking the example of water as a novel emergent of hydrogen and oxygen, he remarked that it would be simpler to regard water as a property of hydrogen in combination with oxygen.[17] However, this is an application of hindsight that is itself obfuscatory. Its *reductio ad absurdum* is that intelligence is a property of hydrogen when in combination with oxygen, carbon, nitrogens sulphur and the rest; an idea that has no more explanatory value than Molière's remark that opium induces sleep because it has soporific qualities.[18] Harrison observed pragmatically that 'it is impossible to develop science wholly from the top down or from the bottom up. The investigator enters where he can gain a foot-hold by whatever means may be available'.[19] C.L. Morgan and Smuts made little attempt to gain a realistic foothold, and embryology's practical approach to a particular integrative level and its ontogenic origin blinkered its investigation to the implications of how that level could have emerged phylogenetically. Nevertheless, the realistic organicism of Harrison and his contemporaries did provide the kind of foothold that holistic evolutionism ultimately needed for its further development.

The biochemist Joseph Needham was a student of F.G. Hopkins, who coined the aphorism, 'Life is a dynamic equilibrium in a polyphasic system', variations of which are still part of the stock in trade of contemporary reductionists.[20] Although he was initially sympathetic to this concept, the chemistry of embryology led him to a more organistic view, but one which was to be couched in chemical terms, without appeal to vitalistic forces of any kind. Organisation was a basic biological problem, not something fundamentally mystical and unamenable to scientific attack. If the organisation and unity of the organism were taken as axiomatic, as J.S. Haldane and E.S. Russell had done, there could be no progress in understanding it. Even so, vitalism had continually drawn attention to the real complexity of biological phenomena, and opposed the mechanistic tendency to oversimplify.[21] Needham found it almost inconceivable that Driesch, having induced the concepts of harmonious equipotential and intensive manifoldness, should have immediately fettered them with an obscurantist entelechy. However, his own work on chemical organisers led him to a premature rejection of nuclear preformation similar to Woodger's. Needham recognised that the organisation of enzymes, or control of their activity, must have a role in development, but without knowledge of the molecular mechanism of differential gene expression the significance of enzyme activity could not be properly appreciated. The most popular metaphors for the physical bases of development were not molecular but physical, with models such as magnetic fields, electrical gradients and vibrations.

The theme of integrative levels was addressed in Needham's 1937 Herbert Spencer Lecture, where he noted that organismic views on mechanicism and vitalism expressed by the Russian delegation to the 1931 International Congress for the History of Science in London came as a surprise to the English audience.[22] Materialist dialectics not only supported organicism but was 'the profoundest theory of natural evolution, the theory of the nature of transformations and the origin of the qualitatively new'.[23] American organicists and emergentists would no doubt have regarded this as a day late and a ruble short. While Herbert Spencer had understood the significance of time in relation to complexity and differentiation Needham confessed that he himself had had to work through years of experience before it finally dawned that time brought higher levels of integration. This was made the introductory theme of a later essay, 'Evolution and Thermodynamics' (1943). Evolution, as it progressed, manifested:

1. A rise in the number of parts and envelopes of the organism and the complexity of their morphological forms and geometrical relations. 2. A rise in the effectiveness of the control of their function by the organism as a whole. 3. A rise in the degree of independence of the organism from its environment, involving diversification and extension of range of activities. 4. A rise in the effectiveness with which the individual organism carried out its purposes of survival, and reproduction, including the power of moulding its environment.[24]

The levels of biological complexity could not be meaningful simply in terms of physicochemical explanation; the living organism was a 'patterned mixed-up-ness. . .a patterned separateness'.[25] How did this evolutionary progress of integration and control stand in relation to the laws of thermodynamics, especially the second law, which dictated the dissipation of energy and the increase of entropic 'mixed-up-ness'? The idealistic view of this could be that a creator wound up the universe in opposition to the second law and it has been winding down ever since, or that there were ghosts in the machine, biological Maxwell's demons, such as entelechy, that held back the tide of entropy. This line of argument is still used in favour of special creation by the 'scientific' neo-creationists. The realistic view of the problem saw two alternatives: either biological organisation was 'kept going at the expense of an over-compensating degradation of energy in metabolic upkeep', or the physicists's concept of order was distinct from biological organisation.

Schrödinger, taking the first alternative, observed that all that was required by the second law was that the

total balance of disorder in nature is steadily on the increase. In individual sections of the universe, or in definite material systems, the movement may well be towards a higher degree of order, which is made possible because an adequate compensation occurs in some other systems.[26]

Eddington also saw the apparent paradox that 'Evolution shows us that more highly organised systems developed as time goes on', and resolved the problem by concluding that the organism has to collect external energy, and if shut off as a closed system would soon come to a state of 'extreme disorganisation'.[27] Bertalanffy had independently proposed that organisms should be regarded as open systems

> whereas in closed systems the trend of events is determined by the increase in entropy, irreversible processes in open systems cannot be characterised by entropy or another thermodynamic potential; rather the steady state which the system approaches is defined by the approach of minimal entropy production. From this arises the revolutionary consequence that in the transition to a steady state within an open system there may be a decrease in entropy and a spontaneous transition to a state of higher heterogeneity and complexity.[28]

The non-photosynthetic organism is actually an entropy generator, since the largest part of the energy that it collects from other systems in the form of complex reduced carbon molecules is squandered as entropic heat due to the inefficiency of conversion. Such an organism increases the rate of entropy production in the biosphere, and man speeds the plough by using non-renewable energy resources. Needham argued that biological organisation was distinct from ordered physical systems like crystals, which did not require metabolism. The antithesis of entropic mixed-up-ness was not organisation but 'separatedness'. 'The world has been moving steadily from a condition of universal separatedness (order) to one of general chaotic mixed-up-ness (thermodynamic disorder) plus local organisation (patterned mixed-up-ness)'.[29] There was no fundamental physical contradiction between physical order and organisation; the distinction was a reflection of the different levels of explanation required by physics and biology. For Eddington physical laws were required to encompass all phenomena, and the definition of the organism as a collector of external energy was all that was necessary. For Needham the concept of physical order was consistent with biological organisation but did not explain it.

Needham's *Order and Life* had been preceded by Ludwig von Bertalanffy's *Modern Theories of Development; An Introduction to Theoretical Biology,* the English translation of which was made in 1933 by J.H. Woodger who was a mentor of both Bertalanffy and Needham and with them a co-founder of the Theoretical Biology Club in 1932.[30] *Modern Theories of Development* was introduced with the realistic assertion that the empiricist tends to forget

> that a collection of facts, be it ever so large, no more makes a science that a heap of bricks makes a house. . .Only if the multiplicity of facts is ordered, brought into a system, subordinated to great laws and principles, only then does the heap of data become a science.[31]

Bertalanffy agreed with his compatriots who held that the consideration of wholeness had properly dislodged the consideration of purpose as a major thrust of theoretical biology. As the mechanist H. Winterstein observed, ' "Purposiveness" is nothing else than a short expression for all the phenomena upon which the maintenance of an observed state or process depends'.[32] From a realistic point of view entelechy could have been treated as an analogous category. The exclusion of descriptive organicism for its unscientific lack of mathematical formality was an inadmissible restriction.[33] On the other hand, the dangers of 'romantic biology' in which organicist descriptions were confused with causality were to be avoided.

Bertalanffy's first formulation of the concept of organism as system was this:

> A living organism is a system organized in hierarchical order of a great number of different parts in which a great number of processes are so disposed that by means of their mutual relations within wide limits with constant change of the materials and energies constituting the system and also in spite of disturbances conditioned by external influences, this system is generated or remains in the state characteristic of it, or these processes lead to the production of similar systems.[34]

Stated in this way it subsumes the phenomena circumscribed by 'homeostasis' as well as development and reproduction, but is still at an immature, intuitive and descriptive stage of its logical development. Bertalanffy was intrigued by R. Goldschmidt's attempt, in *Physiological Theory of Heredity* (1927), to fill the gap between the potential of Mendelian genes and their expression during development and in maturity. The genes were small in quantity, but produced large effects: therefore if, as Bertalanffy believed, the speed of developmental processes was proportional to the amount of gene material, the genes might be chemical catalysts or enzymes, from whose reactions developmental hormones were produced.[35] The genetic and cytoplasmic organ-forming materials had also to be localised by a physicochemical stratification, made possible by the heterogeneity of the 'polyphasic colloidal system' of the cytoplasm. As embryonic development progressed, later products of the gene-catalysts interacted with the appropriately localised primary products to trigger organ formation.

> In every organ the same game of the catalyst proceeds, continually new genes begin to work as soon as their specific substratum has appeared. In such a way, a relatively small number of gene-catalysts and of organ-forming materials may, by their various interactions, yield an infinite number of reactions, and, in consequence, of developmental processes.[36]

This was the simple answer to the old objections to Weismann's biophore theory: by a harmonious interaction of combinations of a relatively small number of gene-catalysts and cytoplasmic organ-forming molecules large numbers of products could be produced. This, as Bertalanffy conceded, was largely hypothetical, since there was no proof that genes were molecules, and the existence of 'developmental hormones' was a speculation based upon limited observations on the amphibian organiser. Although Goldschmidt had worked out an example of how a butterfly wing could be differentiated by different reaction velocities, producing different degrees of chitinisation and colour-pigment deposition, chemical differentiation could not be the whole story, since there remained the problem of how the differentiated materials of the muscle, nerve, bone and blood vessels in a limb, could be finally constructed as a right arm or a left leg; the problem that Driesch had 'solved' with entelechy. According to Paul Weiss's biological-field theory, formative influences belonging to the site of development-determined final form, but while talk of 'forces' or 'fields' named the causes, it did not explain them.[37] Its implications have not yet been fully worked out by embryologists, although the 'old-fashioned' epigenetic view that the local embryonic environment impresses itself upon the way in which the organs take their final form is proving to have more significance than had been anticipated. Goldschmidt and Bertalanffy understood that such theories were relevant to the nature of evolutionary progress as well as to the process of normal development. Goldschmidt observed that while his physiological theory of heredity appealed to physiologists and developmental biologists, its relevance was not appreciated by the Mendelian geneticists.[38]

From a survey of the historical theories of development Bertalanffy concluded that the problem of classical mechanicism, as represented by Weismann, was its 'additive' aspect, and the problem with vitalism was its appeal to transcendent causes. The solution to these problems was to be sought in an *organismic or system theory,* which addressed organismic harmony and coordination without resort to entelechies. At this early stage in the development of system theory, evolution was mentioned only as a unique product of the complex physicochemical and organismic system of the living organism, in contrast to other kinds of wholes. The primary goal was 'to develop biological law as a system complete in itself', proceeding organismically.[39] The first principle to be established was a law of 'biological maintenance', or 'biodynamic equilibrium'. The second goal was a law of 'hierarchical order'. Like the emergentists, Bertalanffy intuitively believed that due to a 'biological uncertainty principle' new laws had to be derived for each new level of organisation. He also adduced a corollary principle of dynamic hierarchical order 'so long as an organic system has not yet reached the maximum organization possible to it, tends toward it'. This principle was to subsume the subordinate aspects of differentiation, organ-information, epigenetic fields, etc., each one to be founded on quantified experimentation. Like Smuts's holism, Bertalanffy's system theory had a Baconian streak, but proceeded to some extent from an empirical foundation and not only from intuition and generalisation. The definitive work on system theory was not to appear until late in Bertalanffy's career; in the meantime a translation of *Das Biologische Weltbild* (1949) appeared as *Problems of Life* in 1952. The opening chapter reveals a strongly sustained interest in organicism as well as an implicit fascination with the vitalism of Driesch. For Bertalanffy and Woodger, Driesch had been the biologist who had provided the knowledge of excess, as in William Blake's aphorism, 'You never know what is enough unless you know what is more than enough'.[40] Although this book predated the identification of DNA as the molecular basis of heredity, Bertalanffy was already focusing on the nucleoproteins as the sites of protein synthesis, and citing research which demonstrated that the substance of the chromosomes was in a constant state of renewal. A crystalline structure, as Schrödinger had suggested in *What is Life?* (1944), was insufficient; it had to be seen as a metabolising aperiodic crystal maintained in a steady state.

Bertalanffy's concept of the hierarchical organisation of the organism now included a physiological *hierarchy of processes,* which paid more attention to the mechanisms of homeostasis. *Segregation hierarchy* was the ontogenic and phylogenic differentiation of epigenetic fields, whereby the primary whole segregated into functionally specialised sub-systems. Phylogenetically, progress was marked by an increasing integration,

> The higher we go in the scale of organisms, the more different is the behaviour of isolated parts from that which they display in the whole; and the poorer it is in comparison with the performance displayed by the whole organism.[41]

The integrative systems of higher animals included the homeostatic *milieu intérieur,* the hormonal system and the nervous system. The progressive differentiation and specialisation and centralisation of control mechanisms was accompanied by a loss of generalised ability — a situation where 'Robinson Crusoe cuts a much poorer figure than Man Friday'.[42]

System theory encompassed not only organisms, but also supra-individual associations, including societies and symbioses. Intracellular symbioses he counted as a higher-order system. Moreover, he wondered if 'the highest unit is the whole of life on earth'.[43] A form of this concept has recently been proposed as the 'Gaia principle' by Lovelock (1979), who argues that the biosphere creates the physicochemical conditions of its own survival and evolution.

Problems of Life began to explore the implications of the epigenetic mutations in relation to their phylogenetic effects. If there were genes whose action could cause profound changes in the organism as a whole as well as those which affected only specific characters, then Goldschmidt's principle of harmonised reaction velocities made good sense. It was possible that mutation could produce rate-changes in gene action, since there were known cases of alterations in allometric growth which apparently were due to a failure of certain genes to produce enough of a particular hormone by a critical stage in development. In addition, the stage in development at which mutagens were applied would be crucial in terms of ontogenic effects, the earlier the stage, the greater the effect: 'A quantitive change, simple in itself, controls a widespread process that leads to profound transformation of the developmental patterns and hence to complicated morphological changes'.[44] Both Goldschmidt's 'hopeful monster' and Schindewolf's parallel hypothesis of proterogenesis represented to Bertalanffy the real possibility of great evolutionary transformations.

The four main problems of evolutionism were: 1) the origin of the multiplicity of forms within a given type of organisation ('type' meaning variety, species and genus); 2) the origin of adaptations to environment; 3) the origin of the types of organisation; 4) the origin of the complex morphological and physiological integration with the organism as a whole.[45] The first two were well within the competence of neo-Darwinism. The latter two were problems of macro-evolution that neo-Darwinism had failed to explain for epistemological reasons, a consequence of the positivist approach. Bertalanffy was bothered by 'useless characters' which persisted, by

morphological peculiarities which form the framework of taxonomy [and] seem to represent 'types' that in themselves have nothing to do with usefulness but can be fitted to different exigencies and habitats much in the same way as churches, town halls, or castles can be built equally well in gothic, baroque or rococo styles.[46]

But then these might be survivors in 'low selection-pressure' environments, or products of genetic drift. That selection theory could offer hypotheses to explain away these useless characteristics illustrated equally that a decisive judgement was impossible. Saltatory effects were not only apparent in terms of fossil record, but occasionally happened even in *Drosophila* (e.g. the four-winged, 'bithorax' mutation), an organism that Bertalanffy felt was stuffed with genes concerned with subtle and, in general, rather insignificant characters. However,

like a Tibetan prayer-wheel, Selection Theory murmurs untiringly: 'everything is useful', but as to what actually happened and which lines evolution has actually followed, selection theory says nothing, for the evolution is the product of 'chance', and therein obeys no 'law'.[47]

His dissatisfaction with selection theory led Bertalanffy to the question of autonomous evolutionary causes within the organism.

While he acknowledged that selection pressure was a limiting condition, and accepted that mutation pressure was a relatively insignificant factor in gene distribution, he used the following analogy which illustrates a fundamental epistemological point: the principle of entropy indicates a limiting condition for physicochemical activities. A knowledge of the principle does not shed any light on particular processes which are thermodynamically allowable: it does not, for example, explain why compounds crystallise in their characteristic forms. For such knowledge, information about the nature of the reacting substances, their rates of reaction, and the lattice forms of the different kinds of molecules was required. Similarly, natural selection could describe the limits of possibility, but did not explain the particular route taken by advantageous characters. This was not a new line of argument; it had been propounded by Mivart in 1871 and pursued by many sceptics.

Bertalanffy was intuitively drawn to the belief that mutations and their effects were not entirely random, but had to fall within certain degrees of freedom. The similarities between the cephalopod and vertebrate camera eyes and the stability of respiratory proteins exemplified such structures. A species represented a state in which 'a harmoniously stabilized "gene balance" has been established, that is, a state in which the genes are internally so adapted to each other that an undisturbed and harmonious course of development is guaranteed'.[48] He agreed with some palaeontologists that the formation of new equilibria both at the species and more radical systematic levels was discontinuous, the gaps in the fossil record representing periods of re-equilibration in relatively small numbers of individuals. Following the establishment of new equilibria, gradual speciation would be the end product. Like Berg and Vavilov, Bertalanffy was fascinated by parallel evolution, which supported his view of autonomous causes producing similar changes within similar degrees of freedom in related types. Parallel evolution could be as simple as parallel mutations of homologous structural genes, or various homologous or analogous epigenetic mutations causing parallel effects in terms of size increase, general pigmentation change and allometric modification.

The limitations of fortuity or randomness drew Bertalanffy to orthogenesis; while some examples could well be the outcome of orthoselection, it was possible that under conditions of 'slackened' natural selection internal shifts in equilibrium could produce not only hypermorphosis, but also favourable results such as increased brain capacity. Dubois had applied such a hypothesis to explain the doubling of brain-size in hominid evolution. While the criterion of usefulness was important, it was not progressive adaptation that brought forth orthogenesis; but orthogenetic trends might eventually create prerequisites for new and higher achievements. What was the 'secret law' at which the chorus of examples hinted? Evolution seemed to be a 'cornucopia of *évolution créatrice,* a drama full of suspense of dynamics and tragic complications', and like Bergson he thought that 'Life spirals laboriously upward to higher and ever higher levels, paying for every step'.[49] Death was the price of the multicellular condition; pain the price of nervous integration; anxiety and possibly self-destruction the price of consciousness.

General System Theory was a

> logico-mathematical field, the subject matter of which is the formulation and the derivation of those principles which hold for systems in general. A 'system' can be defined as a complex of elements and of the relations or forces between them. From the fact that all the fields mentioned are sciences concerned with systems follows the stuctural conformity or 'logical homology' of laws in different realms.[50]

System theory was to make use of models, analogues and homologies for understanding unknown systems in terms of known ones, without prejudice as to mechanistic or organicistic origins, but recognising the limitation of analogues in comparison with homologues. There are innumerable examples throughout the history of biology of physicochemical models or analogues which, despite their inadequacies, advanced our understanding of biological phenomena. More recently, cybernetics has contributed to the understanding of neural systems. The usefulness of such analogies is indisputable. At this point, however, one must ask; is it necessary or advisable that these isolated applications be regarded as part of a natural, unified logico-mathematical system as Bertalanffy hoped; or does system theory suffer from the same limitations and pretentiousness as Smuts's *Personology* or Bacon's *Novum Organum* or Spencer's synthetic philosophy? System theory has the virtue of an encyclopaedia: it centralises — makes accessible — a host of different ideas and model systems. Odd juxtapositions may lead to lateral associations which Edward de Bono has demonstrated as a useful counter-inductive method, but the interrelationships need to be tested case by case.[51] The catholic systems encyclopaedia embraces topics in evolution, general biology, physiology, economics, communications, psychology, psychiatry, general knowledge and education, and there is a large modern literature on its biological applications. However, many critics, such as D.C. Phillips (1976), argue that the system approach has failed to deliver on its original promise, being unable to sustain its objections to the purely analytical method, remaining vaguely defined and failing as a scientific theory because it lacks predictive power.[52] This is somewhat more than Bertalanffy claimed for it, though his camp-followers may be more dogmatic. One of the fallacies of system theory is that a 'photograph' of the problem is an answer. Translating a biological or linguistics problem into the symbols of information theory impresses the uninitiated, and makes the problem more accessible to information theorists without necessarily providing an answer. There is also a good deal of positive feedback among system theorists that generates a community enthusiasm. Is this any more than Bacon's Idol of the Theatre, a peculiar system of philosophy. . .creating fictitious and theatrical worlds? Bacon might have been addressing the system theorists directly when he said, 'Nor do we speak only of the present systems, or of the philosophy and sects of the ancients, since numerous other plays of a similar nature can still be composed and made to agree with each other'.[53] On the other hand, if the system approach is not required to perform like a law of physics but rather as a guide to induction, it can be valuable, and if a self-declared system theorist produces a bright idea we are in no position to judge whether it was the product of the system approach or simply due to the native wit of the individual.

One such idea, that is interesting in the evolutionary context, is Howard Pattee's 'The Evolution of Self-Simplifying Systems' (1972), which addresses the question of how life gets more complicated and how complexity relates to adaptedness. By 'adaptedness' he means 'functional effectiveness', or in other words being good at what you do. Pattee's first generality is that although the best technical devices are the most complicated ones — such as cars and calculators — some artifacts like the toothbrush will survive because they have an effective simplicity. However, these analogues do not tell us how functional complexity arose. Even under the relatively simple conditions of the origin of life from abiogenically synthesised organic compounds, the task of formulating rules for the generation of simple complexes is very difficult.[54] By the calculations of other philosophers of science concerning the possibility of self-replicating machines, Pattee concludes that live evolving systems required a self-description, as well as the ability to reproduce, i.e. first of all they needed both a genotype and a phenotype; in his words 'the separation of the description of the events from the events themselves'.[55] By separating the structure (genotype) of the DNA from its function (phenotype) the protocell has simplified itself in the way that a scientist simplifies analytical problems, or as Aristotle might have said, the cell has separated the offices of heredity and action. Although the model has many imponderables, Pattee has put his finger on one important aspect of simplification relevant to the patterned separatedness of the evolution of function. The same principle applies to evolution by gene duplication. This kind of self-simplification is progressive, conferring the potential of order and control. There is also a regressive self-simplification that allows the organism to escape from specialisation and evolve in different directions. Pattee further argues that random mutation and natural selection are mathematically incapable of generating increased complexity, since the probability of successful random change decreases exponentially, and the probability of lethal effects increases exponentially with progressive complexity. Bertalanffy had an intuitive appreciation of this when he talked about the limited degrees of freedom of harmonious re-equilibration in the evolution of complex organisms. One obvious example of a self-simplifying process is neoteny. Coupled with this is the redundancy of the old adult genetic apparatus, complete with redundant control systems. Understanding the re-tooling of the redundant equipment is a substantial problem for systems or cybernetics theoreticians.

Another influential 'systems thinker' who has expressed alarm at the undisciplined growth of metaphor in biology is the Austrian-born scientist Paul Weiss.[56] His early attitudes were formed in reaction to the reductionism of Jacques Loeb, and through interaction with the gestaltist Köhler, and Bertalanffy. For Weiss the organism had a hierarchical organisation, each hierarchical level having its own set of characteristic organising modes. Weiss's early research used limb transplanting in amphibian larvae to elucidate the role of the nervous system in organisation and development. He concluded that the pre-existing muscles had the epigenetic role of chemically modulating an indifferent portion of the nervous system as a 'selective receiver'.[57] His later research continued to explore the integration of the nervous system by observations on the regeneration of damaged nerve fibres. Contact guidance from the physical territory through which the new axons found their way was the major stimulus for directing the growth of the axons. In the 1950s he opposed the over-simplifications of molecular biology, and was particularly critical of the way in which information theory was abused in its applications to molecular genetics. 'Information', he feels, has been given an over-exalted explanatory role since there is no information about order and its evolution. Confusion arising from the abuse of the metalanguage of information theory has been to some extent removed by Rupert Riedl, who gives particular emphasis to the evolution of order in the context of epigenetics. This and physiological evolution are among the most significant theoretical biological problems for evolutionists.

Genetic Assimilation

While this topic does not dovetail neatly with the other subjects of the present chapter, it is a prerequisite for Chapters 13 and 14 on epigenetics and physiology respectively. To bring out its general implications it is necessary to return to 1896, when C.L. Morgan, H.F. Osborn and J.M. Baldwin independently published explanations of how the genetic fixation of 'acquired characteristics' might be explained in more conventional Darwinistic terms. Osborn was to some extent the odd man out since he was dubious about the role of natural selection in the hypothetical process that came to be called 'organic selection' or 'Baldwin effect'.

By C.L. Morgan's account Weismann had provided the germ of the idea in his 1894 Romanes lectures, which introduced the concept of intra-selection to explain plasticity in organisms. Roux had already suggested that the adaptation of organs resulted from the action of natural selection on competing parts of the body, but Weismann proposed that it was not the particularly adapted form that was inherited, but only a plasticity that allowed a variety of responses depending on the circumstances. For example, if larger antlers had particular selective advantage in deer then ontogenic plasticity would allow for other developmental correlations such as more substantial bone and musculature of the pectoral region to support the additional weight and stress. As a corollary to

Weismann's hypotheses it would have to be assumed that the correlated plasticity would have had to become genetically fixed in a particular mode as time went on, otherwise it would follow that experimental removal of the antlers during the life of the individual would result in general anatomical change. Morgan went on to suggest that it might be shown that particular ontogenic modifications due to plasticity might provide 'the conditions under which variations of like nature are afforded an opportunity of occurring and making themselves felt in race progress'.[58] In modern selectionist terminology it would be said that the plastic ontogenic change altered the selection pressure.

In 'A Mode of Evolution Requiring Neither Natural Selection Nor the Inheritance of Acquired Characteristics', H.F. Osborn noted how environment could have profound ontogenic effects, and suggested that the tendency for the soles of the feet of a human infant to turn in, and the grasping reflex, would be exaggerated if the child were brought up in a tree.[59] This suggested to him an alternative explanation of certain Lamarckistic phenomena which had been cited by Cope:

> During the enormously long period of time in which habits induce ontogenic variations, it is possible for natural selection to work very slowly and gradually upon predispositions to useful correlated variations, and thus what are primarily *ontogenic variations* become slowly apparent as phylogenic variations or congenital characters of the race.

Osborn later took up Baldwin's term organic selection and outlined it as follows:

> ontogenic adaptation is of a very profound character; it enables animals and plants to survive very critical changes in their environment. Thus all the individuals of a race are similarly modified over such long periods of time that very gradually congenital variations are collected and become phylogenic. Thus there would result an apparent but not real transmission, of acquired characters.[60]

Osborn, who admitted former Lamarckist leanings, was conciliatory towards the neo-Lamarckians,

This hypothesis, if it has no limitations, brings about a very unexpected harmony between the Lamarckian and Darwinian aspects of evolution, by mutual concessions upon the part of both theories. While it abandons the transmission of acquired characters, it placed individual adaptation first, and fortuitous variations second, as Lamarckians have always contended, instead of placing survival conditions by fortuitous variations first and foremost, as selectionists have contended.[61]

This compromise was welcomed in some quarters. However, others who accepted the generalities of organic selection disagreed. E.B. Poulton responded,

I do not believe that these important principles form any real compromise between the Lamarckian and Darwinian positions, in the sense of an equal surrender on either side and the adoption of an intermediate position. The surrender of the Lamarckian position seems to me complete.[62]

The other point of disagreement concerned the origins of plasticity. Morgan, in his pre-emergent phase, and Baldwin and Poulton, took the position that plasticity itself was developed by natural selection. Osborn declared himself in agreement with Hans Driesch, E.B. Wilson and T.H. Morgan, in the belief that plasticity was an inherent function of protoplasm rather than a selectively accumulated property. These were over-simplifications: marked phenotypic or ontogenic flexibility is found in plants. Consider the lodge-pole pine of the Pacific seaboard of North America: given good deep soil and a sheltered location it grows as straight as its name implies, yet on the coastal cliffs it resembles a giant ground-creeping ivy. No such magnitude of morphological plasticity occurs in animals. The remarkable ability of regeneration possessed by some invertebrates and lower vertebrates is lost in the higher vertebrates, but is replaced by another kind of adaptability, which is physiological and behavioural, both expressions of physiological homeostasis. The internal environment is not plastic; it can accommodate external and internal changes, but its essential attribute is constancy. Behaviour can be plastic, but the evolution of the nervous system of intelligent animals has the rigid prerequisite of internal constancy. Nevertheless, I concur with Osborn's conclusions:

If we could formulate the laws of self-adaptation or plastic modification, we would be decidedly closer to the truth. It appears that Organic Selection is a real process, but it has not yet been demonstrated that the powers of self-adaptation which become hereditary are only accumulated by selection.[63]

Baldwin was quite happy to accord natural selection the full credit for the evolution of behavioural plasticity, 'a completed view of psycho-physical accommodation requires (first) natural selection, operating upon (second) variations in the direction of plasticity, which allows (third) selective adjustment through the further operation of natural selection upon the organisms' functions.'[64]

He recognised that this form of plasticity was at the core of evolutionary progress:

There is secured a *blanket utility*. . .a general character, through the operation of natural selection which progressively supersedes and annuls many special utilities with their corresponding adaptations, while, at the same time, other special functions having special utilities are given time to reach maturity by variation and selection.[65]

Baldwin admitted the logical difficulty of saying that 'natural selection is responsible for — or is an adequate explanation of — the results which spring from the accommodations of an organism simply on the grounds that the plasticity of the organism has survived by natural selection', and he remarked how impossible it was, 'to put one's finger on anything positive to represent natural selection'.[66] He did not think it through to realise that 'blanket utility' is its own justification and natural selection a supernumerary, without explanatory value. Such is the power of reification.

Baldwin's organic selection was proposed as 'A New Factor in Evolution' in 1896. The choice of the word 'selection' was unfortunate, as it is sometimes difficult to decide whether Baldwin meant a form of natural selection, or selection in the sense of a *choice* on the organism's part. The ability of an animal to choose 'from the actual modes of behaviour of certain modes — certain functions etc.', was an important aspect of organic selection.[67] Baldwin also referred to such choices, as well as automatic physiological responses to environmental change as *accommodations:* 'By securing adjustments, accommodations, in special circumstances, the creature is kept alive', (or 'persists in its own being', or 'gets away with it'). Therefore, 'those congenital or phylogenetic variations are kept in existence which lend themselves to intelligent, imitative, adaptive, or mechanical modification during the lifetime of the creatures which have them. Other congenital variations are not kept in existence'.[68] On the other hand, organic selection was the process whereby, 'certain reactions of one single organism can be selected so as to adapt the organism better and give it a life history'. This he contrasted with 'the Natural Selection of whole organisms', which he suggested would at 'every change in the environment. . .weed out all life except those organisms which by accidental variation reacted already in the way demanded by the changed conditions', i.e. what is sometimes called 'catastrophic selection', which, when the flood comes, sorts out the sinkers from the swimmers.[69] However, Baldwin begged the question; animals do make choices and physiological accommodations, but he smuggled in the genetical predispositions for them, and that is what has to be proved, instead of being used as an axiom.

A similar argument is used by some socio-biologists in their interpretations of the biological and evolutionary phenomena. For example rapists are urged to rape by the 'whisperings' of their phylogenetic history. They have been naturally selected because rape is a successful evolutionary 'strategy', ensuring the reproductive success of the rapist.[70] Rape is 'programmed into the genes'. We may pause and remove our hats at the thought of outraged motherhood, but we do not have to leave our brains in them. Under some circumstances rape, or promiscuous sexual behaviour, may result in the perpetrators begetting more offspring than monogamous males. Proving that there is a genetic predisposition to the particular behavioural structure of rape is an altogether different proposition. Some specific behavioural structures, or stereotyped behaviour, found especially in birds and insects, are innate: it is not surprising to find that some socio-biologists have entomological or ornithological backgrounds. It does not follow, indeed it is a feeble inference, that all forms of behaviour have some degree of genetic predisposition. In mammals there may be a heritable emotional bias based on genetically determined hormonal levels, but an excess of testosterone or adrenaline will not determine in any way the particular action of rape or murder. There is also a semantic confusion arising from the use of 'rape', 'promiscuity', and 'altruism' in ethology and socio-biology as if they were literal rather than metaphorical.

Baldwin's thoughts on behaviour and evolution nevertheless deserve a new airing. Noting that Weismann had been baffled by the problem of how natural selection could explain how the right variations occurred at the right time and place, Baldwin, with reference to the evolution of instinctive behaviour, argued that so long as there was physiological accommodation or intelligent choice of behaviour then natural selection could be held at bay, and 'the species has all the time necessary to perfect the variations required by a complete instinct.'[71] Why does the organism, having the capacity for intelligent choice in the first place, opt for the build-up of a stereotyped response, instead of that intelligent choice? Baldwin would have responded that such a response, like a nervous reflex, is faster and therefore more effective under most circumstances. Baldwin concluded that not only was use-inheritance unnecessary, but organic selection also explained the persistence of certain lines of determinate phylogenetic change, the direction of the orthogenetic evolution being determined by the limits of ontogenic accommodations of the earlier generations.

The adoption of the principle of organic selection eliminated the inheritance of acquired characteristics as the necessary element of evolution, though it did not disprove the possibility of such a mechanism. It emphasised the action of the organism, an important aspect of Lamarckian theory, as an indirect cause of evolution. This was why Osborn had regarded the principle as a compromise between Lamarckism and Darwinism. While Baldwin believed that organic selection would explain away the apparent direct-line evolution found in the fossil record, he did not have enough confidence in the principle to make it explain gaps in that record, but adduced 'sweeping change in the environment' as the hypothetical mechanism that encouraged evolution in a new line. He also developed the concept that among the higher animals tradition, i.e. the young learning from their parents, enhanced organic selection.

In summary, the elements of organic selection discussed by Baldwin, Lloyd Morgan and Osborn included gradual or sudden environmental change, to which organisms responded with one or more of the following: ontogenic morphological plasticity, physiological adaptibility, behavioural flexibility, intelligent choice and guidance from tradition. These responses provided the time to acquire variations which occurred randomly and were selected according to Darwinist principles and were useful in the context of the modified responses. The concept had far-reaching implications for evolution theory, but as G.G. Simpson has observed, its timing was poor.[72] It pre-dated the rediscovery of Mendelian genetics, and by the time neo-Darwinism had reasserted itself organic selection had largely been forgotten. In any case the population geneticists focused narrowly upon single gene effects rather than on the broad prospect presented by organic selection. Furthermore, as a theoretical or even speculative principle it lacked the polemical power required to persuade the many adherents of orthogenesis and Lamarckism.

C.H. Waddington independently developed a similar concept which he called *genetic assimilation*.[73] Since the name is more self-explanatory it is preferable to the misleading term organic selection, and it will be used here to subsume all the phenomena summarised above. Waddington's definition is: 'a process by which characters which were originally "acquired characters", in the conventional sense may become converted, by a process of selection acting for several or many generations on the population concerned, into "inherited characters"'.[74] To demonstrate the occurrence of genetic assimilation it is necessary to demonstrate that non-genetic phenotypic characters which are induced in a given environment will persist when the organism is placed in another environment in which control organisms of the same type had not shown the distinct phenotypic character. In Waddington's early experiment *Drosophila* were given a high temperature shock for 4 hours at 40°C, about 22 hours after puparium formation. Some of the emergent flies showed broken cross-veins in the wings. When the flies with the newly acquired character were interbred the frequency of the trait increased from generation to generation, the heat treatment being applied each time. Samples from each generation which were not heat-shocked did not develop the trait until the fourteenth generation. Flies which now developed the trait without heat shock were interbred, and some of their offspring showed the cross-vein trait. Selection from these produced up to 100 per cent broken-vein types at normal temperatures. The second phase of selection was later found to be necessary only to speed up the process. Later experiments demonstrated the genetic assimilation of larger induced changes such as wing distortions, and the bithorax condition in which the metathorax develops as a mesothorax, contributing a spare pair of wings. Another interesting effect observed in some experiments was the persistence of the assimilated trait after selection was relaxed. Also, for many of the assimilated traits the underlying mechanism was polygenic, sometimes involving all the chromosomes. Inbred stocks of flies showed little ability to demonstrate genetic assimilation compared to normal, randomly bred populations, suggesting that the ability to produce odd phenotypic traits in response to peculiar environmental effects was associated with a genetic predisposition found in small numbers of the normal populations. However, in the case of the dumpy wing and bithorax conditions the traits seemed to be novel, and arose from the experimental conditions, possibly due to the mutagenic effect of the heat. But as Waddington noted, 'It is only because development can be modified by the environment that the acquired character can appear in the first place,' and the selective strengthening of the trait allows it to persist when the inducing stimulus is removed.[75] This property of a developmental process to be somewhat modifiable yet resistant to modification he called *loose canalisation*.

Waddington argued that the appearance of a new trait due only to 'plasticity' was incomprehensible since the plasticity must be an expression of hereditary potential. However, it is clear that Morgan, Baldwin and Poulton were aware of and accepted that implication. Waddington also observed that his own genetic assimilation work, as well as being empirical, focused upon narrow parameters involving the imposition of a single environmental change, omitting the question of what the flies would do if confronted with a choice at a temperature interface.[76]

In the narrowest Waddingtonian sense genetic assimilation is the natural selection of genetic controls for phenotypic characters that were originally induced by the environment, or selection of refinements of such controls. In a broader sense it could be used to mean the selection of a range of alleles that tend to increase the stability of a changed internal environment, the change having been induced in the first place by a change in environment. The terms 'internal adaptation' or 'internal selection' approach this phenomenon, but they lack the sense of preliminary environmental change or organismic action which have altered the internal environment phenotypically. In the loosest sense that 'organic selection' conveyed, an environmental change was again implied, usually through a change in behaviour in the organism, followed by a selective or adaptive adjustment to the new circumstances. The simple term 'adaptation' subsumes these processes, but fails to inform as to the components of the process. I am going to use genetic assimilation as the most general term for selective, genetic adjustments to internal and external environmental changes, and the correlated fixation of such internal changes.

13 EVOLUTIONARY EPIGENETICS

> . . .a change in differentiation may be brought about by changes in the quality, quantity, time and place of formation, direction and speed of transport of the substances in collaboration with an otherwise unchanged general developmental system. Since these substances produce an immense morphogenetic effect when called into action. . .small changes. . .may lead to large results, provided that the general harmony of differentiation is not interfered with.
>
> Richard Goldschmidt 1940[1]

> . . .the attempts to 'explain' genetic and selective processes by all sorts of fancy terms like 'pedogenesis', 'palingenesis', 'proterogenesis', and what not have had a stultifying effect on the analysis. The less said about this type of literature the better.
>
> Ernst Mayr 1960[2]

Many of the biological theories and laws of growth discussed in previous chapters pertained to the essential quality that according to Aristotle kept the organism true to type, and evolutionists were eventually bound to ask how organisms were released from such restraints. The Aristotelian process of the progressive embryonic development of actual features from the potential features of the germ came to be called epigenesis. For William Harvey it was a universal principle of the development of organisation out of chaos, both in biology and cosmology.[3] The doctrine was refined by Caspar Wolff (1738-94) in *Theoria Generationis* (1759) and popularised by J.F. Meckel (1761-1833). Both were embryologists and students of plant morphogenesis.[4] Wolff demonstrated that plant organs such as leaves and roots were differentiated from the meristematic tissues at the growing points of the root and shoot tips, and that the organs of the developing chick embryo appeared from homogeneous tissue. Epigenesis thus implied an emergence of morphogenic novelty from cells which did not originally contain it. This was the counterpoint to preformationism, which had earlier been promulgated by Marcello Malpighi (1628-94), who asserted in *De formatio pulli in ovo* (1689) that a preformed embryo could be discerned in an unincubated egg.

The later preformationists proposed that a complete organism was preformed in the egg and embryonic development consisted only of growth and an unfolding, though some schismatics held that the sperm cells were the containers of the preformed organism. Wolff denounced preformationism and its mechanistic interpretations, offering instead the *vis essentialis,* the vital force responsible for epigenesis, which became the *nisus formativus* in the writing of Johan Blumenback (1752-1840), and persisted in the neo-vitalist literature as Driesch's entelechy and Bergson's *élan vital.* A mechanistic epigenesis appeared in *Über unsere Körperform* (1874) by Wilhelm His, who was attempting to explain the formation of embryos in terms of both differential growth and physical forces, which would, for example, produce the folding and tubulation of embryonic tissues. Wilhelm Roux and his associates, including Hans Driesch, began to study these problems experimentally, and the distinctions between preformation and epigenesis of the earlier debates soon became blurred. For some time the issue was the degree of importance of cytoplasmic organisation. Was it preformed, as it seemed to be in the mosaic eggs, to the extent that the egg cytoplasm was the 'embryo in the rough' as Loeb called it, with the nucleus contributing only some of the details?[5] Was it more susceptible to epigenetic events such as Driesch's experiments, conducted largely with regulative eggs, seemed to indicate, to the extent that a harmonious equipotential, or totipotency, was retained by individual cells until quite late in development? Or did each embryo have the qualities of both self-differentiation due to intrinsic cellular factors and correlative differentiation due to the relation of the developing cells to the embryo as a whole, as Roux was ultimately obliged to conclude?

Spemann and his associates demonstrated that transplants from one part of an embryo to another might either follow a differentiation pattern induced by the surrounding tissue, or might induce the surrounding tissue to change its differentiation pattern to conform to the original embryogenic destiny of the transplant, showing that both self-differentiation and correlative differentiation could occur in the same embryo.[6] Another issue was the role of the nucleus. Weismann's biophore theory had argued that embryonic differentiation was due to a qualitive nuclear division, with the genetical determinants being portioned out to the different cell lines, resulting in characteristic differentiation patterns. Since the germ plasm had to be totipotent, it was necessary also to adduce a totipotent reserve idioplasm to explain regeneration. E.B. Wilson, himself an important character in the drama of classical embryology, provided a retrospective of the period in the third edition of *The Cell in Development and Heredity* (1928). He noted how the focus on cytoplasmic events had broadened, so most embryologists had become content with the idea that many characters were effected by the transmission of a nuclear preformation that found expression in a process of cytoplasmic epigenesis. As early as 1901 Boveri had pointed out that while preformed cytoplasmic features of the egg were important in embryogenesis, the chromosomes must have an epigenetic role, and the cytoplasmic organisation of the egg was itself the product of an earlier process of epigenetic development in which the chromosomes were involved.[7] Experiments with delayed nucleation, first by Loeb in 1894 and later by Spemann in 1914, showed that cytoplasm would not differentiate in the absence of a nucleus, and Boveri's observations on dispermic sea urchin eggs between 1902 and 1907 demonstrated that a particular combination of chromosomes was necessary for normal development, so that by inference the individual chromosomes must possess different qualities.[8] The synthesis of these contradictions did not occur for three decades, and even then there remained the problem of how the epigenetic instructions of the nuclei were obeyed, since light microscopy could not demonstrate the transfer of nuclear material to the cytoplasm. The question was unanswered for two more decades until the role of messenger RNA was elucidated.

The sense that Waddington and Løvtrup now give to epigenetics admits the molecular gene theory, and assumes the presence of genes that have an embryonic role which if altered might have evolutionary consequences. Moreover, the developmental genes act and interact in an 'epigenetic landscape' involving cytoplasmic factors, and repressors and inducers in the extracellular internal milieu and external environment. There remains some confusion about the meaning of epigenetics because of the historical association with epigenesis and because the term suggests a branch of genetics. The sense of 'origination' or 'production' in the *genesis* root has been superseded by the sense of 'heredity'. However, epigenetics involves not only genes but also their differential expression, events, interactions and environments which cannot be predicted from a knowledge of the genes and their Mendelian combinations, and the momentum given by Waddington and Løvtrup to the term epigenetics to describe these concepts justifies its continued use.

During the period of classical embryology, from His to Spemann's later work, there were few attempts at interpreting the evolutionary significance of epigenesis, there was considerable discussion of the associated recapitulation or biogenetic theory, as S.J. Gound's *Ontongeny and Phylogeny* (1977) illustrates. I intend to disentangle the two and concentrate upon the question of how evolution occurs through embryonic changes, rather than how ontogeny reveals evolutionary history.

As H.F. Osborn discovered in his search for the unknown factors of evolution, one of the most important epigeneticists was Geoffrey, who had understood that the earlier in development a change occurred, the greater the evolutionary saltation might be. His associate Étienne Serres believed that deformities in mammalian embryos were caused by local retardation of the action of the formative force, serious retardation creating teratological monsters.[9] Kölliker, also a saltationist, argued that the probable mechanisms were epigenetic changes or changes in the behaviour of larvae. Heterochronous changes, i.e. alterations in the relative timing of developmental events such as acceleration and retardation, were regarded as important causes of variation by the neo-Lamarckians Hyatt and Cope, and neoteny was classified by Mivart as a saltation. Driesch concentrated on epigenetic stability, but from harmonious equipotentials the next intuitive step was to consider the harmonious re-equilibration that would be necessitated by epigenetic change in order to achieve evolutionary progress: Bateson's 'accommodatory mechanism'. Bertalanffy and Goldschmidt were among the first to ponder evolutionary epigenetics from a realistic standpoint.

Richard G. Goldschmidt (1878-1958) studied zoology under Richard Hertwig in Munich. Appointed as a lecturer he developed a course in genetics, and in 1910 he began to study the gypsy moth *Lymantria,* extending these studies in 1913 to Japan, where there was a strain that produced curious sex determination effects when crossbred with the European variety.[10] Stranded in the United States by the outbreak of World War I, Goldschmidt continued his *Lymantria* research with the assistance of T.H. Morgan, E.B. Wilson and Ross Harrison. His research was mainly conducted at the Bussey Institute at Harvard, where he came in contact with E.B. Castle, W.M. Wheeler, E. East, Sewall Wright and E. Sinnott. The *Lymantria* work continued in Germany and in Japan until 1930, and in 1938 he emigrated to the United States. Goldschmidt's early views on genetics were neo-Darwinist, and research on melanism in *Lymantria* led him to the conclusion that natural selection must be more important than mutation in effecting population changes. Subsequent studies of the varieties of the gypsy moth, however, led to the conclusion that, 'geographic variation is a blind alley, leading only to micro-evolution within the species, and not the source of real evolution'.[11] The only alternative was the existence of macro-mutations, which in rare cases could affect early embryonic processes so that through embryonic regulation and integration a major step in evolution could be accomplished.

Goldschmidt had already considered the equilibration of the rates of embryonic processes in *The Physiological Theory of Heredity* (1927), and it was a simple step to consider the effects of a radical change in those relative rates. The result was the 'hopeful monster', a character whom he 'half-jokingly' introduced in a lecture given at the World's Fair in Chicago in 1933. Monsters have appeared frequently in the history of epigenetics and saltatory evolution; Caspar Wolff left an unpublished manuscript on the generation of monsters.[12] The 'hopefulness' of Goldschmidt's monster was a recognition of the limitations of teratology for evolution theory as well as the recognition that a monster would need unusual circumstances to survive, and to reproduce. Six years later he returned to this theme in his Silliman lectures at Yale, which were expanded into *The Material Basis of Evolution* (1940). His discussion of macro-evolution opened with the role of chromosomal mutations. Their possible evolutionary significance might be that they were simply insignificant, freakish occurrences; that they represented significant changes in numbers of genes and gene mutations; or that they affected the organism without involving change at the gene level. Since Goldschmidt had long suspicions concerning the reality of the genes and their mutations he was inclined to the third option, postulating that 'position effects' in the chromosomes were radical factors in evolution. Changes in the structure and patterns of chromosomes could be 'sub-threshold', without discernible effects until a new stable pattern was reached and a new phenotype of even new species emerged, separated from the old type by an apparently bridgeless gap: ' "Emergent evolution" but without mysticism'.[13] This view of macro-evolution could not be conceived of on the basis of accumulation of micro-mutations, but large phenotypic changes could be effected by small 'systemic', chromosomal mutations. Evolution implied the 'transition of one rather stable organic system into a different but still stable one', signifying a changed process of development which could only have evolutionary significance if 'the subsequent different process of development are again properly integrated to produce a balanced whole, the new form'.[14] Was such a change possible in terms of one or a few macro-evolutionary processes rather than many micro-evolutionary ones? A number of works, including Kammerer's, proved that slight environmental changes during development could produce marked phenotypic changes. His own research on phenocopying had shown that a range of phenotypic changes known to result from micro-mutations could be evoked in a non-heritable form, by, for example, subjecting embryos to temperature changes. If these large viable changes could be produced by quite small external stimuli, then small systemic mutations could produce the same phenotypic effects, but on a hereditary basis. Contemporary work with hormones, both embryonic and adult, suggested that

> a change in differentiation may be brought about by changes in the quality, quantity, time and place of formation, direction, and speed of transport of the substances in collaboration with an otherwise unchanged general developmental system. Since these substances produce an immense morphogenetic effect when called into action. . .small changes. . .may lead to large results, provided that the general harmony of differentiation is not interfered with.[15]

The experimental production of neotenic amphibian larvae, research in which he had participated, was another case of large phenotypic change induced by small epigenetic change, and a case known to exist in nature as a heritable condition. Harms had induced metamorphic changes in the tropical mangrove-swamp fish, the mudskipper *Periophthalmus,* by feeding it thyroid extract. The effects included changes in the pectoral fins which made them more limb-like, reduction in the gill cavity, and drying of the skin.[16] Human foetalisation and other variations such as gigantism and dwarfism were other examples of large phenotypic changes effected by simple alterations of the endocrine system, which in these cases obeyed Mendelian principles. Following a similar line of argument, Stockard (1931) had suggested that *Homo sapiens* had evolved in a single step effected by hormonal changes.[17]

In response to Fisher's rejection of macro-evolution on the grounds that it could not produce adapted organisms, Goldschmidt proposed that phenotypic and behavioural flexibility buffered the organism after epigenetic changes; adaptability allowed the organism to 'get away with it' under a variety of circumstances. Seeing the final outcome in terms of reproductive success, the selectionist attributes the cause to natural selection, and an expression of adaptability in response to a particular change is interpreted as adaptation. This is masked by the circularity of the concept of adaptation: the organism exists in this environment, therefore it must be adapted to it, since no organism can survive in an environment that it is not adapted to. Adaptive specialisation may possibly evolve exactly as the neo-Darwinists suggest. Goldschmidt's error was to work his hypothesis to the point of collapse by trying to subsume both integrative evolution and adaptive specialisation in the same hypothesis. Goldschmidt compared the epigenetic disengagement of integrated systems with changing gears in an automobile — it could be achieved by breaking the motor shaft, or it could be done harmoniously by declutching, so that the disengaged systems were in a position to enter new combinations — a good system analogue, though Goldschmidt seems to have been unaware of Bertalanffy's early thoughts on the subject. Goldschmidt embraced D'Arcy Thompson as a tacit macro-evolutionist, and welcomed Julian Huxley's conviction that allometry would go part of the way to explaining 'non-adaptive' changes, thereby lightening the explanatory burden of natural selection. [18]

For Goldschmidt a Manx cat was 'just a monster' but *Archaeopteryx* was a 'hopeful monster', because the mutant feathered tail could be useful for flight. [19] A compressed fish with both eyes on one side of the head was hopeful because it could begin to live like a flounder. Both the monster and the hope are misleading anthropomorphisms. Man has bred the achondroplastic dachshund to chase badgers into their setts, but the breeders see no monstrosity. A similar natural process could have produced otters and seals and whales. There was an element of 'hope' in as far as the emergence of such macro-mutations would be aided by a suitable environment. A dachshund without a badger sett, or an otter without a stream would be *incongruous* in Uecküll's terms. But is the unfulfilled hope of a new environment such a let-down? Homeostatic emergences may be self-sufficient, and novel anatomical harmony of form its own reward. Could convergent and parallel terms not be coincidences of harmonious epigenetic change? E. Bonavia, in 1895, had anticipated Goldschmidt by pointing out that monstrosities such as *Archaeopteryx* could in some circumstances be evolutionary saltations.[20] Goldschmidt was already regarded as unorthodox for his views on the gene; reaction to *The Material Basis of Evolution* was stronger: 'I had certainly struck a hornet's nest. The neo-Darwinians reacted savagely. This time I was not only crazy but almost a criminal.'[21] The crime was that he was too close to the truth for comfort. He could not be dismissed for freakish associations such as Larmarckism or vitalism. He had been inclined towards neo-Darwinism and indeed, he had contributed to the success of neo-Darwinism with respectable experiments in population genetics and a fifteen-year study of *Lymantria* varieties. The results drove him to alternatives unacceptable to the majority of the orthodox, reductionist congregation. He was not alone in doubting the existence of the corpuscular gene, and other geneticists were uneasy with the molecular model as late as 1951, though not prepared to support Goldschmidt's assertion at the Cold Spring Harbor Symposium session on the theory of the gene that point mutations were simply sub-microscopic chromosomal rearrangements. In the same session Barbara McClintock, who had worked briefly with Goldschmidt in 1933 in Berlin, presented her paper, 'Chromosome Organization and Genic Expression', describing the regulation of chromosomal mutation in maize. The experimental demonstration of a regulated 'position effect', with clear implications for cell differentiation and hence for evolution was grist to Goldschmidt's mill, and he heartily endorsed it as a vindication of his own opinions. The epistemological implications of this event are discussed by Evelyn Keller in her biography of McClintock, *A Feeling for the Organism* (1983). The assembled geneticists lacked the vocabulary needed to understand gene transposition, which, though based on firm experimental data had a strong intuitive content which obscured its meaning. Goldschmidt's enthusiasm was a further strike against McClintock, and resulted in a similar reputation for 'madness'. Moreover, the Phage Group, which had been meeting annually at Cold Spring Harbor during the 1940s was swinging reductionist opinion into line with the physicochemical interpretation of the gene. Within two years of the gene-theory symposium the group's crowning success, Hershey and Chase's identification of DNA as the molecule responsible for phage reproduction, together with Watson and Crick's description of the double helix, guaranteed the dominance of the molecular biological paradigm.

Aromorphosis was the name given to homeostatic emergence by A.N. Severtsov, the Russian evolutionist, in 1927. It was an early ontogenic event: 'a fundamental structuro-functional change which increases the general efficiency of organisation and renders the animal capable of further adjustment and adaptation to changing environmental conditions'.[22] Severtsov also included *katamorphosis* in his new evolutionary lexicon; this implied regressive changes which simplified the connections between the organism and the environment, including neoteny.[23] K. Beurlen, writing about the evolution of the Crustacea in 1930 and comparing their differentiation with that of the Mollusca and the vertebrates, concluded that periods of 'explosive' evolution gave rise through sudden epigenetic change to these new major types, which were not immediately advantageous but could undergo adaptive evolution.[24] Like Eimer, Beurlen believed that even adaptive evolution had an orthogenetic component. Similar views were held by Schindewolf concerning saltatory archetypal evolution;

> Through big evolutionary saltations (*Entwicklungsspringe*) there arose first the major structural plans, and thereafter followed, within their limits, a descending differentiation of types of lower order. . .These first exemplars of a new typal organisation connote the starting point of an increasing diversity of form, of a further elaboration of the typal characteristics[25]

The new types must have arisen epigenetically, by *proterogenesis*, though all of its products would not be viable. Like Beurlen he rejected Lamarckian interpretations. The most radical epigenetic hypothesis of this period is found in Austin H. Clark's *The New Evolution: Zoögenesis (1930)*. The complete absence of any intermediate forms between the major groups of animals, which was one of the most striking and significant phenomena brought to light by zoology, had been ignored, or at least overlooked. To explain this condition he applied embryological evidence to construct an epigenetic saltation theory of *eogenesis*, which explained archetype formation. 'Mutations' were minor saltations within particular typal phyletic lines, and *zoögenesis* was the complete history of animal evolution, which subsumed orthogenesis as an evolutionary process. Clark regarded a gastrula-like organism, close to the coelenterate level of evolution, as the common ancestor of animals. From this gastruloid type different lines evolved due to divergences of cell relationships in time based on changes in the adhesive properties of the cells. The four main lines of evolution were: (1) types which became linear colonies by budding; (2) types which budded internally within the original unit; (3) types which remained solitary; (4) types which were colonial, but the individuals were independent.[26] Between any two types there could be intermediates which combined the characters of both. Clark proposed that these major types and the viable intermediate forms produced seventeen major divergent groups of animals, plus the original form, each group having appeared more or less spontaneously, and these were the only forms capable of 'meeting successfully the conditions of existence'. He ruled out the planktonic larvae of the marine invertebrates as transitional ancestral forms since he regarded them as later specialised distributional adaptations. There was, of course, room for great diversification within a viable type, by a combination of eogenesis which could suddenly produce the classes, together with orthogenesis and adaptation. But since the conditions of existence had always been the same, especially in the sea, adaptation to environment could not have played a major part in zoögenesis. Because the same fundamental cellular properties were called on time and again in different combinations or different modes of interaction it was not surprising to Clark (nor Berg nor Vavilov) why convergent and parallel evolution would occur. As in Willis's differentiation theory, over-specialisation would not be an obstacle to progressive evolution since the specialised forms were and would remain the terminal products of evolutionary lines. Clark recognised that completeness of structural, and presumably physiological balance was an important quality of animals, but his thesis did not generate any explanation of how this complex, organised completeness could come into being by a random process of differential cell adhesion; it simply asserted that it had to have done so. This probably explains why Clark's eogenesis is missing from an otherwise comprehensive discussion of epigenetic saltationist hypotheses by E.S. Russell in *The Diversity of Animals* (1962). Russell adopted the idea of aromorphosis, confronted the problem of the evolution of complex functional morphological organisation, and for his pains was regarded as being somewhat of a vitalist. The problem that he found with saltation or macro-mutation was how the complex goal-directed activities could be put together meaningfully. If throwing bricks in a heap one at a time was no way to build a house, neither was throwing the bricks all together at once.

> While it seems probable that the formation of new types or sub-types has come about by early ontogenic variation, however caused, it seemes highly unlikely that new types appeared in this way fully formed, that for instance, the transition from reptile to bird was achieved in one single step.[27]

While Russell conceded that typogenesis must occur by 'abrupt deviations from the normal course of development occurring at an early stage, switching development into new lines and producing rapidly a new or altered type', he was unwilling to accept genic macro-mutation as the mechanism of such an event since he believed macro-mutations to be non-directive and likely to disrupt the normal harmony and coordination of developmental processes.[28] All that Russell was able to suggest as an alternative was that, 'in addition to the Darwinian and the Lamarckian factors, a mode of evolutionary change which is directive and unified, occurs at an early stage in ontogeny, and is cumulative from generation to generation'.[29]

Two other pre-molecular biology writers whose work had some relevance to evolutionary epigenetics, but who were much closer to the mainstream of Darwinism, were Gavin De Beer and I.I. Schmalhausen. The importance of embryology to neo-Darwinism was enunciated by De Beer in *Embryos and Evolution* (1930) and *Embryos and Ancestors* (1940). The latter remains a standard reference for the classification of heterochronous evolutionary changes, although the simplified scheme provided by Gould's *Ontogeny and Phylogeny* may prove more popular in the longer term. De Beer agreed with Julian Huxley's views on the significance of allometry in general and as an explanation of orthogenesis in particular, but he regarded the cleansing of the Darwinist stables of heterodox accretions, especially the recapitulation theory, as a major duty.

I.I. Schmalhausen was a student of Severtsov at the University of Moscow, and ultimately became Director of the Institute of Evolutionary Morphology.[30] Like other members of the Severtsov school Schmalhausen stressed the importance of progressive changes in self-regulation, both in developmental events and in the physiology of the mature organism. In *Factors of Evolution* (1949) he wrote:

Evolution has on the whole been a liberation of the developing organism from accidental environmental changes. It has been accompanied by the elaboration of internal regulating mechanisms controlling processes of individual development. Liberating the organism from the *determining* influence of environment involves establishment of a system of internal factors of development which determines the specific course of morphogenetic processes.[31]

This then was the means whereby animals progressed away from the ontogenic process that left plants responsive to the environment at every stage of development: useful environmental stimuli were internalised and brought under control, 'canalized into the most profitable forms of reaction'. Non-adaptive modifications or reactions might arise physiogenically under novel environmental circumstance, and then be canalised or genetically assimilated. Although Schmalhausen was unaware at the time of Waddington's parallel thinking on this subject, he noted that C.L. Morgan and Baldwin had been on the right track, and thought it regrettable that the success of neo-Darwinism and the stress on adaptation had resulted in the neglect of these concepts. The 'functional activity' of the organism, both behavioural and physiological, needed to be accounted for. All really new reactions of the organism were never adaptive, since they were the results of accidental mutations, 'becoming adaptive only in the course of historic development in certain environments.'[32] Progressive evolution of regulatory mechanism could proceed by duplicative mutations whose products could then diversify to provide a variety of differentiation. Multi-cellularity and segmentation were among the most important historical duplicative events. The morphogenic substances involved in differentiation appeared initially as the products of specific meta-bolism. These *attendant conditions* then became incorporated as *necessary conditions* of morphogenesis.

Dobzhansky assessed *Factors of Evolution* as 'the missing link in the new synthesis', because it started from embryology and the mechanics of development.[33] However, despite the fact that Schmal-hausen got there first with the most men in the sense that he touched on a larger number of important evolutionary issues, and anticipated the ideas of later authors like Ohno, Løvtrup and Riedl, the similar theories of C.H. Waddington were more sophisticated and based on more experimental evidence, and have therefore prevailed in the West.

C.H. Waddington (1905-75), according to his own account in *The Evolution of an Evolutionist* (1975), derived his interest in epigenetics from early studies of the ammonites, fossil cephalopods whose shells bore the record of their entire development.[34] He had an early interest in the metaphysics of Whitehead, and in the neo-Darwinist genetics of J.B.S. Haldane, but settled into the study of embryology with the feeling that the 'potencies' and 'organizers' popular in the 1920s and 1930s should be expressible, as T.H. Morgan had hinted, in terms of the activities of genes. He considered that Goldschmidt had made the most prominent contribution to the question of biological organisation in terms of genetics, but disapproved of Goldschmidt's rejection of the gene as a significant locus of evolutionary change.[35]

Waddington's best known work, *The Strategy of the Genes* (1957), emphasised the importance of epigenetics for a comprehensive theory of evolution. Published just after the elucidation of DNA and the first research on the genetic code, the book was not quite ready to assess the significance of molecular biology in relation to epigenetics, and the relative importance of the 'material gene' and chromosomal qualities was still undecided.[36] A theory of evolution involving only the selection of fortuitous variations of the material genes was inadequate to explain changes in orderly systems of development. Studies of population genetics and individual genes were 'tactical' questions for evolutionists, but Waddington was concerned with the 'strategic' question: 'How does development produce entities which have Form, in the sense of integration or wholeness?'[37] It is interesting to observe how selectionists have subsequently debased the more holistic metaphor of strategy and applied it to what Waddington called genetical tactical situations. Waddington's model of the epigenetic landscape is a successful metaphor for the fate of a differentiating cell in relation to the external epigenetic influences, depicting a ball rolling down a channelled slope, like a water eroded hillside (Figure 13.1). Minor deviations in direction at the beginning of the downhill journey produce major changes by the end of the journey; the choices of route before the ball at any point on the journey are limited by choices already taken. *Homeorhesis* is the epigenetic equivalent of physiological homeostasis, signifying the return of the ball to its normal path after some dislodging alteration in the topography of the landscape, or in more direct terms, the restoration of normal development after some disturbance.[38] Homeorhesis is a stripped-down version of entelechy, divested of its transcendental implications.

Figure 13.1: The Epigenetic Landscape

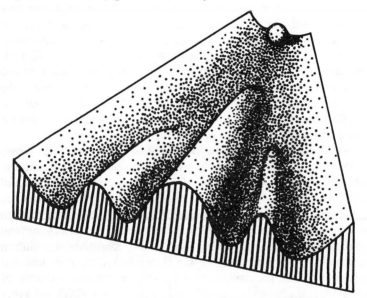

Source: After Waddington 1957. The diagram is a metaphor for the alternative routes available to a differentiating cell, as represented by a ball running down an eroded slope. The depth of the channels can be taken to represent the rigidity of canalisation. The shallower the channel the simpler it is for an environmental or internal change to alter the route.

Evolution had a 'ticklish task — to tighten canalisation against deleterious changes and loosen it in favour of adaptive ones'.[39] Also, if the homeorhetic mechanisms produced a high degree of developmental stability they would resist both adaptive and grade evolution. Waddington doubted that natural selection would be able to produce adapted organisms that were also adaptable. Experiments that varied selection pressures did not produce a loosened canalisation. A genetically adaptable population, i.e. one containing allelic diversity, could respond positively to a variety of selective pressures, to the extent of becoming polymorphic. But how was individual adaptibility, either homeorhetic or homeostatic, to be explained? And how did the organism evolve from simple stable states to complex ones? Waddington concluded that the aromorphic cause must be environmental. 'Abnormal' environments could induce acclimatisation in ontogenesis, resulting in changes that the homeorhetic mechanisms could not bring back under control. Combining this with the concept of genetic assimilation it could be argued that a novel ontogenic state imposed physiogenically by an unusual environment at a crucial time in development might be genetically assimilated or fixed by natural selection. For example it was known that the larval newt raised in oxygen-poor conditions developed larger gills with thinner walls, which allowed more oxygen to be absorbed, while general growth was slowed so that less oxygen was required.[40] Waddington regarded this as a loosening of canalisation whose end result was still the development of a normal newt, and it is not difficult to extrapolate from this to an organism that might exhibit paedomorphic loss of adult features, or alternatively have an adult with larger or more efficient lungs. Under these circumstances natural selection is a *post hoc* event. The role of environmental change and behavioural response in this loosening of canalisation has been well illustrated recently by R. Matsuda.[41] He notes that some terrestrial amphipods have compensated for a move to a drier environment by seeking a moist micro-climate. Such a micro-climate is necessarily dark, and the lack of light produces a neotenic form through hormonal effects. In this case a change in behaviour results in a change in environment, requiring the further choice of a micro-environment, resulting in changes in development. Similarly, temperature has been shown to be an agent of neotenic change in several species of salamander. The changes may be completely reversible or may be genetically fixed, i.e. not alterable in the offspring if they are restored to the old environmental temperatures.

Waddington doubted that theoretical population genetics had any major claims to originality and novelty. At the time of writing *The Strategy of the Genes* he thought that embryologists such as Goldschmidt, Schmalhausen and Dalcq were asking the most significant questions. Dalcq had also understood that disequilibration was necessary to move evolution from a conformist heredity to the novel, and he adduced the term *onto-mutation* to designate the saltatory appearance of a new type of egg organisation necessary to insure the beginning of a new phylum.[42] In his later writing Waddington continued to develop the 'post-neo-Darwinist' epigenetic themes of genetic assimilation, viral transduction of DNA and the problem of palaeontological gaps. He believed that genetic drift might provide a degree of freedom for the organism from the exigencies of the environment and from natural selection; occasionally the race might not be to the swift nor the battle to the strong. The conventional statement that 'the raw materials of evolution are provided by random mutation' appeared hollow to him when the events and interactions responsible for the building and alteration of the epigenetic landscape were taken into account.[43] Evolution might proceed for some time according to orthodox conceptions, but it might

suddenly produce a form which has a stability, or a capacity of response, of a different order of magnitude to anything which has been demanded of it in the past. It may either meet an old challenge so very efficiently that it rapidly and completely takes over from all the old and much less efficient phases that have led up to it; or it may be capable of being used in some radically new way which allows the ancestral forms to survive only in much simpler contexts where their ancestral relation to the novelty may be almost unrecognizable.[44]

This principle of archetypal evolution that he thought might apply particularly well to the emergence of language in man has strong affinities with the emergentism of Morgan and Alexander.

Before proceeding to more recent hypotheses, I wish to reiterate the problem whose solution would seem to lie in the area of epigenetic and molecular biological research. As Britten and Davidson have pointed out, 93 per cent of the enzymes bacteria can synthesise can also be synthesised by man; most organisms are fundamentally similar at the biochemical level;[45] the differences between bacteria and man must be largely interpreted in terms of how these organisms put their abilities to synthesise proteins to use, during the course of their development. To use the perennial image a number of bricks may be thrown into a meaningless pile, or may be put together to make a pigsty, or a palace if there are enough of them. The difference between the pile, the pigsty and the palace is a matter of organisation. Similarly, the same cells that may form an unorganised clump in tissue culture may form an integrated organism if they are given the correct conditions and organised so that the right things are in the right place at the right time, as Driesch stated so clearly. Neo-

Darwinists have been mainly concerned with minor modifications in the bricks and their consequences in terms of architectural style. If the size or surface configurations of the bricks are changed the style of the pigsty can change from functional to baroque. Without a fundamental change of plan or organisation the building always remains the same, structurally and functionally. The problem is: how is the organisation changed in such a way as to produce a structuro-functional advance, remembering that meaningless piles are a common product of organisational dysfunction? I have been discussing proteins as 'bricks': the code for proteins is in DNA; but knowing what bit of DNA represents what protein does not solve the problem; unorganised DNA just brings us back to meaningless piles. All the same, a sound molecular gene theory is clearly a prerequisite for a sound epigenetic theory, when we consider how the lack of one nullified much of the good sense of Woodger, Goldschmidt and Russell.

Of the early pre-DNA commentators Uexküll seems to have understood the problem most clearly:

> It helps us to understand this, if we imagine the genes to be keys of a piano, only waiting to be struck for all manner of tunes to sound forth. We may consider the notes of the score, which arrange for us the laws of a possible phenomenon, or we may listen to the playing, when the keys give out the sound according to the law prescribed. The aim of descriptive biology must be to set down, by means of a kind of musical notation, the laws according to which the genes in various animals sound together in succession, this notation is imparted to use from the beginning by the laws controlling the sounds emitted by the genes in any animal, the genesis of which from the germ, we are attempting to observe.[46]

He defined the gene as a 'latent ferment', i.e. enzyme, united with an 'activating impulse'. He further supposed that new genes might not arise at all, and that,

> It is only the melody of the impulse-sequence that changes. If we compare the genes to the keys of a piano it is obvious that all tunes can be played with relatively little material substance. If we assume that in the germ of the first living creature were present all the ferments necessary to effect all the changes in form and substance that we observe in the development, we might maintain

that the difference between the forms of animals from that time until now depends merely on the fact that only a limited number of ferments were used by the primitive impulse-sequence of the first organism. In course of time, the impulse melodies became richer and more intricate so as to create at last the symphony of the Mammalia. At the same time, perhaps, in consequence of the splitting off of species the originally complete keyboard lost more of its notes, so that in animals at the present day the possibility of shaping new melodies are diminished as the melodies are developed.[47]

Unfortunately, Uexküll's belief that the environment might directly cause changes in the 'melodies', together with his attribution of ineffable transcendental qualities to the 'activating impulse', assured the trashing of his insights. He has been criticised particularly for treating 'conformity with plan' as one of the laws of nature, but his 'conformity with plan' had a strong epigenetic significance and was not simply unreconstructed progressionism. As Woodger pointed out, differences between organisms are largely differences in cellular differentiation, and in order to understand how such differences arise, the process of ontogenic differentiation is what must first be understood. In modern terms that means the comprehension of the ontogenic regulation of gene expression, a fundamental component of epigenetics.

The range of evolutionary epigenetic thought has been extended recently by Søren Løvtrup, Professor of Zoophysiology at the University of Umeå. The neglect of epigenetics by Darwinism gives him a dim view of the creed:

When the history of evolutionary thought in the first century after Darwin is to be written by future historians of science, the closest analogue they will find will be the medieval times of scholastics, with Darwin assuming the role of Aristotle.[48]

He calls for a comprehensive theory of evolution that would assign to natural selection a causal role of imposing order on the disorder of random variation through the increase of useful traits.[49] However, the focus of his thought is epigenetics and the consequences of mutations of the genes that regulate development. *Epigenetics* (1974) is a comprehensive amalgam of classical embryology, cellular and molecular biology and developmental genetics, together with a historical and philosophical overview. In the sense that the preformed structures of the zygote and the predetermined information of the DNA interact with each other and with the environment, and since these interactions become more complex as embryogenesis proceeds, modifying the expression of the preformed and predetermined characters, all of the post-fertilisation events are defined as epigenetic. In a narrower sense an epigenetic character is one, 'which is causally dependent upon a preceding [ontogenic] stage'.[50] Therefore most ontogenic characters are epigenetic. Løvtrup uses 'cell diversification' as a more general term than 'cell differentiation'.[51] Cell diversification includes cellular transformations, processes that determine the cell type, and cellular differentiation, to which he gives the narrower meaning of the modification of the developing cell line. Cell diversification therefore occurs because cells with the same genome have produced different patterns of protein, according to the repertoire of genome and the ability of the ontogenic process to select among the range of protein-synthetic possibilities. This was understood by Jacob and Monod as a consequence of their studies of bacterial gene expression and was implicit in McClintock's earlier interpretations of the regulation of cellular differentiation in maize.[52] Løvtrup points out that the products of cellular syntheses include 'essential proteins' which participate in general cell function, and 'diversification products', i.e. proteins and other compounds which are the products of differentiation enzymes. As Holtzer and Abbot (1968) have pointed out, the diversification products might appear to be unessential or even harmful in terms of the general physiology of the cells, but are epigenetically and homeorhetically essential.[53] The output of different spectra of diversification products in different cells must, since the cells have the same genome, be the result of pre-existing polarities, or dissimilarities in the cytoplasm of the parent cells. Polarity, and associated concepts of cytoplasmic fields and gradients, have been dominant themes of embryology. Løvtrup narrows the problem to the mechanisms involved in the nucleo-cytoplasmic interactions, which underlie cellular diversification. It must be remembered that some of the relevant cytoplasmic qualities are the result of influences external to the cell, such as the nature of surrounding cells, hormones or inducers in the body fluid, and physicochemical changes in the external environment. Since the differential replication of DNA is rare the differentiality must come down to DNA transcription. The now classical model for this process is

Jacob's and Monod's lac operon (1961), which describes how an operator gene controls the action of structural genes which are the coding cistrons for ß-galactosidase. This utilises the milk sugar, lactose. The *operator gene* is itself controlled by a *regulator gene*, through the production of a *repressor* molecule. A second low-molecular weight molecule, the *effector*, may participate by activating or inactivating the repressor. Enzyme substrates may be the inhibitory effectors allowing the operon to produce the necessary enzyme. The products of this synthesis, if they accumulate, may in turn act as the repressor activators, providing a feedback control of unnecessary synthesis. In the case of epigenetic diversification this feedback repression would not be operable. Some proteins are known to be non-repressible, and their synthesis goes on all the time at maximum rate. In Løvtrup's view embryonic induction 'involves nothing but the induction of new diversification patterns through a mechanism of permanent derepression, the stabilization being accomplished by positive feedback'.[53] The origin of induction may be spontaneous and endogenous to the particular cell, or might be exogenous, arising outside the cell. Endogenous inductor molecules might remain latent until certain earlier events have occurred and their metabolism exhausted. As a model for this the bacterium *Escherichia coli* metabolises glucose but not lactose if both are present in the incubating medium. Only when the glucose is exhausted do the cells begin to synthesise ß-galactosidase and use the lactose as an energy source.

Løvtrup's achievement is to make us first of all discover our ignorance of evolutionary epigenetics, and then to focus our attention on the particular unknowns that must be investigated. The first phylogenetic problem is the control of cell division and adhesion. Unicellular organisms do not require this, but organised multi-cellular organisms do if they are to produce the characteristic size and form. Secondly, the emergence of diversified structural, physiological and regulatory cell types adds new dimensions to the phylogenic possibilities. Thirdly, the novelty produced by non-lethal mutations of structural genes is minor, compared to novelties involving the expression of the structural genes.

We have seen that in widely different taxa the same, and rather few elements are responsible for elaborating quite different body forms, a result explainable primarily by changes in the pattern of interaction between the substrate elements in time and space.[54]

Løvtrup admits that this implies changes in some of the structural genes coding for diversification products, but the diversity of these is exceedingly small. Moreover, Løvtrup concludes, as Uexküll had also inferred,

> the secret of phylogenetic progression does not seem to lie in the elaboration of a more and more complicated starting point, as represented by the fertilized egg, but rather in establishing the preconditions for a more and more complicated and extended series of causally connected epigenetic (morphogenetic) events. [55]

Every constituent of the egg may be concerned with epigenesis, not just the heredity material of the nucleus. Løvtrup argues that geneticists have largely ignored epigenetic factors, because of the limitations inherent in their experimental approach, which confines them to intraspecific studies of structural genes. Extra-nuclear components of the zygote are not subject to Mendelian inheritance, and Løvtrup suggests that since the slightest deviation from the normal extra-nuclear constitution is likely to be deleterious, experimental studies of their inheritance are difficult. He would require a more holistic approach that would enlarge the concepts of genotype and phenotype along the lines of Waddington's suggestion that all the hereditary material of an individual be called the *epigenotype*, and the form of body produced by the epigenotype through the epigenetic process be called the *epiphenotype*. 'The epigenotype is identical with the embryogenetic substrate, i.e. the fertilized egg, while the epiphenotype is the adult animal in toto.' [56]

Like Waddington, Løvtrup finds the concept of the archetype that originated in the work of Cuvier and Owen a valuable one, though this has brought down upon him anathema for idealism. [57] On the subject of mutation he requires no radical conceptual changes; mutations include gene mutations, chromosome mutations and genome mutations. These subsume position effects, in which the locus of the gene in the nucleus has some bearing on how it is expressed, and gene duplication (DNA repetition). Løvtrup accepts Ohno's argument that novelty can be effected by the redundancy of duplicate genes which are not required for the original function. He regards the mutation of control genes as of particular phylogenic significance, considering the conservativeness of the structural genes, though point mutations of structural genes for diversification products might have a strong impact. A change in the amount of a particular cell constituent due to a change in a control gene (or by gene duplication) might be phylogenically significant, but would also require an added morphogenic control. For example, epidermal cells might produce large quantities of keratin, but the forms that it can take — feathers, scales, calluses — require epigenetic regulation.

In Løvtrup's view saltatory evolution is a strong possibility, in the context of molecular mechanisms already acceptable to orthodox evolutionists. Gross ontogenic changes may be produced by early epigenetic changes or by later changes that are epigenetically amplified. Co-operative effects among apparently trivial mutations may be significant. Løvtrup is one of the few modern evolutionists prepared to accept the evidence of their eyes and treats 'sports' as a contradiction of the selectionist assertion that macromutations cannot survive. Referring to Goldschmidt's contention that the various sports and types in the dog species would be given supra-specific status if found in nature, he remarks, 'the fact that the responsible mutations can survive only through the protection of man shows that they may be of little advantage under natural conditions, but this does not alter the fact that the mutations exist'.[58] Moreover, by neo-Darwinist arguments, nature seems to be superior in finding utility in variations, and isolation of dissimilar races of dogs in nature would have further given rise to fertility barriers. More recent studies show that large phenotypic effects result from point mutations of single genes, such as the formation of clawed digits on the wings of chickens and changes in intraspecific characters in arthropods.[59] Løvtrup concedes that epigenetic mutations are more likely to be deleterious than advantageous, since their modification is amplified through interaction with other epigenes. Nevertheless, rather than natural selection, 'The real Creator is indeed genetic — and epigenetic — mutation, operating with the assistance of isolation and interspecific competition.'[60] Simple change in the patterns of cleavage, cell division and diversification products can effect drastic morphological changes. The presence or absence of mucopolysaccharides in the body cavity, for example, has a radical effect on the form of the coelom, or may dictate its absence. The adhesiveness of muscle cells, the timing of the onset of production by shell glands, the hardening of the outer coats, the acquisition of a notochord, can largely be viewed as all-or-nothing mutations which, while trivial at the chemical level, have had very profound morphogenic effects. Conceptually, this borders on the simplicity of Clark's eogenesis, and offers no account of how such radical changes, having occurred, were accommodated by the organism to retain its integrity as a functioning whole. This is not to say that Løvtrup (or even Clark) is wrong. These phylogenic events are known to have occurred, and a gradualistic hypothesis is inadequate. An animal can no more be a little bit coelomate than a little bit pregnant.

The Phylogeny of Vertebrata (1977) continues the epigenetic theme, as well as considering systematics and presenting a strong challenge to the conventional view of the origin of the vertebrates from the deuterostome/echinoderm line: it argues that a protostome/mollusc origin is more likely. In the vertebrates the possession of a notochord is an essential epigenetic characteristic, and again Løvtrup stresses that it is not the ability to synthesise the chemical constitutents of the notochord that is significant, but the co-operation of two cell types to secrete the chemical constituents in the right place at the right time as an organised structure. As an illustration of the relative simplicity of the embryogenesis that determines the vertebrate form, Løvtrup notes that in amphibian development, by the end of the intermediate embryonic stage consisting of gastrulation, neurulation and tail-bud formation, the essentials of vertebrate structure have been determined, yet hardly any tissue differentiation has occurred; hardly any of the structural genes that produce the components of the completed organism have been called into action. Relatively few enzymes are synthesised in the early crucial stages, and significant syntheses, which include the production of hyaluronate, chondroitins, glycosaminoglycans and collagen are relatively few in number. Thus, Løvtrup concludes that it is a mistake to think of specific genes as being responsible for carrying out distinct specific embryonic phases, and these events must be understood in terms of the physical interactions of the early embryo. Genetic events and changes are still important since the physical properties of the early embryo are to some extent hereditary. This does not however justify a common response to Løvtrup's thesis: 'But it's still all genetics isn't it?', inferring that Mendelian genetics together with molecular genetics subsume epigenetics. They are insufficient, since they fail to predict that certain kinds of mutation can produce qualitatively different physiological and morphological consequences than others, and that in such cases 'mutation pressure' can be more important than 'selection pressure'. Furthermore, they fail to explain non-nuclear organisation, and the interaction between DNA and epigenetic physicochemical factors in the cellular or extracellular environment.

On the subject of regulators in the higher organisms, the earlier theories of Britten and Davidson (1969, 1971) are relevant since these were submitted in an epigenetic context. For example, taking the vertebrate pancreas, an organ which has the two fundamental functions of the production of enzymes for digestion and the secretion of hormones for the control of blood sugar, they pointed out that the same enzymes and hormones are to be found in the invertebrates. Therefore, the novelty of the pancreas was the consolidation of these activities in a single specialised organ. This novelty was superimposed upon the older phylogenic innovation of the organisation of the intracellular organelles for the synthesis of the appropriate secretory products. To account for this organisational complexity, Britten and Davidson proposed a model that parallels Jacob's and Monod's operon mechanism: *producer* genes (structural genes) with related functions are organised in batteries that can be usefully activated at the same time; but in this model the members of a particular battery may be separate, or even reside on different chromosomes, since each has its own *receptor* (operator) which responds to a diffusible regulatory molecule, and thus all are activated simultaneously. Britten and Davidson describe some examples of gene duplication as *saltatory replications.* For example, in *Mus musculus,* the common mouse, there have appeared a million copies of a particular DNA sequence, a phenomenon of replication that is absent from two related mouse species, but is present to a much lesser degree in the rat.[61] They note that it is not a random replication since it only occurs in particular sequences. This could be regarded as an orthogenetic event, an argument that Britten and Davidson tacitly support with their statement that dissemination of the replicative trait in a population could be either by 'association with a favorable genetic element or simply because of their multiplicity'; also, 'repetitive sequence families originate accidentally. We assume that, later on, some or all of their member sequences become useful to the species, and only at this stage does selection pressure become important'.[62] The process was relatively sudden, since if it had occurred slowly throughout evolution the older sequences would have drifted detectably, and such drift is not observed in the examples investigated in mice and rats. Some repetitions are ancient in origin, but the greater the phylogenic separation of the two groups the fewer the repeated sequences they have in common. Britten and Davidson proposed that organisational changes in regulation might occur though chromosomal mutation, for example by the insertion of a foreign integrator gene into an existing set, with the result that a new combination of producers might be activated, and the morphogenesis of the developing organism altered. Repetition of regulatory genes thus potentiates a 'large number of possibly new regulatory relationships'.[63] As Ohno also argued, the repetition and mutation of regulatory DNA could take place out of the sight of natural selection, producing, for example, a complex spectrum of antibodies that would provide the immune system with the ability to deal with antigens that it had never encountered in its evolutionary history; or, for example, a spectrum of recognition molecules that would allow the organised complexification of an evolving nervous system.[64]

In *Order in Living Systems* (1978) Rupert Riedl argues that since early developmental regulation became established early in evolution, it cannot be altered in the higher organisms because of the disruption of so many later epigenetic events, and further evolution involves modification of less radical, later ontogenic stages. Riedl is concerned to define the limits of epigenetic evolution as well as its potentialities, by the application of information theory. While he regards himself as an adherent of the synthetic theory as far as it goes, Riedl is also a proponent of the Germanic tradition of evolutionary biology that is rooted in nature philosophy and has always stressed trans-specific or macro-evolution. He argues that the emphasis of neo-Darwinist empirical research on micro-evolution has been that whenever macro-evolutionary difficulties have arisen they have been swept under the carpet, or

> Instead of feeling excited when contradictions and traps appeared in our basic theory, people have tended to be disappointed. They have treated these difficulties as gothic or baroque ornaments unfitted to the plain architecture of modern biology.[65]

Order is, 'a meaningful connection between independent quantities according to internal laws'.[66] Organisational simplification of an evolving system, 'a decrease in the number of interpolated decisions required to produce the system,' is a necessary feature of evolution since it decreases reproductive costs and removes mistakes and adaptive difficulties that accompany complexity.[67] He underlines the problem of the biophore theory, noting that the human zygote, which starts out with 10^6 bits of information, matures into an organism with 2×10^{28} bits of information. This spectacular increase arises from the fact that the same laws have been repeatedly applied, or the same structural genes have been repeatedly used in different combinations in different specialised cells, so that the key to understanding evolution is therefore in organisation. The old entelechy solutions, despite their vitalism, at least recognised these problems and gave a sense of direction to the evolution of order, and Riedl says the time has come to move beyond comforting unanimity in evolution theory and the pseudo-explanations of such axioms as 'nature does not make jumps'. Riedl favours a model of epigenetic regulation which involves switching mechanisms whose operation, like their electrical analogues, instruct a system to act until switched off. The switches operate standard parts; in the organism this standardisation is in the protein-synthetic apparatus as well as larger anatomical parts. In the latter an initial phase of proliferation of parts is followed by the subordination of the parts to over-ranking systems, then reduction or individualisation of parts.[68] The same argument would apply at the level of gene repetition. Moreover, these switches stimulate a sequence of related events, so that the result is automatic instead of requiring an on-off decision at each stage in the sequence. This prevents 'long-windedness in the sequence of commands and allows the dismantling of hidden redundancies'.[69] However, repetition and long-windedness are inevitable initial qualities of an organised system, because novelty is accidental and evolution has to accommodate it whenever it can get it. Also, if a character was adaptive to a former environment the adaptation may no longer be relevant; hence the importance of self-simplification. Like Schmalhausen, Riedl believes that the internalisation of epigenetic stimuli stabilises order: 'Order, developed on the basis of accidental decisions, is the easier to modify adaptively the more it can be excluded from undesirable accident.'[70] This is analogous to Bernard's axiom that the control of the internal milieu was the secret of independent life.

Riedl does not believe that order *per se* has selective advantage. However 'order' is a morphological view of function frozen in time, and function, whether it be adaptive or adaptable is useful. Riedl argues that the possibilities and impossibilities of the molecular organisation specify a limited range of advantageous patterns, recalling the 'inner principle' or 'internal selection' that have so often been adduced to explain evolution.[71] Implicit in his thesis is that failure of order is detrimental, if not lethal, in a complex organism. As he points out, the probability of lethal failure in development for any zygote in some species may be as high as 90 per cent, and this failure is epigenetic, occurring before the organism has fully entered the game of adaptation to environment. Thus is it *super-selection* (internal selection) that is the most powerful process in the life of the organism. It is powerful because of *epigenetic burden*, which is 'genetically specified by the number of subsequent decisions that depend on a preliminary decision or by the number of single events (or features) dependent on a preliminary decision, or on a fundamental event (or feature)'.[72] Accordingly, early epigenetic events have a high burden and need to be features of maximum fixation or canalisation. For example, in the vertebrates, the fourth aortic arch, the optic chiasmata and hypophysis are parts which have been preserved in spite of their obscure anatomical position, and remain anatomically fixed, though their functions may have changed in the course of evolution. Therefore, to Riedl's way of thinking, Goldschmidt's hopeful monster would likely be hopeless, due to epigenetic burden. As a further illustration of this point Riedl observes that in a building under construction the position of the elevator shaft in each floor must be made to correspond with its position in adjacent floors. A mistake in position in the lowest floor would make the function of the elevator impossible. This architectural metaphor is a little too rigid: a homeorhetic elevator might have the option of moving horizontally as well as vertically, and a homeorhetic architect has the option of changing his plans at each floor to make sure that whatever mistakes might have been made earlier are minimised by later accommodations. In the development of an insect wing a vein may provide a useful road to the destination of a developing axon of a nerve cell. But if the road is removed by epigenetic accident the axon may take a short-cut across country to find an alternative route.[73]

Novel features added in late stages of development may give an organism a large measure of adaptibility which may later become fixed and specialised. For example, the low-burden, vascularised pharyngeal sacs of a transitional fish could become lungs or a swim bladder, whereas gill modifications might have more drastic consequences. While Riedl acknowledges that one of the main tenets of the synthetic theory has been the rejection of saltations in evolution, and considers that this is somewhat justified by the concept of epigenetic burden, he feels that the macro-evolutionary consequences of epigenetic change must be further explored. He describes three traditional hypotheses for such saltatory events. First is the 'simultaneity hypothesis', exemplified by Goldschmidt's hopeful monster, produced by a large chromosomal mutation. This is rejected by Riedl on the grounds that some well-known cases, such as aristopedia in *Drosophila* and the three-toed condition in horses, have been shown to have 'mistakes' in the neural and vascular anatomy.[74] These mistakes are similar to those made by an army that gets the date of its battle wrong; the internal organisation is there but the army is in the wrong place at the wrong time. Second is the 'coincidence hypothesis' which argues the possibility of several small mutations occurring simultaneously; as many critics have pointed out this is astronomically improbable. Third is the 'storage hypothesis', whereby neutral mutations might persist and accumulate until a subsequent organisation mutation brings them together in a novel functional harmony. This was proposed by Julian Huxley in *Evolution: the Modern Synthesis*. Riedl makes the orthodox neo-Darwinist objection that alleles do not spread nor even persist if natural selection does not increase their frequency. Even if they are not disadvantageous they rapidly disappear by attrition. However, this would not be true if an orthogenetic mechanism were operative. Whatever macro-evolutionary hypothesis is applied the epigenetic burden concept is relevant. Thus, the macro-evolutionary novelty must be added either at a late stage in development where the burden effect is minor, or at an early stage only if the burden effect can be evaded or done away with. One obvious way of achieving this is through a neotenic process where old differentiation pathways are lost or permanently repressed: 'Enormous selective advantage can be gained by dismantling the redundant decisions of a hierarchical pattern'. This might also increase the likelihood of success of an early epigenetic novelty. One means whereby novelty in organisation can be added to pre-existent order is the evolution of interdependence between two formerly independent organs. An arboreal monkey's forepaw and tail may become functionally interdependent as grasping hand and prehensile tail, illustrating the behavioural element required for functional interdependence. Riedl's line of reasoning parallels that of the early organic selectionists, but he adds that ultimately such interdependence becomes genetically assimilated at the level of synchronised regulator genes and the system loses it original flexibility, becoming firmly canalised: 'in the last analysis the external environment decides. . .but only according to the possibilities which the internal environment permits'.[75] Although some of his exposition provides only photographs of the problems raised by other epigeneticists, Riedl's rules of order provide a satisfactory formal treatment of the subject, one that both enlightens and confirms historical intuitions, even in the absence of a complete theory of epigenetic regulation.

To leave Barbara McClintock's contribution to epigenetics to the end of this chapter flouts a strict chronological progression, but the general impact of her research was not felt until the late 1970s. Her early research on corn had illustrated the position effect of chromosomal mutation on pigmentation: that is, the way in which the genes responsible for the colour of the corn kernels were expressed depended upon the chromosome in which the genes resided, and translocation would alter the expression.[76] At the 1951 Cold Spring Harbor Symposium McClintock described an epigenetic regulation mechanism that had come to light through the study of chromosome breakage. A genetic factor was found to effect breakage at its chromosome locus, and the factor was under the control of a gene at a distant locus on the same chromosome. The controlling gene was also found to affect several other loci. The genes that it regulated were occasionally transposed to different chromosome loci, resulting in some changes of gene expression in the affected cells. The controlling gene could also be transposed, and its action was also affected by the consequent change in its genetic milieu. McClintock argued that these events were not abnormal, but 'an example of the usual process of differentiation that takes place at an abnormal time in development.'[77]

A second regulatory system, effecting gene suppression, activation and transposition was described in 1956, and at this point McClintock proposed that this kind of regulation might be found in other organisms. In 1959 Jacob and Monod confirmed this prediction with the first description of the lac-operon system in bacteria. Although McClintock's 1961 paper given at Cold Spring Harbor pointed out the parallels between bacterial and maize regulation, it was not taken seriously since the transposition component had not been seen in bacteria, and did not fit the preconceptions of a scientific audience conditioned to think in terms of the molecular genetics of micro-organisms.[78] Recognition had to wait until bacterial genetics caught up, a decade later, with the discovery of bacterial transpositions that affected gene expression, and the exchange of genetic material in the form of extrachromosomal fragments, called plasmids, from one bacterial cell to another via a phage vector. This research had been hotly pursued because the mechanism was responsible for the rapid spread of antibiotic resistance in populations of pathogenic bacteria. The penny finally dropped at the 1976 Cold Spring Harbor Symposium, and a 1977 publication by Nevers and Saedler detailing the maize genetics in the context of bacterial gene expression brought the subject to a larger scientific audience. Subsequently, similar systems have been found in yeast, and transposed genes in *Drosophila* have been found to be responsible for large epigenetic changes such as the bithorax condition and aristopedia.[79] After two centuries the mortmain of Leibnitz's *natura non fecit saltum* has been lifted sufficiently to allow the application of the expression 'jumping genes'.

McClintock now argues that internal and external environmental changes may be important triggers for the evolution of responsive, adaptable regulatory mechanisms, and the introduction of such novel mechanisms might be integrated into the normal epigenome, with possibly rapid evolutionary consequences.[80] This would then have a direct bearing on the significance of physiogenesis, phenocopying and genetic assimilation, and further justify the Lamarckist stress on the responsiveness of the organism to its environment, without resort to the inheritance of acquired characteristics. It also reinforces the ideas of Schmalhausen and Riedl which focused upon the necessity for the internalisation of epigenetic environmental cues in the acquisition of hemeorhetic stability. A century of theorising, in the face of positivistic, reductionistic and gradualistic resistance, is giving rise to a new synthesis of evolutionary epigenetics.

14 PHYSIOLOGY AND EVOLUTION

> There has been no evolution of function. All living things have certain fundamental 'metabolic activities'. . .this fact seems to have been partly responsible for the lack of interest in evolutionary ways of thinking among physiologists.
>
> J.H. Woodger 1929[1]

Evolutionary epigeneticists give full measure to developmental physiology, both in its internal expressions and in relation to the environment. However, physiologists, trained in the holistic approach for over a century, have given little regard to either developmental physiology or physiological evolution. E.S. Russell complained:

> When we read the books of Cannon or Haldane or Sherrington, for example, which unveil part of this amazing complexity of inter-related function, do we pause to consider that all this is built up afresh, in the course of individual development, from a single microscopic cell, with the help of the most elaborate structural and physiological provision in the body of the mother for the protection and nourishing of the developing embryo?[2]

This lack of synthesis was due to the historical morphological bias of embryology, which ignored the adult functions once the adult morphology was attained. The modern generation of comparative physiologists who consider themselves to be evolutionary physiologists, devote most of their time to the investigation of adaptive specialisations, and the balancing of selection pressures in terms of energy budgets, and the question of how the homeostatic ordering mechanisms were established phylogenetically is ignored. Comparative physiology may have got off on the wrong foot because its original purpose was the clarification of human physiology, and Woodger's argument that the fundamental similarity of all organisms at the chemical level discouraged physiologists from evolutionary thinking may be partly true.

Some biologists were not so narrow in their thought. C.M. Child's *Physiological Foundations of Behaviour* (1924) discussed the evolutionary significance of physiology both in development and in relation to the external environment, in realistic-holistic terms:

> The problem of the organism as a whole is the problem of the origin, development and maintenance of the mechanism of integration in their relation to origin, development and maintenance of the individual. This is first of all a problem of physiology, not of heredity, because. . .heredity does not account for the individual, but merely for the potentialities some of which are realized in the individual.[3]

The wholeness of an organism was an expression of the physiological correlation of parts that differed from one another, but the problem of understanding wholeness by focusing upon the individual organism and its activities in relation to its environment had never been adequately explored. The physiological gradients found in differentiated organisms, in part expression of the protoplasmic gradients in the zygote, could not change autonomously; they had to be imposed physiogenically, i.e. as a consequence of the physical or chemical nature of the environment. Physiological evolution was a progressive process of fixation which had

> standardized for every species both these [physiological and behavioural] mechanisms and the range of conditions to which they are likely to be exposed. The result is on the one hand what we call the normal in nature and on the other the capacities of the organism for 'return to normal', 'adjustment of internal relations to external relations', or regulation.[4]

E.S. Russell began his study of the epistemology of physiology with *Form and Function* (1916), a history of the early debates about which caused the other. He called himself a functionalist, but understood that it was an over-simplification to separate the causal roles of the two. His first thoughts on the directiveness of physiology and behaviour were the subject of an address to the British Association in 1934, and later developed in *The Directiveness of Organic Activities* (1946). Here form and function were unified and the time factor stressed: 'the organism is essentially a spatio-temporal process, or a dynamic pattern in time.' The activities of maintenance, development and reproduction were, when considered objectively, directed towards the end of completing the life cycle.[5] This gave biological wholes a quality that inorganic ones lacked:

> The unstable and self-regulating organisation reached in the
> development of a living organism is therefore something totally
> different from the stable equilibrium, which is the natural end-
> state of an inorganic system. Life is a dangerous venture,
> inorganic processes tend toward stability.

Thus he agreed with Cannon that there was an 'almost miraculous'
quality to the survival of a system which was open to the vagaries of
its environment.[6]

To illustrate directiveness Russell described the complex process
of wound-healing in insects, which involved thigmotaxis and chemo-
taxis by epidermal cells, with the goal of restoring the continuity of
epidermis, replacement of cuticle and basement membrane, and the
restoration of normal cell density. He also considered insect
behaviour as directive activity. Although much of it was instinctive,
or consisted of fixed action patterns from which they did not deviate,
the larvae of the caddis fly *Molanna* showed at least six different
kinds of response to the removal of the posterior portion of the
larval case. The larva dealt with unusual contingencies which were
unlikely to have occurred in its evolutionary history, so that its
power of effective response was fundamental and primordial, not to
be accounted for by selection. The plastic behaviour of *Molanna*
indicated that the specialisation of instinctive behaviour was not
universal, but rather a secondary development, a limitation of the
general power of adaptive response which all organisms possessed.[7]

The goal of complex directive actions was related to a biological
end, usually maintenance, development and reproduction. But goal
and end were distinct: the goal of wound-healing was the closure of
the wound; the end was the maintenance within viable limits of
physiological parameters such as the volume of body fluids, or the
survival of the organism. The goal could be experimentally achieved
by covering the wound with an artificial membrane and the directive
activity brought to a halt. By the use of the word 'goal', it was not
implied that that the agents concerned were conscious of it before it
was reached; their action was directive but not purposive. In goal-
directed activity the action ceased when the goal was reached; if the
goal was not reached, the action usually persisted [continued nest-
building and egg-laying in birds whose nests or eggs have been
destroyed]; if the goal was not reached by one method other methods
might be employed, [*Molanna* case renewal]. Where the goal was

normally reached by a combination of methods, deficiency of one method might be compensated for by increased use of other method [e.g. blood pressure control by heart rate; blood vessel sphincters; hormonal action and nervous activity; or the case of hypophysectomised rats which having lost the physiological ability to maintain body temperature build themselves warmer nests]. The same goal might be reached in different ways, and from different beginnings; the end-state (= goal) being more constant than the method of reaching it. Goal-directed activity was limited by conditions but not determined by them. The drive to completion dominated the needs of individual organs, for example, a regenerating planarian would mobilise and resorb individual organs to become a whole animal. A migrating salmon mobilised its muscle tissues to build its gonadal tissue, and sacrificed its wholeness to reach the high goal of reproduction.[8]

G. Sommerhoff in *Analytical Biology* (1950) gave directive activity a more formal analysis in the Woodger tradition. A number of other biologists and philosophers also attempted to distinguish between 'goal-directed' and 'goal-intended' activities, and Morton Beckner has commented that these attempts to come to terms with purposive or directed activity can be reduced to the following: '(1) There is a goal. (2) The system shows persistence in reaching that goal. (3) The system shows sensitivity to the conditions which support and/or impede the reaching of the goal.' He also points out that emphasis on goals alone can give an over-simplified view of the directive activity, since

any activity that subserves one biological end. . .also subserves an indefinitely large number of other ends, but the goal of an activity is a special case. It is discovered not by an analysis of the adaptive significance of an activity, but by application of the criteria of purposive behaviour, namely, persistence and plasticity, to the action.

He concludes that the analysis of this phenomenon should be considered independently of adaptation, although 'purposive' actions are often also adaptive and adaptive actions are often also 'purposive'.[9] Sommerhoff introduced the term *directive correlation,* meaning the accommodation between the activity of the organism or a population and its immediate circumstances, which allows a particular goal, to be reached. Directive correlation subsumes both physiological adaptability and genetic adaptation.

In *The Diversity of Animals* (1962) Russell returned briefly to the question of evolutionary advance in physiological adaptability:

> It is clear that the mammals and the birds as well, show a mode of evolutionary change which we do not find to any great extent in the Decapods for instance, namely marked advance in physiological behavioural efficiency, a stepping up or heightening of vitality. This mode of change which raises life to a higher level of intensity and efficiency may be termed *grade evolution*. [10]

While it does not follow that understanding genetic adaptation alone gives any insight into physiological adaptability, this assumption seems to be made by some physiologists and biological theoreticians, with the result that adaptability is discussed as genetic adaptability and physiological adaptability is ignored. C.H. Waddington agreed with Russell that this was due to the morphological bias of phylogenetic studies, and the fact that structure is, generally speaking, not modifiable. Waddington believed, however, that physiological adaptability could be related through epigenetic adaptability to genetic adaptability. The logical flaw in this argument is that genetic adaptability is a property of populations, the consequence of which is the *adaptation* of individuals, not *adaptability* of individuals. The real problem here is the confusion of two words, based on the same root and prefix but which signify quite different biological phenomena. If physiological adaptability is to be related to the process of adaptation by natural selection it must be shown that adaptability is the cumulative result of different adaptions to different conditions. It is possible, but not axiomatic, that homeothermy, for example, evolved as an accumulation of adaptations to warm climates and cold climates.

Waddington perceived that there had to be a compromise between the extremes of loose canalisation of epigenesis and rigid homeorhesis. Nevertheless, evolution has produced both types in their most extreme forms, i.e. plants and the higher animals. The difference between the two types of organisms may be related to the degree of genetic fluidity that arises from gene transposability. Nature, in Darwin's sense of the word, does not choose between plants and animals, and any intermediate forms such as colonial cnidarians or amorphous sponges do not seem to have done very well for themselves by the compromise. Waddington added that as a result of the striking of the balance 'homeostasis' was achieved. But what he meant by homeostasis in the context was simply survival — a steady state of persistence in being — not a system of physiological regulation. Just as there are two fundamental epigenetic directive correlations — loose canalisation and strict homeorhesis — that achieve the goal of survival and reproduction by the adult, there are two types of physiological directive correlations; and again there is little compromise to be made between the two. In some cases it seems that opting for the one shuts the door on the other. The first type is physiological conformity with environment in which the internal medium conforms approximately to the external environment: Bernard's 'fluctuating life'. In the second type of physiological directive correlation there is homeostatic adjustment to keep the internal milieu as constant as possible — Bernard's 'constant life'. To reach a physiological end, such as healing, maintenance or reproduction, a certain degree of flexibility is required, so that interference with the normal sequence of physiological events activates unusual responses that bring the sequence back on track so that the goal is reached. There seems to be no physiological burden equivalent to Riedl's epigenetic burden whereby a change in activity early in a sequence is likely to have a more detrimental impact than one occurring late in the sequence; which suggests that Riedl might be over-emphasising the significance of epigenetic burden. However, specialisation has the same effect in physiology as in epigenetics, that of reducing flexibility, of reducing the manouevring space of each directive correlate.

Another author who recognised that physiological evolution was more than an epiphenomenon of natural selection was L.L. Whyte, a specimen of the broadly educated, non-specialised lay holist that Mayr considers to be more perceptive than the myopic specialists. A theoretical physicist and philosopher of science, Whyte was an early member of the Theoretical Biology Club, and an associate of many of the biologists who contributed to the growth of theoretical biology. *Internal Factors in Evolution,* published in 1965, introduced the concept of *internal selection,* a process of 'selection of mutants at the molecular, chromosomal, and cellular levels, in accordance with their compatability with the internal coordination of the organism. The restriction of the hypothetically possible directions of evolutionary change by internal organizational factors'.[11] It was to be distinguished from external Darwinian selection, which was viewed as differential reproduction of individuals with different adaptive fitness resulting in the restriction by external factors of the evolutionary possibilities permitted by internal selection.[12] Thus it can be seen to be equivalent to Reidl's later 'super-selection'.

As a physicist Whyte was interested in quantum mechanics and ordered systems, and was stimulated by Whitehead's opinions concerning organisms. Disturbed by the simplistic reductionism of the physiologists of the time who regarded cells as a bags of chemicals, Whyte was intuitively convinced that cells were hierarchically ordered systems. In *The Unitary Principle in Physics and Biology* (1949) he surmised, 'all mutations to new stable patterns may necessarily possess favourable or unfavourable properties in relation to the self-stabilizing organization of the system'.[13] Whyte was dismayed that the new discoveries of molecular biology had rapidly become commonplace, while their implications for evolution theory had been largely ignored. For this reason he decided to enter the lists and challenge the 'over-orthodox' neo-Darwinists, emphasising, like Woodger, the need for a formal logico-mathematical treatment for organisation, integration, unity and correlation. His *coordinative conditions* are

> the general algebraic conditions expressing the biological spatio-temporal coordination, the rules of ordering which must be satisfied (to within a threshold) by the internal parts and processes of any cellular organism capable of developing and surviving in some environment.[14]

The term thus implies the conditions that allow the organism to 'persist in being' or to 'get away with it'. Whyte was not content with the over-simplification that haphazard mutations caused evolution, since they could have no directive influence *per se*; their meaning had to be understood in the context of the pre-existing coordinative conditions. Nor could mutations be so different from the original forms that the transition to accommodate the novelty would be prejudiced. Whyte saw the stability of coordinative conditions as the prerequisite for the Darwinian test of relative reproductive success, and one that had to be viewed in isolation from adaptation-to-environment to be understood. He pre-empted the criticism that internal selection is a special case of natural selection with the assertion:

Nothing could better display the harm of intellectual fashion. It is good style today to equate the internal and the external environments; therefore there can be only one universal type of natural selection. Thus contemporary habit tends to inhibit the recognition of a distinction which is indispensible to clear thinking. For the internal and the external environments are not alike; life is not a vague property diffused through the universe; the nature of life lies partly in the fact that it is concentrated in units whose interior is subject to an ordering principle which does not apply to the inanimate spaces between organisms. These units of organization do not lose their distinguishing form of order because they are in perpetual give and take with the environment; the [coordinative conditions] do not operate through open spaces, they are only effective inside boundaries.[15]

The achievement of internal selection was not differential reproduction but improved coordination. These distinctions are not mutually exclusive, but they do narrow the focus of attention on an aspect of biology neglected because of the historical focus on the external. As Riedl later pointed out, most of the important events in an organism's development are aspects of internal selection, and have occurred before the question of adaptation to the external environment arises. A successful improvement in the coordinative conditions has intrinsic value, but because of external conditions numerical reproductive increase may not result. For example, the early mammals can be assumed to have been of a higher grade of coordinative conditions than reptiles, but in competition with and under predation by physically stronger reptiles their reproductive success was less than impressive. Similarly, extreme physicochemical conditions often favour organisms of a lower grade of coordination.

Whyte's views are relevant both to sequential and simultaneous complex events; both to epigenetics and to physiology. The major unknowns of evolution were: (a) If internal selection was important in phylogeny is there any direct evidence of it in systematic or biochemical or cellular biological data? (b) Is internal selection involved in macro-evolution or more important in adpative evolution? (c) Do the causes of major evolutionary changes reside in chromosomal, cytoplasmic, or cell cortical organisation, or by interactions between them? (d) Could the topological (helical) arrangement of polynucleins and proteins be of special significance? (e) Are pulsating or vibrating molecules, or electromagnetic polarisation important factors, and are they altered by genetic mutation? (f) Is there an inherent ordering tendency related to the physical laws? (g) Can advances in coordinative conditions occur by internal selection?[16] The historical roots of a number of these questions are immediately recognisable. Some of them are relics of the molecular gene theory's sojourn in the wilderness, and of the ideas of other physicists like Whitehead, Bohr, Delbrück and Schrödinger (questions d, e, f). The concept of molecular vibration was discussed by Haeckel, Mivart, Morgan, Spencer, Bateson, Woodger and others; and more recently the significance of oscillatory enzymatic and cellular systems have been reconsidered.[17] Question (c) can be answered with 'yes to all alternatives', but the relative importance of each cannot be fully determined, as seen in the previous chapter. The following rephrased questions will be dealt with in this and the following chapters, (i) Is there evidence from comparative biochemistry and cellular biology that sheds light on internal selection in particular and physiological evolution in general? The question of the physicochemical basis of biological order will also be referred to here. (ii) How does progressive change in the coordinative conditions occur; or, how does homeostasis evolve?

Comparative biochemistry is of post-Darwinian origin, and has therefore always had phylogenetic or evolutionary implications, especially in the pioneering work of Joseph Needham and Ernest Baldwin. The traditional comparative physiologist's view of biochemical evolution was that it was additive. As organisms evolved from the marine environment and encountered new conditions in fresh water and on the land, they would 'evolve' new enzymes that would allow then to enter those environments and thrive in them. Take the example of how the evolution of excretory processes was explained: ammonia is a toxic product of the metabolism of dietary amino acids; in small aquatic animals its toxicity is not dangerous because ammonia diffuses rapidly into the surrounding water. In larger organisms which are not wide open to the environment, where the integument is impermeable, it is advantageous to be able to detoxify ammonia in the body as urea and excrete it with water as urine. Thus, the acquisition of the enzymes of the ornithine cycle, that produced urea, would be selectively advantageous. Then, under very dry conditions where water is scarce, uric acid would be most useful as an excretory product since it can be excreted as a dry solid. So the acquisition of a new set of enzymes for the synthesis of uric acid would give an even greater selective advantage. As Needham pointed out, the advantage of particular modes of excretion have to be interpreted in terms of the needs of embryos.[18] For example, mammals and birds both inhabit deserts, and water conservation is desirable. But mammals excrete urea, losing water in the process, while birds excrete uric acid. The mammalian embryo is buffered by the physiological capacity of the mother through the placental connection, while the bird embryo is confined to an egg where urea would not be accumulated without ill effect. Although the comparative physiology of excretion is rarely presented as simplistically as this, Baldwin's *Introduction to Comparative Biochemistry* (1937), a text that influenced the thinking of generations of zoologists, remarked, 'The rest of this important mechanism [the orthine cycle] must presumably have been added later in the process of evolution as an adaptation to limitation of water supply, which. . .called forth the production of urea.'[19] A much more complex story is told by J.W. Campbell in his contribution to C.L. Prosser's *Comparative Animal Physiology* (1973). The biochemical pathways capable of producing urea and uric acid, together with the enzymes which could degrade them, urease and uricase, were all present in the ancestral eukaryotic unicellular organisms as nutritive pathways. That is, they were concerned with the synthesis of amino acids such as arginine, and the purine bases of the nucleic acids. With the appearance of the heterotrophic habit of eating other organisms and digesting their proteins to relase the amino acids, the nutritive functions of these ancestral pathways could be usurped, and metabolic intermediates such as urea and uric acid used for osmoregulation, acid-base regulation and ammonia detoxification. In contrast to the earlier belief that new enzymes were added *de novo* as new environments were encountered during the course of evolution, the most striking feature of excretory biochemical evolution has been its regressiveness, the loss of key enzymes. For example, the loss of uricase seems to have been a prerequisite to the utilisation of uric acid as a metabolic end-product in birds. The loss of arginase, which converts arginine to urea and ornithine, is another characteristic of uricotelic animals, although the functional alternative to losing part of the pathways, by their temporary or permanent repression, is occasionally found.

The concept of regressive evolution was developed by André Lwoff, whose studies of the synthetic abilities of micro-organisms indicated that their physiological evolution could be defined as, 'a degradation, a regressive orthogenesis'.[20] As they evolved morpho-logically, they lost the primitive ability to synthesise amino acids, vitamins, chlorophyll and other compounds. The results of these studies and their general implications for physiological evolution were discussed in *L'Évolution physiologique* (1944). Marcel Florkin had expressed similar thoughts on metabolic evolution and nutritional requirements in the *Introduction to General Bio-chemistry* (1943), but he initially believed that excretory pathways had evolved adaptively by the acquisition of new enzymes. Lwoff was convinced that the physiological regression that accompanied morphological progress would be found to be a universal pheno-menon among animals, and that this would apply also to the excretory system. He argued that the loss of synthetic abilities in differentiated tissues was a partial expression of regressive evolution, and cited neoteny as a regressive phenomenon. Man had been particularly prone to regression in the form of foetalisation which involved, the simplification of bony anatomy and dentition. Also, in comparison to the lower primates, man has a number of regressive or vestigial organs. In Lwoff's view enzymatic novelty could occur only through the mutation of a pre-existing enzyme.

The 'loss' of an enzyme can occur in a number of different ways, all of them involving gene expression. The literal loss or deletion of the entire cistron through a mistake in recombination or chromo-somal mutation is the first possibility, but is likely to have lethal consequences. A point mutation could result in the wasteful synthesis of an inactive enzyme or a nonsense protein, whereupon the deletion of the cistron or its permanent repression would have positive advantages. The conventional wisdom of selectionism is that even permanently repressed genes are likely to be weeded out of the genome unless some selective advantage can be identified. However, some cases of these *dormant genes* have long phylo-genetic histories. For example, birds retain their 'teeth genes', but never develop teeth.[21] It is only very recently that the DNA recombination techniques that are capable of discovering unsuspected dormant genes have been developed, so it is not yet known how common they are in eukaryotes. The postulated immediate advantage to the loss of an enzyme is the saving of cellular energy, which is particularly significant in prokaryotes. In the case of the excretory system the loss of an enzyme such as uricase may also potentiate the ability to reproduce under unusual circumstances, as well as providing additional associated physiological bargains. Sometimes utility is difficult to identify. Baldwin, for example, puzzled over the loss of uricase by human beings where the loss allows the debilitating condition of gout, due to the accumulation of uric acid crystals in the joints. Perhaps there is no hidden advantage, and gout has simply not been sufficiently detrimental to affect human survival and reproduction. It has been suggested that hyperuricaemia in the higher primates has had the epigenetic effect of inducing rapid brain expansion. Although the concept is contentious it does provide a hitherto missing utilitarian explanation for the universality of the phenomenon.[22]

The example of regressive evolution that most interested Lwoff was the loss of the ancestral ability to synthesise certain amino acids and vitamins. In human terms the loss appears to be detrimental considering the number of individuals who die of malnutrition. But this has also to be seen in terms of cultural stupidity. *Homo sapiens* is the only animal that subjects itself to the difficulties of having to pay for nutritious food when the food may be plentiful and the exchange medium scarce; to the choice between food and junk when money is available; to the monoculture of crops that lack dietary essentials; to living in cities, prisons and concentration camps; or to going on long sea voyages on a diet of salt meat and hard tack. Other animals survive quite well by obtaining the essentials in their normal diets, unless those diets are manipulated by man. It is usually argued that if the essentials are in the diet and the synthetic pathways are not called upon they become genetically labile.

The ability to digest certain kinds of food also appears to have been lost in some animals. It was tempting at one time to view the evolution of digestive functions as a history of encounters with new diets and the *de novo* appearance of appropriate enzymes. However, virtually all digestive enzymes found in the higher vertebrates are present in the lower invertebrates. A phylogenetic relationship between the protein-digesting enzymes trypsin and chymotrypsin has been established in terms of both active site function and primary amino acid sequences. This suggests that some novelty is possible by the duplication of the original structural gene and consequent mutation and functional divergence. But one did not evolve from the other in response to new diets, or if it did, it was before the diploblastic grades of animal evolution, since both are found in cnidarians. Pepsin is a gastric enzyme found only at the vertebrate level and used to be regarded as a 'new enzyme; but it probably derived from one of the low pH optimum intracellular cathepsins whose phylogeny goes back to the early eukaryotes and prokaryotes. The changeover in activity from intra-cellular to extra-cellular digestion was a necessary step in the physiological evolution of most metazoan phylogenetic lines, and may have required adaptive changes, but not necessarily new types. New diets have of course appeared from time to time in evolutionary history. In cases where the diet is new in the sense of different proportions of the carbo-hydrate, fat and protein components, adaptation involves only change in the proportions of the appropriate enzymes. One interesting case of a new diet was the appearance in land plants of large quantities of cellulose, a compound which had been scarce during the long period of aquatic evolution. Most animals, if they ever possessed cellulases in their early history, had lost them by the time the new food came on the market. The ability to process cellulose was not entirely lost. Some molluscs seem to have an endogenous poly-ß-glucosidase that can digest simple cellulose polymers but not the native variety found in plants. Some inverte-brates also possess cellobiase, an enzyme that hydrolyses the dis-accharide breakdown products of cellulose digestion. Some animals, the tunicates, can even synthesise cellulose. However, most animals that adopted cellulose diets did not re-invent cellulases: they borrowed them from symbiotic bacteria and protista, which may originally have been free-living in plant material ingested by animals. Despite the failure of herbivorous animals to synthesise cellulases, they have gone to considerable lengths in the production of morphological novelty to accommodate their symbionts, in the form of special gut compartments that allow efficient activity.

If regressive evolution, whether adaptive or as a prelude to grade advance, is so common, and real novelty in enzymatic evolution so rare, where and how did the original complement of ur-enzymes and proteins arise? This problem is addressed by Émile Zuckerkandl in 'The appearance of new structures and functions in proteins during evolution' (1975). He argues that with the exception of certain periodic proteins such as silk, collagen and keratin, *de novo* evolution of new protein classes has not occurred since the earliest phase of the evolution of protocells. All 'modern' proteins are derivatives of perhaps less than five hundred ur-proteins which appeared even before the familiar transcription-translation mechanisms for protein synthesis were perfected. Taking a number of cases of proteins that were traditionally believed to be evolutionary late-comers, he argues, for example, that the immunoglobins involved in the vertebrate immune system were most likely proteins responsible for the specificity of cell surface interactions in older animals. Haemoglobin is often referred to as a protein which has evolved independently in a number of groups, the ease of this process being made possible by the universality of iron-porphyrin haem which is also used in electron transfer in cellular respiration. Even this molecule may be monophyletic and have been lost a number of times.

Zuckerkandl defines the origin of life as the co-emergence of two molecular types, a polynucleotide controlling the structure of a polymerase, thus combining the qualities of reproduction with the ability to make new macro-molecules.[23] Some kind of protocellular bag to hold them together so that they did not dissipate in the medium, would be necessary, and various demonstrations of the abiotic production of such membrane systems have been made. Utility of these original macromolecules would be in terms of stability and survival, but with the addition of a primitive self-replication any useful quality would be preserved and extended. Randomly, abiotically synthesised polypeptides have a variety of weak catalytic activities, and even a single one of these polypeptides may show several such enzymatic qualities. By the random coalescence of protobionts a greater profusion of these inefficient enzymes, together with their primitive synthesis mechanisms could be brought together in a single cell. The variety of functions brought together by this random symbiosis would result in a functional diversity or adaptability in this primitive cell. Thereafter there would be advantages accruing from the co-evolution of specialised single-function enzymes with greater catalytic efficiency, and a more exact transcription-translation mechanism: 'Each progress in functional efficiency in proteins caused progress in fidelity of translation and vice versa'.[24] Once this functional efficiency had been attained it would be impossible for new inefficient proteins to compete. 'Thus, for proteins, evolution soon became a family affair. For them living systems rapidly turned into a nearly closed club with few new members admitted without family ties to old members.'[25]

It must be stressed again that Zuckerkandl is discussing the evolution of protein classes, those groups of homologous proteins which have been identified through amino acid sequencing work by molecular biologists. Within classes enzyme functions may vary, or change radically, or even converge functionally with enzymes belonging to different classes. For example, in a strain of *E. coli* which had lost the ability to synthesise ß-galactosidase, an enzyme appeared which could carry out the same function but was twice the size of the original and showed no immunological cross reaction. Zuckerkandl regards this as support for his assertion that it may be possible to find in any organism, 'protein molecules that are nearly competent for any thermodynamically possible function that had not been carried out before'.[26] The matter of thermodynamically possible functions comes back to Whyte's question of how order relates to physical laws. Zuckerkandl points out that proteins have the *general functions* of physical parameters such as overall solubility and charge, determined by similar qualities in the constituent amino acyl residues. A large number of different primary sequences would satisfy a given general function, but the general function would have partly defined the coordinative con-dinative conditions of the protocells. Functions like high charge might not be allowable because of reactivity and tendency to electro-phoretic concentration. *Specific functions* are determined by particular portions of the protein molecules and molecular topology: the primary amino acid sequence and the three-dimensional secondary and tertiary structures. Specific functions include local hydrophilic or hydrophobic regions, and the active site in enzymes, where specific amino acyl side chains push and pull at their substrates to effect a reaction. They also determine inter-molecular reactivity and binding such as is found in the sliding filaments of muscle cells, and in microtubules. It is inconceivable that an active proteolytic enzyme such as many organisms need to recycle their structure and acquire amino acids and energy from protein food, could have been permitted to work freely in the proto-cellular environment. The specific functions of these enzymes therefore demanded the co-evolution of a protective membranous organelle such as a lysosome or a mask for the active site. Thus, specific functions further defined the coordinative conditions of the cell. According to Zuckerkandl the potential evolutionary versatility of enzymes was due to three factors. First, different chemical functions may be fundamentally related so that simple modifi-cations of active sites may be adequate for a change of function. Most enzymes deal with a variety of substrate arrangements of the atoms carbon, hydrogen, oxygen and nitrogen, and these arrange-ments are not unlimited. Some proteinases that lyse carbon-nitrogen peptide bonds can almost as easily deal with carbon-oxygen ester bonds without any modification of active site at all. Secondly, the product of a biochemical reaction usually retains part of the structure of the original substrate, so that a slight modification of the enzyme through mutation of a duplicate of the structural gene may be able to further process the product of the original reaction. N.H. Horowitz suggested that nutritive biochemical pathways were built up as compounds which were formerly available by abiotic synthesis in the original 'dilute soup', became limiting factors, eliciting the evolution of 'new' enzymes that would synthesise the limiting compounds.[27] Zuckerkandl's thesis indicates how such 'new' enzymes could have been formed. Thirdly, 'old' enzymes already had general functions that integrated them with the whole cell, so that the 'new' modified enzyme had the same compatibility. Another way in which the complex metabolic spectrum that characterises the eukaryotic condition would be built up is by endo-symbiosis. The association of a host prokaryote that excreted ammonia with a symbiont that fixed ammonia for amino acid synthesis would be immediately beneficial to both, as well as providing a classic case of emergence.

While the evolution of radical novelty in protein type may well have been terminated in primitive prokaryotes, biochemical evolution is not characterised entirely by regression and stasis. 'New' enzymes, with new functions can appear from old enzymes, certainly in prokaryotes. Zuckerkandl believes that even the vertebrate herbivores could eventually have evolved cellulases if it had not been for the swifter expendient of symbiosis with prokaryotes. The modification of 'old' enzymes to alter their functions in a manner appropriate to changing internal or external exigencies belongs largely to the realm of biochemical adaptation. Great increases in the proportional synthesis of given enzymes can be effected by gene duplication or by increases in the volumes of the specialised secretory organs. Enzymes may undergo changes in their temperature and pH optima and allosteric sensitivity through point mutations. In the same adaptive way respiratory pigments may have their oxygen-binding properties altered so that they function more efficiently in changed oxygen and carbon dioxide concentrations. The substrate affinity of an enzyme may change to make it compete more effectively for substrate than another formerly dominant enzyme, resulting in the emphasis of alternative pathways. Refinements in the timing of regulation may be adaptive to environment as well as satisfying the coordinative conditions of developmental physiology: a frog tadpole is ammonotelic, but switches to ureotelism at metamorphosis. Gene duplication is a safer route for biochemical evolution since the original function can be retained while another is provided for experimentation. This can produce a spectrum of responses for varying conditions. For example, Peter Hochachka has discovered that salmonid fish-muscle kinases, enzymes which are concerned with energy conversions, exist in a variety of isoenzymatic forms which have appeared by duplication and mutation. Each has a different temperature optimum. A drop in environmental temperature induces synthesis of the isoenzyme with the appropriate kinetic qualities, and consequently the fish can go on swimming at its normal speed, unlike simpler poikilotherms which slow down in the cold.[28]

Thus, comparative biochemisty and molecular biology reveal a great deal about adaptation, and about biochemical conservatism: the limitations of novelty at the protein level. They elucidate the significance of regression as a prelude to some important physiological developments, and the importance of gene regulation changes, and improvements of the coordination of regulators. This satisfies a number of Whyte's questions. The second major question is how do the coordinative conditions evolve, or how does the homeostatic condition evolve?

15 THE HOMEOSTASIS PARADOX

If homeostasis is characterised as constancy, and evolution characterised as change, how did the homeostatic condition evolve?

Evolutionary progress is synonymous with the acquisition of advanced states of epigenetic and physiological coordinative conditions, implying greater flexibility of response to a greater range of disequilibrating environmental variables, leading to greater internal stability. The most extensive expressions of functional-morphological adaptive radiation are found in organisms at the highest levels of grade evolution because they are able to enter and thrive in the greatest range of environments.

Changes in form are the most striking features of adaptive radiation. However, at the organ level there is remarkably little morphological innovation. Once they were differentiated the major morphological changes that occurred in them were rather general ones such as volume and surface-area elaborations. The functional evolution of the organ systems has largely been concerned with improvements of their internal coordination and integration with the rest of the organism through improvements in the regulatory endocrine organs and nervous system. Morphological adaptive radiation is at its most striking in the locomotory and feeding structures. Consider the forelimbs of mammals for example: flukes in whales, wings in bats, clawed paws in cats, hooves in camels, hands in humans: contrast these remarkable differences with the underlying physiology. One might expect that since diving is one of the most exceptional and characteristic functions of whales there might be some unusual physiological features such as a blood system with an extraordinary capacity for oxygen, or a special air-storage system. This is not so; there is simply an exaggeration of some of the functions found in all mammals: the ability to shunt blood away from the peripheral circulation and to the brain; an improved tolerance for blood carbon dioxide and lactic acid levels, some increase in myoglobin. Even the relevant morphological features, with the exception of the locomotory ones, involve only small modifications: changes in capacity of particular blood vessels, increase in

295

the muscularity of blood vessel sphincters, and the elaboration of a retial system. In the bat the anatomy related to flying and echolocation for food capture is striking, but the only notable features of their general physiology is their diurnal torpidity and tolerance for the ammonia generated in their dormitory caves. The cat has a proportion of proteolytic digestive enzymes appropriate to its high protein diet, and an acute sense of smell, both minor adjustments of existing qualities, not innovations. The camel lives in the extremely hot and dry environment of the desert. It used to be believed that its hump was a water store; but the hump is a fat depot, remarkable only for its small contribution of metabolic water. The camel's physiological desert survival kit consists of an ability to tolerate dehydration, a humidity barrier in the form of sweaty hair, and a tolerance for fluctuating body temperature. By departing from its homeostasis and letting its temperature rise above normal, instead of sweating or panting to keep it at a fixed level, an average camel saves about four litres of water a day.[1] Since the body temperature is lowered and the camel is in a semi-torpid state during the cold desert nights it saves energy that might be otherwise expended in shivering. This probably accounts for another notorious characteristic of the camel — its reluctance to get started in the morning. The human species, the generalist among mammals, can develop surprising individual abilities and approach the feats of diving mammals in terms of deep free diving if not in dive duration. The legendary hardy prospector Pablo walked 120 miles at an air temperature of 30°C over a period of eignt days in the Arizona desert without water, tolerating a body-weight loss of 25 per cent in the process, and providing an archetypal subject for cartoonists.[2]

Thus, while function evolves progressively, from Bernard's fluctuating life to more constant life, function itself does not undergo much adaptive radiation. Adaptation of function in the mammals includes slight modifications of pre-existing abilities, or improved tolerance of environmentally imposed changes, even the relaxation of the constant life mechanism in some cases. The adaptive radiation of the locomotory and feeding structures is possible as a result of two aspects of animal physiology: the ability to survive in different environmental conditions provided by the homeostatic mechanisms — *la vie libre* — and the possession of finely tuned hormonal and nervous integrative systems that make possible the behaviour for which particular morphological adaptations are especially advantageous.

The paradox is this: a high evolutionary grade is characterised by the constancy of the internal milieu established by complex interactions of physiological mechanisms and behaviour, but evolution implies change rather than constancy. Indeed, the internal milieu has changed during the course of grade evolution. It has not simply involved refinement of the original coordinative conditions: fundamental alterations in osmotic pressure, temperature, ionic composition, organic content, oxygen content, chemical messenger content and defence mechanisms have all occurred in the internal milieu, and these changes have not necessarily been gradual. Therefore, homeostatic evolution must be considered in terms of change as well as in terms of constancy. The paradox is a tacit element in holistic evolutionary writing, from Galton, Mivart and Spencer, to Severtsov, Schmalhausen and Waddington. Their concept of an existing equilibrium being upset and re-established at a higher grade, Severtsov's *aromorphosis,* is the clue to the resolution of the paradox.

For the sake of simplicity the evolution of the coordinative conditions might, as Whyte suggested, be considered apart from the environmental conditions, and be entirely pertinent if only the refinement of an existing system were at issue. But in considering aromorphic change the impact of the environment must be brought in. The direct influence of the environment takes us back to Buffon, Geoffroy and the neo-Lamarckists. E.D. Cope's *physiogenesis,* the physiochemical imposition of internal change in organisms by the environment, is a particularly important aspect of aromorphosis. He also concluded that the imposed changes would eventually become heritable, through their direct impact on the germ plasm. However, when F.W. Hutton remarked that such forces were beyond the control of the organism, which had to depend ultimately for its evolution on the natural selection of random mutations, he neglected the crucial point that to some extent the physicochemical forces are controlled by the behaviour of the organism. By entering or leaving an environment that imposes change the organism has altered the 'selection pressures'. Future selection has to conform to the new rules of the game. Schmalhausen regarded the genetic assimilation of coincidental environmental influences that evoked useful epigenetic change as a feature of evolutionary advance. He used the term physiogenesis to imply not only the physicochemical changes caused by the environment, but also the individual's ability to respond effectively to the changes, a complex of causes and effects

that he also called *accommodation*.[3] Physiogenic modification
included changes in the surfaces of leaves caused by intensity of
illumination; changes in pigmentation in animals caused by sunlight
or air temperature; increase in erythrocytes at high altitudes;
responses to temperature change in homeotherms; responses of bone
growth to mechanical stress, and compensating growth in an organ
such as the kidney when one of the pair had been removed. Schmalhausen was clearly more interested in the responses to physicochemical changes that had already been partly genetically
assimilated. These processes were facets of his *stabilising selection,*
which he used in the sense of attaining stability as well as maintaining
it. What the organism did or chose to do was paramount: 'the origin
of new functional differentiations is always based upon the vital
activity of the organism itself'.[4] Thus the choices available to the
organism as a result of its adaptability were important, especially if
the organism lived near an interface between two environments,
where there were powerful disequilibrating forces. I prefer to restrict
physiogenesis to the limited sense used by Cope: the physicochemical
changes imposed by the environment. Accommodation is a useful
general term for the responses of the organism able to survive such
changes, whether these be simple conformity, tolerance of a degree
of change in the internal milieu, or a fuller homeostatic response to
stabilise the internal milieu. Pietro Omodea (1975) has argued that
adaptability, which he too equates with evolutionary progress, is
selected in the face of contradictory environmental conditions,
which may be encountered all at once or over the course of the life of
the organism. This is on the right track, but the full sequence of
events must include some or all of: (1) Behaviour that leads to
physiogenesis, or the random imposition of physiogenesis by
environmental fluctuation. (2) Accommodation of the physiogenic
change. The organisms may simply tolerate it physiologically, or
move backwards and forwards through the environmental interface,
balancing the detrimental effects of physiogenesis with some
advantage such as food or safety from predators. (3) Acquisition and
enhancement of homeostatic mechanisms that improve tolerance or
accommodation. (4) Adaption of internal factors such as enzymes to
the physiogenically imposed new coordinative conditions. (5) Genetic
assimilation of the new coordinative conditions through the
selection of stabilising physiological controls, even if the external
environment continues to fluctuate. (6) Regression of adaptibility,
leaving the organism as a specialist for the new environment.

A brief survey of the physiological evolution of the marine invertebrates and the early vertebrates reveals what a strong influence physiogenesis has been. Modern vertebrates have body fluid concentrations and ionic compositions which differ from those of their invertebrate ancestors. The geologist Macallum hypothesised in 1926 that the cellular environment and the internal milieu reflected the condition of the external marine environment at the time when the internal systems were closed off from the sea. The high potassium and low sodium of cells suggested an origin in an ancient sea that had similar proportions of ions. The low salinity of vertebrate body fluids suggested that they appeared before the sea had attained its present salinity. Modern geological opinion is that metazoan evolution began at a time when the oceans had the same ionic content and salinity as they have at present. Nevertheless, there is a logical parsimony similar to Macallum's in the belief of most physiologists that the dilute body fluid of fish reflects a sojourn in fresh water. It would be very interesting in this regard to know what the ionic composition of the prebiotic 'dilute soup' was. Did the general and specific functions of proteins arise in a high sodium or high potassium environment, or did the high potassium of primitive cells appear as a refinement of their coordinative conditions due to the acquisition of a selective sodium/potassium membrane pump? This mechanism is present in almost all cells, and maintains the intracellular composition of cells exposed to the external environment despite the general leakiness of cell membranes. Whatever the prevailing primitive conditions potassium now has an important role as the co-factor of many enzymes, whereas sodium is relatively unimportant in this regard. Moreover, the ionic differences between the interior and exterior of the cell created the weak electropotential that was the prerequisite for a neural resting potential and the evolution of the nervous system.

The acquisition of a relatively impermeable integument, with ionic diffusion and selective permeability limited to the gut and gill surfaces made possible the partial isolation of the internal milieu, so that it could act as a buffer medium between the cells and the vagaries of the external enviroment. Specialisation of ionic pumps in the respiratory epithelia and kidney tubules reinforced a trend towards reduction of some divalent ions and allowed the increase in the organic content of the body fluid. The next step in vertebrate evolution was reduction in the body fluid salinity. With hindsight we can see that only in animals which underwent this change did a full flowering of homeostasis occur. But while certain immediate advantages may have accrued from dilution, such as a reduction in the load on tissue membrane pumps and more efficient blood pigment and enzyme function, this does not seem to have been 'pressure' enough for it to have evolved in the higher animals remaining in the marine environment, such as most of the decapod Crustacea, the Cephalopoda, and primitive marine agnaths such as the hagfish. A physiogenic imposition of dilution by the environment seems to have been necessary. The permeability of the respiratory surfaces may also have been variable and controlled by hormones, permitting entry into fresh water. The shore crabs *Carcinus* and *Hemigrapsus,* perennial subjects for comparative physiology students, illustrate perhaps the early responses in the entry into the fresh-water environment. When they are in sea water their body fluid salinity is the same as sea water, isotonic. As the external environment is experimentally diluted their body fluid salinity declines linearly until it is about 60 per cent of the original concentration. Then the crabs begin to osmoregulate by absorbing salts through the gills as fast as they are lost, and by controlling the output of the antennal glands. When the blue crab *Calliactes* enters fresh water its response is an interesting example of physiological serendipity that Mangum and Towle (1977) call *enantiostasis.* The ammonia released into the blood for osmoregulatory purposes has the useful coincidental effect of raising the body fluid pH and making the haemocyanin a more effective oxygen-transport molecule, thereby compensating for the dilution which decreases haemocyanin efficiency. Since haemoglobin does not show this detrimental dilution effect the vertebrates were not restrained by this obstacle. The refinement of hormonal controls of permeability, and an accumulation of biochemical adaptations to the new internal condition, dictated the homeostasis of body fluids at the diluter level when the fish returned to the sea. This was achieved by increased gill secretion of salts in teleosts and by the use of urea as a means of maintaining osmotic equilibrium together with low salinity in the body fluids of the cartilagenous fishes.

The relatively impermeable integument acquired as an early accoutrement of homeostasis in aquatic environments was also one of the preconditions of emergence from the water onto land. The crustaceans have rarely acquired the watertight cuticles that their insect relatives possess, with exceptions like the coconut crab *Birgus.* While the insect's cuticle is regarded as an adaptation to life on land,

Beament (1961) has shown that its impermeability is just as useful when the insect returns to water because it keeps water out as efficiently as it keeps it in. Beament raised cabbage white caterpillars under water through several moults. The waterproof integument is an example of a property that eludes conventional adaptationist interpretation. The waterproof integument of terrestrial animals was a 'pre-adaptation' possessed in fresh water and a 'self-adaptation' to marine metazoan homeostasis. It falls into the class of characters that are intrinsically beneficial under a variety of environmental circumstances.

The acquisition of an appropriate respiratory apparatus was another precondition of emergence on to land. This gave Darwin a problem because he and his contemporaries thought that the lung was evolved from the teleost swim-bladder, and how a swim-bladder could have been acquired gradually from nothing was problematical. The conventional view now is that fish with swim-bladders evolved from fish with lungs, and that lungs were acquired as adaptations to a stagnant, low-oxygen, fresh-water environment. Many fish take air into the buccal cavity under low-oxygen conditions. Everyone is familiar with the sight of a large goldfish gulping at the surface of a small bowl. This intrusion into the atmosphere was probably accidental, when primitive fresh-water fish sought higher oxygen levels in the water at the interface of the water and air, and contact with the aerial environment would have the immediate physiogenic result of a raised blood oxygen by its diffusion through the buccal epithelia. Morphological and physiological adaptations to this new feature of the internal milieu followed. Many fish have vascularised epithelia which may have appeared as an ontogenic response that was genetically assimilated. The tropical mud skipper *Periophthalmus* respires almost entirely through its enlarged, vascularised pharynx. A lung can be regarded simply as a surface-area increase of a vascularised pharynx, and also acts as a flotation device reducing the muscular energy required for keeping the fish at the interface. The transitional fish *Gymnarchus,* which lacks a lung, inflates its buccal region and floats at the surface by means of this flotation-cum-respiratory organ. R. Rosen (1973) calls the notion that a number of potential functions co-exist in a given organ and can be called forth by functional adjustment the *principle of function change,* and argues that such multiple potentials are essential qualities for the evolution of new organs.

The transition from water to air is by no means as simple as the acquisition of a surface-swimming habit and the acquisition of a lung. An adjustment to the physiological trigger for ventilation and functional morphological changes were co-requirements. Most aquatic animals pass water over their gills faster if oxygen levels in the blood drop. The effect of doing this in combination with taking up oxygen from the air would be the immediate diffusional loss of that oxygen through the gills to the stagnant environment, because the buccal circulation goes directly to the heart, which sends the oxygenated blood to the gills. As Johansen (1968) has demonstrated, the ability to shunt blood to the organ that is receiving oxygen, found in the electric eel *Electrophorus* and in some lung fishes, is a useful transitional ability. The transitional fish also needs to inhibit the low-oxygen trigger for increased ventilation, and at the same time procure an anatomical adjustment in the gill circulation that will permit a direct flow circulation through the gills, with closure, or degeneration of the gill capillaries. At the same time the improved availability of oxygen attained by breathing air, plus gill closure, results in a build-up of carbon dioxide, with possible drastic effects on blood acid levels and haemoglobin function. Therefore, a co-requisite is increased sensitivity to blood carbon dioxide which becomes the primary trigger for lung or buccal-cavity ventilation. Cheap and easy access to oxygen, even without specialised respiratory mechanisms, was the key to future living at high metabolic rates in the terrestrial environment. Gravity would also have a physiogenic effect as the early tetrapods hoisted themselves out of water and on to the land, their limbs and other organs subjected to new mechanical stresses. Again, physiogenesis would be followed by the genetic assimilation of appropriate functional morphological, ontogenic responses.

Looking back over the major physiogeneses of early vertebrate evolution, we can see that the sequence and timing of these events is significant for further grade progress. Consider the Cephalopoda, such as squid and *Octopus,* which have undergone no dilution physiogenesis like the fish, but whose morphological and neurological plasticity has allowed them to successfully co-evolve with the fish, acquiring remarkable convergent features such as image-forming eyes and flotation systems in the process, as Andrew Packard (1966) has demonstrated. However, having evolved their sophisticated nervous systems in a sea-like internal milieu, there is almost no possibility that these animals could tolerate dilution for the length of time necessary for accommodation to the new conditions. The inability of the Cephalopoda to enter fresh water has denied them the conditions necessary for the development of a terrestrial respiratory system and its metabolic potentialities. The less advanced Gastropoda have been able to enter fresh water and invade the land by virtue of the shell, a portable micro-environment that permits accommodation, and time for genetic adjustment.

The acquisition of a regulated body temperature, the homeothermic condition, was the next requirement for animals that were to realise the higher evolutionary grades. A core body temperature higher than the ambient temperature is not the exclusive preserve of birds and mammals; large size is enough to ensure this in some cases. Some sharks and tuna have a core-muscle temperature that is maintained by a heat-recycling vascular exchange system, so that they can maintain high swimming speeds in cold water.[5] Another way of maintaining a relatively constant body temperature is by behavioural change. Many reptiles have a 'preferred' body temperature which they sustain by hiding in holes in hot conditions and exposing themselves to the sun when their body temperatures drop. Behavioural compensations are also common in mammals when physiological adjustments are inadequate. In birds and mammals lowered body temperature is compensated for by increased shivering, adrenaline secretion, heart beat and food consumption, non-shivering thermogenesis and the fluffing of feathers or fur. Excess body heat is dissipated by increased ventilation and the associated evaporative cooling, increased subcutaneous circulation and sweating, and reduced food consumption. These responses are regulated by a neural thermostat and hormonal controls. Hypotheses concerning the evolution of temperature homeostasis are largely speculative, but there is some room for legitimate inference. A long sojourn in a stable, warm climate would physiogenically impose a temperature-stable internal milieu, redefining the coordinative conditions that internal selection would work around. For example, a spectrum of isoenzymes with different temperature optima would be redundant. However, under the circumstances of prolonged constant high temperatures, poikilotherms are not disadvantaged; there are no temperature drops to slow them down. Bennett and Ruben (1979) have argued that under these conditions the evolution of thermoregulation *per se* would have been of no particular advantage, but the acquisition of improved aerobic respiration and the consequent ability to sustain high locomotory speeds, through improvements in pulmonary and vascular efficiency, and increases in mitochondrial volume, would be distinctly advantageous. Animals committed enzymatically to high stable body temperatures would then require the ancillary mechanisms of neural and hormonal thermoregulation and insulation to survive a climatic cooling trend, and to be able to start fast on a cold morning. The advantage would not be all theirs since the cost of temperature maintenance is high at all times, while poikilotherms, though vulnerable at low temperatures, have low fuel bills. Behavioural adjustments, especially cooling behaviour, would be required in the process of acquiring temperature homeostasis. Sweat glands, useful as they are for cooling, are not essential for life in warm climates. Some warm-climate mammals either do not use them, or lack them altogether, and birds get by very well without them. The additional means of staying warm in cold climates, erective fur, hair and down, might have been adaptive refinements of a system most of whose features existed before the protobirds and mammals entered such climatic conditions, since there is evidence that hair had evolved in protomammals by the time the dinosaurs were emerging.

Much controversy has been generated by the theory that some of the dinosaurs were homeothermic. Robert Bakker (1975) has argued that on the basis of predator-to-prey ratios calculated from fossil remains there were fewer predators than there should have been had the predators been poikilotherms. The bone structure of the putative homeothermic reptiles has the blood vessel pattern more characteristic of large mammals than of modern reptiles: there is also some evidence of the presence of air sacs, suggesting an efficient bird-like ventilation system, and a four-chambered heart. However, most palaeontologists argue that the case for dinosaur homeothermy is circumstantial. A dinosaur might well have high core temperature that would allow it to survive a cold night, but several cold nights and days would be enough to drain the internal heat store completely and leave the animal unable to fend for itself. If asked why some small dinosaurs did not adopt the behavioural compensation of burrowing or nest-building and parental care of the young, Bakker would reply: 'Some did: the birds.' The small cryptic mammals had probably acquired these habits long before and taken them to the lower temperature extremes for further adaptive refinement, and then were in a position to advance into dinosaur territory when the freeze came. Other mammal acquisitions included the secretion of milk for nurturing the young, first present in egg-laying mammals as little more than a source of moisture from enlarged sebaceous glands. The egg-laying habit suggests an initial warm climate initially, but the use of parental body warmth in nests and burrows followed by the sharing of temperature homeostasis in marsupials and placentals finalised the temperature independence of the mammals. These refinements also made possible the evolution of large-brained mammals with critically narrow homeostatic requirements which then underwent the most impressive explosion of adaptive radiation in evolutionary history. They also provided a survival kit for the kinds of astronomical climatic catastrophes that are presently being discussed as periodic events through biological history.[6]

In preparation for concluding this survey of homeostatic evolution an exercise of holistic simplification is appropriate: i.e. the construction of a model that retains the characteristics of the original complex whole, but is sufficiently simple that the significance of the parts and their relationships can be comprehended. Once this is done the acquisition of the parts and their regulatory mechanisms can be explored. This requires no new ground-breaking, since W.R. Ashby's *Design For a Brain* (1952) provides a model of an electro-mechanical 'homeostat' which reduces the complexities of an organism with a learning capacity to comprehensible physical properties and cybernetic principles, without leaving gaps for vital factors to hide in. Ashby asked: if the operation of interrelated specific homeostatic mechanisms maintained essential variables such as temperature and blood sugar in the internal milieu of the higher animals, what were the basic characteristics of such mechanisms? For one thing, the essential variables had to be allowed to fluctuate to some extent since if they were absolutely fixed the flexibility or adaptability of the whole system would be jeopardised.[7] How, for example, could an animal have a carbohydrate meal if a temporary increase in blood sugar were impermissible? 'That a whole dynamic system should be in equilibrium at a particular state it is necessary and sufficient that each part should be in equilibrium at that state, in the conditions given to it by the other parts.'[8] By extrapolation, the homeostatic mechanisms must have reserve capacity to deal with fluctuations in essential variables rather than to be in all-out activity all the time, which might preserve the desired equilibrium of the whole, but would leave it vulnerable to further change. A marine fish subjected to dilution physiogenesis might accommodate through the all-out use of a limited ability to hormonally decrease its permeability. An improvement in this mechanism would bring the regulation into greater equilibrium, as well as saving energy. This could simplistically be called 'adaptation to fresh water', but holistically it has the significance of restoration of stability.

Ashby called the theoretical model derived from his experiments with the homeostat the *ultrastable system.* Disturbances to the equilibria of this system come directly from the environment, e.g. by temperature and salinity changes, indirectly from the environment, increase in blood sugar after a carbohydrate meal, or infection. There are disturbances generated by the organism, e.g. build up of lactic acid, carbon dioxide, ketosis, but these are usually relieved by interaction with the environment. Thus, the environment is a primary component in the diagram of the ultrastable system (Figure 15.1)[9]

As the organism goes about its business in its environment it receives sensory input and responds to that input through its motor system, e.g. by flight, pursuit or feeding. The portion of the system that responds thus is R. The environment acts directly on some of the essential variables (V), such as body temperature, and these may vary until they reach survival limits. Ashby represented this with the symbol of a meter whose pointer oscillates between high and low limits. When the limits are reached there must be a feedback to the organism so that it can take action through a conscious response or through a regulatory mechanism. In either case parameters in the organism, which may be unvarying under the initial environmental conditions, become variable, with an effect on the organism. These constitute the homeostatic functions that Ashby called *step-mechanisms,* since they switch suddenly to a high output from a state of inactivity or low output. To the naïve organism the environment is a 'black box' whose contents can only be ascertained through its responses to trial and error input from the organism. Persisting with the example of heat, a drop in the temperature of the blood entering the brain brings a neural parameter to a critical state where it then functions as a thermostat, a step-mechanism that triggers a shivering response. If a naïve animal touches a fire the step-mechanism of sudden activation of pain causes an immediate motor response. Ashby discovered that the 'fully joined' system of his electro-mechanical homeostat inhibited fast learning.[10] His analogy was a learner-driver who steers the car off track when trying to change gears; that is, if both tasks are given equal concentration at the same time they interfere with one another and hinder the learning process. Therefore, the 'har-

Figure 15.1: The Organismic Ultrastable System

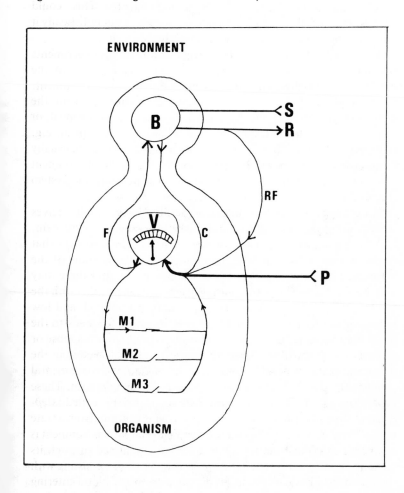

Source: Based on Ashby 1952.

B: brain
C: central control of step mechanisms
F: sensory loop to central control system
M1, M2, M3:   correlated step mechanisms
P: physiogenic (physicochemical) impact of environment on essential
                variables
R: behavioural response to S
RF: behavioural feedback control of essential variables
S: environmental stimulus received by organism
V: essential variables

monious equilibrium of the whole' requires a certain degree of independence for the parts. A diving mammal, for instance, takes advantage of this independence by closing down its peripheral circulation, and the oxygen supply to organs that can tolerate anoxia, while maintaining blood pressure and oxygen supply to the brain.

Any step-mechanism will be brought into critical state at a fixed threshold, and in an ultrastable system such as an organism is likely to have an inhibitory feedback loop to prevent it from pushing the essential variable too far in the opposite direction. Its adaptability is therefore limited to the on and off modes, e.g. activation and inhibition of enzymes. It may also be said to be adapted to a particular aspect of the internal and external environment. If there is a physiogenic change the step-mechanism's threshold may adapt to that change through gene mutation, but it thereby loses its ability to function at the original threshold. Therefore, as Ashby argued, in order to broaden adaptability it is necessary to acquire accessory step mechanisms, each with a different operational range, so that if the functional limitations of one is reached another can step in.[11] The array of muscle kinases with different temperature optima in salmonid fish which allow efficient swimming over a temperature range fits this model of a battery of accessory step-mechanisms serving the same function. The evolution of this arrangement will most likely involve gene duplication and diversification by mutation, the molecular mechanism discussed by Ohno, which is essentially a variation of Cope's Law of Repetitive Addition and Bateson's Rule, which were established on the basis of morphological evolution.[12]

With a few exceptions body temperature is not an essential variable that has been brought under control at the fish grade of evolution. Blood osmotic pressure is however an essential variable in modern fish, regulated at levels different from those of the external environment. The stenohalinity of most fish, or their inability to regulate flexibly if the external environment changes, is probably a regression, permissible because of the consistency of the aquatic environments. In the migrating salmonids the essential variable of osmotic pressure is allowed a limited range, with step-mechanisms controlling each end of the range. An acclimatisation period is required during migration to allow the appropriate molecular step-mechanisms, ion pumps and other gill-cell membrane properties to be restored. The salmonid is actually functioning with a bimodal osmotic pressure equilibrium, which may represent the intermediate 'accommodation state' in the evolution from a fixed 'adapted state' to the adaptable condition. In full homeostasis multiple step-mechanisms produce a single equilibrium that is protected from physiogenic effects. This is the limitation of Ashby's concept of ultrastability. It explains the refinement of the homeostat, and the expansion of adaptability at a given level of organisation, but does not explain how advances in organisational grade are achieved. Nevertheless, he provided a clue by suggesting that a change of goal might bring potentially disequilibrating events to ultimate advantage. A kitten attempting to play with a falling ember quickly learns the goal of avoidance. However, in human history the goal of avoiding fire has been changed to one of using fire, a pivotal step in the emergence of civilisation. A change of behavioural goals took the primitive marine agnaths into fresh water, triggering physiogenic disequilibration and the subsequent re-equilibrations that provided the physiological foundation for subsequent tetrapod evolution. Ashby also pointed out that the nature of the variables in the environment was of great importance to evolving organisms. An 'easy' environment had only a few independent variables; a 'difficult' environment had many interdependent variables. Thus, the original transitional fish moved into a terrestrial environment with a food supply, a constant temperature, and no predators, so that the acquisition of suitable new step-mechanisms was relatively simple. This can be compared to the difficult environment of modern fish, where the variable 'worm' may be dependent on the variable 'hook'.

Ashby's mechanistic approach made no distinction between the physiochemical ultrastable system and the living one, other than a difference in complexity. The ultrastable system was one that adjusted itself to environmental change and thus survived. Reproduction was smuggled in as another step-mechanism that added to regulation and coordination. The environment would always 'select' ultrastable systems. Therefore,

> the development of life on earth must thus not be seen as something remarkable. It was inevitable in the sense that if a system as large as the surface of the earth, basically polystable, is kept gently simmering dynamically for five thousand million years, then nothing short of a miracle could keep the system away from those states in which the variables are aggregated into

intensely self-preserving forms.[13]

He then went on to express a selfish-gene concept of life, with the primitive genes building around themselves the 'active defences' of their bodies and extensions such as nests and tools. This is in marked contrast to Russell's and Cannon's view of the 'almost miraculous' quality of survival of a living system open to the vagaries of the environment. However, despite the intolerant abstraction implicit in Ashby's interpretation, it is a forceful answer to the fundamental metaphysical question of whether life is inherent in the physicochemical nature of the universe, and Ashby represents the positivist ideal of the astronomer who gets on with spectrophotometric analysis while others gaze at the stars with their mouths open.

In addition to the improvement in the regulation of particular essential variables the evolution of homeostasis has involved the organisation of its regulatory mechanisms into a hierarchy of controls. The existence of such a hierarchy in the nervous system was pointed out by J. Hughlings Jackson in his 1890 lectures on convulsive seizures. Sherrington demonstrated this hierarchy by dissection of various parts of the nervous system: spinal section of a mammal caused the loss of the ability to stand and walk due to the loss of control from the cerebellum. Decerebrate animals could stand with an unusual posture, but were unable to walk. Thalamic animals had normal postures, and could walk; animals with only the cortex cut off from the rest of the nervous system could thermoregulate and show emotional responses, but had lost the ability to respond with appropriate behaviour to new circumstances, and to learn. Thus, there is a rough correspondence between the morphological evolution of the brain, with its anterior progression, and the addition of the new hierarchical levels.[14] The advantage of hierarchical control in the nervous system has often been compared to the function of a telephone switchboard: instead of each subscriber being directly linked to every other, switching at a local exchange greatly simplifies the wiring, and national and international exhanges allow the entire world to be integrated into an intercommunicating whole. However, when telecommunication involves only a few functional units, direct wiring between each is a satisfactory arrangement, as is found in simple reflexive components of the nervous system.

In considering the evolution of the hierarchical complexity of the nervous system, not only the intrinsic advantage of organising many units into a coordinated whole that must be taken into account: the circumstances under which the new units were acquired must also be examined. In addition to its role in the regulation of the essential variables of the internal milieu the nervous system receives internal and external stimuli, integrates these, and orders appropriate muscular responses. Sensitivity to the physical stimuli of light, gravity and vibration, and to a wide variety of chemical stimuli, are properties of many cells in the most primitive organisms. An electropotential across the cell membrane, which is characteristic of all cells, is increased to become the neuronal resting potential, the physical starting condition for the propagation of nerve impulses. All of these properties are the inevitable consequence of the physico-chemical conditions in any living cell, enhanced by the biological nature of the selectively permeable cell membrane. They cannot be considered adaptations to any specific element in the external environment, but rather define the primitive coordinative conditions that both permit the cell to persist in its being, and also potentiate cellular adaption or specialisation once the multi-cellular state has been achieved. Cells with long cable-like axons, which are specialised for impulse propagation, and reflexive circuits of these cells are found in the simplest and most primitive metazoa, and form the functional-morphological units that are built up into large arrays and ultimately brought under a number of hierarchically organised levels of control as evolution progresses. It is the conventional wisdom of neo-Darwinism that both relative increase in size and the progressive complexification of the brain have been wrought by 'selection pressures', largely the need to obtain food and escape from predators, which is neither better nor worse as explanation than Lamarck's assertion that the animal, feeling the need of an organ, goes about the business of evolving it. Centralisation of the nervous system into a ganglionic complex or brain at the front end of the animal is a consequence of having a distinct region which is the first part of the animal to enter an environment, and where it is useful to concentrate the sensory organs (or vice versa). The proximity of the integrating centre to the sensory organs reduces the lag time between stimulus and response, is a convenient arrangement for the future evolution of short, fast-acting interneuronal con-nectives, and can be easily protected by a compact cover such as a skull. This centralisation of brain and central nervous system has evolved independently a number of times in the animal kingdom. Some brains can be properly described as specialising for a particular kind of environment. For example, among the fishes there are proportionate increases in the sizes of different brain regions associated with the integration of the most important stimuli in the life of the fish. Sharks have 'olfactory brains' and turbid water fish have cerebellar expansions related to lateral line, vibration sensitivity. In the case of electric fish which can detect changes in the electromagnetic fields that they generate in the turbid water, there is an enormous cerebellar expansion. There appears to be a good correlation between the proportion of stimuli of a given type and the amount of brain tissue needed to process the incoming information.[15] However, as brain evolution progresses there are some innovations that defy the adaptationistic interpretation. Various mammals with different modes of life have, for instance, evolved colour vision. For what could this be said to a specialisation? As intelligence evolves, there is an associated increase in the amount of cerebral integration and in the ability to respond flexibly to changes and inconsistencies in the incoming stimuli, in other words, as Le Gros Clark and others have argued, adaptability becomes the major attribute of progressive brain evolution.[16] These reservations concerning the strict neo-Darwinists interpretation of brain evolution are expressed by H.J. Jerison in *Evolution of the Brain and Intelligence* (1973). He also sheds some light on the question of the relative importance of adaptation and adaptability in mammalian brain evolution, with hypotheses based largely upon the gross brain anatomy of living animals and the endocasts of the brains of fossil tetrapods. He begins with the eye and brain of the reptile ancestors of the mammals, which integrated light stimuli in the retinae: although they could respond selectively to signals, their responses were largely reflexive, and there was little or no expansion of the visual integrating region of the brain itself, and therefore no mental visual pictures of external events, such as we can produce in great variety simply by closing our eyes. The first major step in the progress of mammalian brain evolution was hypothetically due to the development of nocturnal habits in protomammals in the late Triassic. Since vision is relatively useless in the dark, great advantage would accrue from improvements in the olfactory and auditory organs. Selectivity or discrimination among the sound stimuli depends upon an increase in interneuronal connections, but the ears, or tympanic organs, unlike the retinae do not have the neural capacity to develop in their own immediate vicinity, and so require an expansion of the mid-brain. Along with a refined ability to locate sound it is also useful to be able to identify specific sound sequences, such as we do when we listen to speech or music, and a neural, short-term memory is a minimum requirement for this ability. That is, some mid-brain circuits must receive signals, connect them with the appropriate motor responses, and at the same time remember the signal and be able to integrate it with earlier sound memories to produce a crude sound-picture, or mental auditory model of the environment.[17] Since these animals did not live in total darkness all the time the visual apparatus would be occasionally useful, and further brain progress could be made by a closer integration of visual and auditory messages, a refinement that would allow the animal to both see and hear prey and predators and potential mates, and to recognise that both types of stimuli were generated by the same source in the external world. A nocturnal habit is not entirely necessary for these arguments since the same reasoning would apply in a habitat consisting of dense vegetation, where hearing and olfaction might be more important than vision. So the acquisition of defined hearing confers adaptability to a variety of environments. All in all, the consideration of the quality of a genetic or epigenetic variation, in terms of what range of benefits might accrue, seems to come more directly to the problem than casting about for 'selection pressures' that cause the variation to appear.

Jerison argues that a visual 'time-binding' capacity, able to hold past optical signals and integrate them in sequences, might have evolved in the mid-brain in a parallel manner once there was a return to a more open diurnal environment. The coordination of these auditory and visual capacities would now contain a dynamic perceptual audiovisual 'map' corresponding to the environment and with the potential to make predictions about events in the immediate future.[18] The underlying processes of such evolution involve epigenetic changes in the relative rates of brain development, with duplicative production of neurons as well as a genetic duplicative mechanisms that would increase the number of dendritic 'recognition' molecules such as Ohno suggests would allow the functional differentiation of an expanded mass of nerve tissue.[19] The external impacts of the environment and the sequence in which they occur constitute the other major element. In the dark, olfactory and auditory stimuli are paramount. Moreover, by Jerison's hypothesis the cephalisation of the visual mode requires the precondition of an expanded mid-brain auditory function. This aspect of homeostatic evolution is also linked to temperature homeostasis: if nocturnal cooling adversely affected a nervous system upon which the greatest coordinative demands were made at night a means of heat generation and insulation, to maintain a nervous system already enzymatically adapted to a stable, high ambient temperature, would be of the essence.

It was about the end of the Mesozoic, after the great reptile extinctions, that the most successful mammalian adaptive radiation occurred, and it is plausible that the new audiovisual-modelling brain, functioning in a homeothermal milieu, was largely responsible. As an analogy, consider the advantages of the modern warroom, with computer-modelled situation-maps displayed on video display terminals, over the trench-warfare company commander who had to respond directly to what he could see or hear in front of him, with only occasional hints of strategic requirements coming in by runner or over a field telephone. Nevertheless, from the point of view of either, the overnight extinction of the enemy's heavy artillery and armour would be a distinct advantage. If, as a number of biologists and astronomers now argue, the biosphere has been occasionally devastated by catastrophic astronomical events such as direct hits by comets, meteors and asteroids, or pole reversals caused by near misses, small, burrowing, nocturnal, omnivorous homeotherms with rudimentary intelligence would have the upper hand during the period of ensuing chaos, whether it was cold, dark and dusty, or characterised by harsh diurnal radiation.[20]

In the further progress of the mammalian brain-size, increase through flexuring and cortical surface-area increase through fissurisation were important epigenetic changes. Brain enlargement Jerison puts down to the 'selective pressures' of co-evolving prey and predators, although he also cautions against the interpretation of progressive changes in the direction of intelligence in terms of species-specific adaptations in specialised niches. Moreover, brain enlargement conferring behavioural flexibility would have a general advantage to animals ranging widely and exploring new environments.[21] From the example of the small-brained lemurs, which have a number of anatomical arboreal adaptations, it can be seen that regardless of potential the 'selection pressures' of the arboreal habitat do not invariably cause brain expansion in the primates.[22] Cortical expansion of the brain can be correlated with the development of manipulative skills in primates, but while this can be loosely described as 'adaptation to diet' its advantage as adaptability to a variety of diets is a fuller explanation. This may again be contrasted with the condition of the lemurs, whose expanded finger pads can be unequivocally called 'adaptations for climbing'. The final event in progressive brain evolution was the elaboration of a language function which allowed individuals to share their mental situation-maps, with all of the consequences for food acquisition, protection from predators, mating, tool-making, education, art and mythification that make the most hardened neo-Darwinists throw up their hands at the problem of identifying 'adaptive zones' and sequences of 'selective pressures'.

In conclusion, while the traditional approach to evolutionary physiology has been to call every novelty an adaptation: adaptation to environment, internal or self-adaptation; adaptation to changing environments, and while I have been antithetically urging Whyte's approach to interpreting physiological evolution in terms of the internal coordinative conditions, there ultimately has to be a synthesis of the two. A spectrum of biological functional change can be discerned: at one end is the simple effect of a point mutation that alters the structure of a protein in a way that is coincidentally adaptive to external or internal conditions. This is equivalent to an adaptive morphological change, such as in the feeding apparatus. Next, if the modified protein is an enzyme its synthesis and activation may be of the nature of a step-mechanism. That it can be switched on and off depending upon the condition of internal essential variables makes it not only adaptive to a particular circumstance, but also confers the first level of adaptability, i.e. of homeostatic regulation, upon the organisms. Theoretically this acquisition will have an effect on the other coordinative conditions resulting in maximum flexibility of the whole system. This low-grade adaptability can then be improved by gene duplication and diversification to produce a range of controlled responses. The step-mechanism adaptability is developing in an organism that even at the simplest cellular levels has the ability to respond to certain stimuli by contracting away from the stimuli or making stronger contact with them; another kind of on/off adaptability. In plants this may amount to no more than an ontogenic alteration in growth pattern, so that a shoot will grow towards light, a climbing plant will wind round a support, or a tree exposed to prevailing winds will have an appropriate profile. In animals this evolves in conjunction with the endocrine and nervous systems into complex behavioural responses to the environment, which are linked through the alteration of the internal essential variables to the homeostatic step mechanisms. The highest grades of adaptability involve: the ordering of control mechanisms into a hierarchy; the development of an ability to model the environment in the central nervous system, with a memory to match new and old models, and a logical faculty that by-passes trial-and-error responses; and finally the ability to communicate both the models and the deductive and inductive inferences based upon them. These highest levels of adaptability are fundamentally made possible by evolution of homeostatic coordinative conditions which have come into being largely as a result of a unique sequence of environmental encounters, each of which has imposed disequilibrating physiogenic changes upon the internal milieu, which have been brought into new equilibria through genetic assimilation. Thus, although Whyte's arguments are justified from the point of view of holistic simplification, the whole lacks its unavoidable links with the environment, and we must agree with Rosen when he writes, with reference to the evolution of metabolic organisation:

> The important novelty at this level, which distinguishes this kind of argument from the simple kind of parameter adjustment previously invoked to account for function change, is that, in effect, the parameter adjustment must take place in the system *environment*, or better, in the coupling between the system and its environment rather than just in the system itself.[23]

These conclusions appear to be in stark contrast with those expressed by E. Jantsch in *The Self-Organizing Universe* (1980). Although he would not exclude physiogenesis as a disequilibrating cause he ignores it on the grounds that self-organisation is an autonomic or autocatalytic process, and the system is 'self-referential with respect to its own evolution'.[24] A self-generating disequilibration comes into his scheme; as his mentor, the chemist I. Prirogene argued with regard to non-living dynamic systems, 'non-equilibrium may be a source of order, or organization, that became the foundation of a non-linear thermodynamics of irreversible processes now permitting the description of phenomena of spontaneous structuration'.[25] Although Jantsch notes that although this new *order through fluctuation* has only been clearly recognised since 1967 it is implicit in most holistic evolutionary writings. However, a partial resolution of some of these contradictions becomes available when Jantsch goes beyond the organismic system to biological co-evolution and ecological niche complexification, emphasising that the behaviour of the organism in relation to these factors not only allows it to 'determine within relatively wide boundaries to which natural selection it subjects itself', but also to participate in further evolutionary progress, through responding to new 'stresses' which arise from its new activities.[26]

Lamarck, Cope, C.L. Morgan, Osborn, Baldwin, Child, Schmalhausen and Waddington all recognised the significance of some of these aspects of functional evolution, and Herbert Spencer (and Jantsch) had it half right when he argued that the evolution of organismic complexity was due to the evolution of environmental complexity.[27]

16 ALARMS AND EXCURSIONS

> . . .the emergence of new theories is generally preceeded by a
> period of pronounced professional insecurity.
>
> T.S. Kuhn 1962[1]

According to Løvtrup (1982) there are four complementary theories
of evolution.[2] The first is evolutionism itself, the 'theory of the
reality of evolution', to which both Lamarck and Darwin made
major contributions. The second is the theory of the history of
evolution, which deals with the course of evolution, its progressions
and regressions and fits and starts: the realm of phylogeneticists,
systematists and palaeontologists. The third is the 'ecological
theory' which accounts for adaptation, and which neo-Darwinists
usually call the 'synthetic theory'. Finally, there is the 'epigenetic
theory' of macro-evolution, that Løvtrup argues is the most
important element needed for a wholly comprehensive theory.

The most entertaining aspect of modern discussions of evolution
is that every one of these four theories is presently a polemical
battleground. The first is the arena for the debate between the so-
called 'scientific creationists' and evolutionists. Scientific
creationism is a contradiction in terms since its axioms are the
received word of the Judaeo-Christian deity, as expressed in
Genesis. The 'science' is a quasi-materialistic sophistical critique of
existing evolutionary theories, with the simplistic tacit assumption
that somehow if Darwin got it wrong it had to be God that got it
right. Met at first with a disdainful silence by biologists, neo-
creationism shifted into the educational, political and judicial
arenas, thereby demanding a vigorous response from evolutionists
and resulting in a discernible loss of equanimity and occasional wild
assertions that came close to surpassing those of the creationist pro-
vocateurs. Some of the arguments of the neo-creationists concerning
the putative contradiction between evolution and the second law of
thermodynamics were answered by Bertalanffy and Eddington half
a century ago, and criticisms of the accuracy of dating techniques
arise from axioms that make no more sense than P.H. Gosse's
nineteenth-century assertion that God put the fossils in the rocks to
test man's faith.[3] D.J. Futuyama's *Science on Trial* (1983) marshalls

the pro-evolutionistic arguments, and provides a useful crib-sheet for naïve biologists who might be called upon at short notice to defend the faith. In the epistemological context it can be seen that creationists share the perplexity of some biologists concerning explanations of the process of grade evolution in terms of random point mutation. But lacking an appreciation of the independence of the historical theory from the theories pertaining to mechanism, they falsely assume that if Darwinian selection theory has flaws then the entire structure of evolutionism is in question. The politics of this conflict are ably described by Dorothy Nelkin's *The Creation Controversy* (1982).

Within the historical theory there are two internecine struggles going on, the first of which is over the pace of evolution. In 1972 a palaeontological study by Eldredge and Gould concluded that speciation was in some cases surprisingly rapid, but followed by long periods of stability. These 'punctuated equilibria' were proposed as 'an alternative to phyletic gradualism'. 'The history of evolution is not one of stately unfolding but a story of homeostatic equilibria, disturbed only rarely (i.e. rather often in the fullness of time) by rapid and episodic events of speciation.'[4] The very suggestion of a non-gradualistic interpretation of the fossil record was enough to provoke apoplexy among orthodox neo-Darwinists, and several meticulous and exhaustive studies of very small segments of the fossil record have since demonstrated a gratifyingly gradual mode of change — at least in the context of those very limited segments. However, as Løvtrup has pointed out, neo-Darwinism has nothing to fear from the theory of punctuated equilibria, since the hypothetical causes of these phenomena are no more than alterations in selection pressure, presumably brought about by environmental changes, which not even the most orthodox of neo-Darwinists would dispute.[5] Mayr's 1959 statement, cited in Chapter 11, concerning 'explosive evolution' was the more revolutionary since it was couched in epigenetic terms.

The sanctity of gradualism developed not only from the power of Darwinian authority, but also, as Himmelfarb suggested, through the identification of gradualism with naturalism.[6] In contrast, transcendental evolutionary axioms tended to produce saltatory interpretations. This sanctity is also outraged by another tea-cup storm, the debate over the application of cladistics in the current evolution exhibits at the British Museum of Natural History. Cladists attempt to establish phylogenetic or systematic relationships by comparing known, quantifiable biological parameters, and the debate over its usefulness as a tool of systematics has raged back and forth across the pages of *Systematic Zoology* for the last decade. That the museum has stuck to the cladistic facts and not made assumptions based on Darwinistic interpretations of the course of evolution is, as Hitching puts it, 'what has made traditional Darwinists so angry. When the oldest and perhaps the most prestigious natural history museum in the world starts beating a retreat from Darwinism, a funeral march is on the way'.[7] *Nature* asked, 'Can it be that the managers of a museum which is the nearest thing to a citadel of Darwinism have lost their nerve, not to mention their good sense?'[8]

Although cladists do not take the continuity of gene flow as axiomatic, they make no prejudgements about the gradualistic or discontinuous nature of the pace of evolution. Nevertheless, some critics have decided that cladism is opposed not only to gradualism but also the authority of Darwin and his living philosophical heirs. Since discontinuity is the essence of dialectical materialism, cladism, by failing to adhere to gradualism, must be Marxist subversion.[9]

Charges of Marxism are more appropriately applied in the perennial conflict over the ecological theory of evolution, which attempts to elucidate the causes of adaptation, and, as the synthetic theory, subsumes all evolutionary causes under the adaptationist selectionist thesis. The Lamarckist alternative of adaptation by the inheritence of acquired characteristics was not liquidated by Kammerer's bullet.

Popular interest in Lamarckism was revived in 1927 by the experiments of the psychologist W. McDougall, who trained rats to use a particular maze route by administering electric shock, and then found that the offspring of the trained rats needed fewer lessons than control rats to learn the route.[10] In 1936 F.A.E. Crewe found that rats' ability to learn is partly hereditary, and he surmised that McDougall had been unconsciously selecting for intelligence and had produced a line of smart rats.[11] A prolonged series of experiments by Agar seemed to support McDougall's initial interpretations, but then disclosed that the control animals were also learning faster due to a putative improvement in the health of the entire colony.[12] Rupert Sheldrake's *A New Science of Life* (1981) has an explanation for this. The control rats were in fact subjected to the 'morphogenetic fields' of the experimental animals and so absorbed the results of the training. This pot was kept boiling until 1954. In the meantime another pot had been coming to the boil, unnoticed by many biologists outside Russia. In 1928 Trofim Denisovich Lysenko interpreted the vernalisation of wheat as evidence of the inheritance of acquired characteristics.[13] Over the years Lysenko's influence and the importance of the form of Lamarckism that he called *Michurinism* grew. The first suggestion of an ideological battle for control of the conduct of genetics in Russia came with a false rumour about the arrest in 1936 of I.N. Vavilov, Director of the Genetics Institute of the Academy of Sciences of the USSR, and the cancellation of the Congress of Genetics scheduled for 1937. Vavilov was actually arrested in 1940 and died two years later in prison.[14] It was only in the late 1940s that the fate of Vavilov and a number of other Soviet geneticists became generally known, through a number of articles and translations which were compiled by Conway Zirkle as *Death of a Science in Russia* (1949). In the same year Lysenko's 1948 inaugural address as President of the Lenin Academy of Agricultural Sciences was published in English.[15] He agreed with Engels that,

> The entire Darwinian teaching on the struggle for existence merely transfers from society to the realm of living nature Hobbes's teaching on *bellum omnium contra omnes* and the bourgeois economic teaching on competition, along with Malthus's population theory. After this trick. . .has been performed the same theories are transferred back from organic nature to history and the claim is then made that it has been proved that they have the force of eternal laws of human society.[16]

Up to this point I disagree with neither Lysenko nor Engels. However, he then condemned Bateson, who had been sympathetic to the scientific policies of the revolutionary regime, as an obscurantist, Mendel and Morgan as debasers of Darwinism, and Weismann as a mystic since he had postulated that the germ plasm was immortal. On the other hand, Lamarck, who had rarely been mentioned before 1948 in the Michurinist literature, represented 'materialistic' evolution, and the true synthesis was between creative Soviet Darwinism and Michurinism.[17] According to this synthesis the whole organism and its experience was supposed to be distilled down into a somatogenic cellular quintessence, ensuring the inheritence of acquired characteristics. The deviants Schmalhausen,

Vavilov and 'the Morganist-Weismannists' had supposedly slandered Lysenko for the suppression of non-Michurinist trends. The 'truth' was that he had neglected to provide adequate facilities for Michurinism, effectively allowing the opposing metaphysicians and scholastics to have their own way. This statement was published in English two years after Darlington's account of the 'restriction' by liquidation or death in prison of ten Soviet geneticists, and so it is quite understandable that Darlington and anyone else familiar at the time with the fate of Soviet genetics should despise Lysenko's cant, intrigue and fanciful ideas about heredity.[18] It is less comprehensible why the burden of blame should be loaded on to Lamarck and why he should continue to be reviled and misrepresented. This is the very sophistry that Zirkle and Darlington were at pains to expose in Lysenkoism. The Lamarckian concept of the inheritence of acquired characteristics lends itself just as well to racism as it does to Marxist ideology, but so does natural selection, and no one blames Darwin for social Darwinism.

One prominent supporter of Lamarckism after 1930 was F. Wood Jones. His *Habit and Heritage* (1943) had an unfortunate introductory note on 'the outstanding work of Lysenko and Michurin'.[19] However, the full account of the Lysenko affair was not common knowledge until 1949, and Wood Jones, like J.B.S. Haldane, cautioned that the account of Lysenko's work available at the time in translation was inadequate to allow proper judgement. Wood Jones was anxious to disembarrass Lamarckism of the burden given it by Eimer and Weismann:

> Man has cut off the tails and ear of his domesticated animals, he has removed the prepuces, pierced the ears and noses of his fellows; but the fact that those mutilations do not become perpetuated by heredity has nothing whatever to do with the question of the inheritance of *acquired* characteristics as these things are properly defined.[20]

Trends of Life (1953) discussed Russell's concept of directive activity as excluding the necessity of natural selection acting on fortuitous variation, but simplistically assumed that a Lamarckist mechanism caused the evolution of directedness. The calcareous and horny opercula and epiphragms of gastropods were the products of morphogenically directed activity that could not have evolved by natural selection. These he compared with the use of the head as a stopper by the burrowing toad *Bufo emphurus* and the trapdoor of the spider of that ilk: 'the needs created by any well-defined ecological situation are likely to be met by all living things subjected to them by directive responses of a similar kind'.[21] These 'needs', he thought could be satisfied by the growth of one part in harmony with that of another. It does not, however, follow that morphogenically directed activity or analogous behaviour falls to Lamarckism by the rejection of the Darwinist position.

Wood Jones raised a number of interesting cases that he thought confounded Darwinism, that are amenable to epigenetic interpretation — such as the sharp claws that baby kangaroos use to climb into the maternal marsupium, which are then lost, to be regrown at a later age; also the loss of the swim-bladder by species of the Indian loach *Nemachilus* which inhabit torrential mountain streams. When some of these ascend to alpine lakes they develop new swim-bladders but not from the vestiges of the old ones. The same phenomenon also occurs in some carp and catfish.[22] The extra prehensible thumb of the panda seemed to him another example of a novelty meeting a need. Jones cited the example of the way in which the bones of the foot in man may be modified by the style of squatting. He observed that these modifications of the foot anatomy appear in the embryo before any habitual squatting had been able to induce the morphological change. Interesting though these examples are, they can be largely 'explained away' as quasi-Lamarckist phenomena using the arguments of organic selection or genetic assimilation as Waddington did with he callosities of the ostrich. Proof of somatogenic induction was still missing.

The Lamarckist banner was also carried on by H. Graham Cannon, whose *Lamarck and Modern Genetics* (1959) was published on the 150th anniversary of the *Philosophie zoologique*, coinciding with the hoopla of the Darwin centenary celebrations. Cannon was convinced that Lamarck's disrepute was due to distortion of, and omissions from what Lamarck had really said, and to the Kammerer issue. Cannon had been MacBride's research assistant at the time of the Kammerer visit to London, and he supported Bateson's view of Kammerer's fraudulence. Like Wood Jones, Cannon believed that Lamarckist experiments with mutilations, and other abnormal interferences with organisms, had been a waste of time, and that more significant was 'the experiment in which the disturbing condition is one which the animal might expect to meet, or at least would not be surprised to meet, in its normal

activities'.[23]   Cannon   regarded   Kammerer's   colour-change
experiments with salamanders as meeting these conditions. The
focus of Cannon's attack was contemporary genetics. There was no
point to 'all the soul-destroying mathematics that have been
expended in analysing the power of natural selection', since
variations were simply ontogenic.[24] Nor could genes bring novelty:
'Genes can *alter* things but they cannot produce new things.'[25] To be
given major evolutionary significance, gene mutation had to be
shown to be capable of generating novelty. For example, the
novelties of the egg could not be the 'products of gene
complexes. . .the white of egg, the shell, and the egg-opener, these
are entirely new. They were wanted and therefore they appeared'.
There is little difference between this and the optimism of 'selection
pressure'. Cannon did not dismiss the gene theory out of hand
admitting (and this was in 1959) that 'It might even be said that we
are beginning to study their molecular anatomy'.[26] But he felt that
each bit of a complex organ must have its own genes, thereby forcing
gene theory into the absurd position in which Weismann put it.
Finally, since he believed that bacteria had no chromosomes it was
'sheer mysticism. . .exceeding the entelechy of Driesch' to attribute
the power of bacteria to adjust themselves to changing conditions to
undifferentiated lumps of chromatin.[27] To opponents of the gene/
chromosome theory the existence of prokaryotes which apparently
lacked chromosomes had been a comforting delusion.

Cannon asserted that only Lamarck's second law, response to
need, was really important. The fourth law of inheritance of
acquired characteristics he inferred was redundant. But smuggling
the fourth law into the second does not get over the problem that
Lamarckism in any of its forms stands or falls by the proof of
somatogenic induction. Like Haeckel, Cannon believed '"organic
invention" is a property of protoplasm'.[28] The physicochemical
equilibria established in non-living systems was acquired by the
earliest protobionts and became more sophisticated as evolution
advanced. This is the kind of over-simplification that Bertalanffy
and others had been at great pains to show was based on false
analogy. Cannon believed that the concept of homeostasis illustrated this point, but did not appear to realise that regulated
equilibria in the internal milieu were quite different from the
physicochemical equilibria that would be established if regulation
was withdrawn. By this error and by the more fundamental one of
the rejection of 'neo-Mendelism' and the gene theory as central

concepts of genetics Cannon undid the good that he might have done in obtaining restitution for Lamarck.

Alister Hardy, though a supporter of selection theory, echoed Cannon's claim that Lamarck had been much maligned and deserved better. In *The Living Stream* (1965), based upon his Gifford Lectures, Hardy stressed the importance of behaviour in animal evolution, as an additional degree of flexibility in ensuring survival. Organic selection, which he equated with genetic assimilation, was the mechanism whereby behavioural 'pre-adaptation' not only brought about genetic fixation of the new behaviour, but also triggered physiological and anatomical departures. Lamarck was closer to appreciating the organism's activity as an agent of evolution than the Darwinists, and organic selection explained many aspects of evolution for which simple selection theory was inadequate. Nevertheless, there remained some phenomena for which organic selection was also inadequate. Eyeless *Drosophila,* if inbred, could produce eyed offspring by some alternative epigenetic route. Furthermore, the embryonic segments that produced the homologous limbs in vertebrates were not always the same.[29] D'Arcy Thompson's demonstration of coordinate trans-formations and Russell's directive activities also seemed to defy the cumulative selection hypothesis. These phenomena make more sense in the context of epigenetic regulator models. However, selection theory does not even generate Hardy's questions, far less their epigenetic answers. Therefore, Hardy invoked a kind of collective unconscious or racial memory similar to the hypotheses of Samuel Butler and Carl Jung:

> If it is established that impressions of design, form and experi-ence. . .can occasionally be transmitted by telepathy from one human individual to another, might it not be possible for there to be in the animal kingdom as a whole, not only a telepathic spread of habit changes, but a general *subconscious* sharing of a form and behaviour pattern — a sort of psychic 'blueprint' — shared between members of a species?[30]

Setting aside a positivistic scepticism about telepathy, there remain two major problems with these ideas. If the whole group could benefit from the experience of a few individuals why has evolution been so slow? The second point was raised by Lamarck himself, although it rarely appears in the work of his disciples. The business of adaptation by the inheritance of acquired characteristics is an obstacle in the way of grade evolution: both somatogenic induction and natural selection take a flexible, adaptable organism and lock it into a pattern of rigid specialisation. Organic selection fixes freedom of choice into inevitable stereotyped behaviour patterns. Therefore, if we want to understand progress we need to stop paying exclusive attention to adaptation to environment. The same criticisms apply to Rupert Sheldrake's similar concept of morphogenetic fields.

Hardy's thesis also recalls aspects of Albert Vandel's *L'Homme et l'évolution* (1958), which stressed psychism as an important factor in the later stages of progressive evolution, and dismissed adaptive evolution as an insignificant Anglo-Saxon obsession. Vandel felt that migration was important in providing newly progressed forms with the opportunities to realise their potentials. He agreed with Lamarck's emphasis on the response of the organism to its environment, and the importance of progressing to a neurological level where psychism was an effective force, but rejected the eclecticism of the neo-Lamarckists. Sympathy with Lamarckism has understandably remained strong in the country of its origin. It sometimes forms an amalgam with Bergsonism, as is evident from the title of P. Wintrebert's *Le Vivant, créateur de son évolution* (1962). Wintrebert argued that protoplasm has creative properties, and the organism has a role in its own evolution, initially through a chemical intelligence.[31] Taking somatogenic induction as axiomatic, he proposed that the real creativity was the ability of the protoplasm to respond to its needs in the first place by creating adaptive genes by 'mutation', through the combination of an 'adaptive antibody' with a nucleoprotein. This variant of the chemical-Lamarckism theme seemed to be a better alternative to the fortuity of neo-Darwinism and divine intervention which both removed from the organism the capacity of ordering it own life.

The range of thought expressed in P.-P. Grassé's *Evolution of Living Organisms* (1977) makes it difficult to classify him easily; however, he is fundamentally a Lamarckist. Grassé's views on the chronology of morphological evolution are analogous to those that I have expressed with regard to physiological evolution. For example, he notes that, 'The evolutionary "value" of any variation depends upon its time appearance.'[32] In the evolution of the functional morphology of snakes the crucial variation is multiplication of vertebrae. This is embryogenic, nerves and blood vessels being formed epigenetically under the stimulus of the presence of new vertebrae. The loss of limbs is to be regarded as a contribution to the production of a harmonious whole being, in contrast to 'those brutal mutations which pertain to the pathology of genes and thus belong to teratology'.[33] While he recognises that evolution consists of coordinated multiple changes effecting bones, muscles and nerves, 'not a mosaic of random variations effecting just anything at any time', he nevertheless dimisses Goldschmidt's concept of systemic mutations as 'pure fantasy'.[34] Although the organism must intervene in the genesis of control systems, models such as Britten and Davidson's are flights of fancy that may be relevant to normal gene activity but not to evolution. Grassé also gives short shrift to the belief that redundant DNA might have a regulatory role. This must be 'pure fiction', because 'genic regulation does not seem to be any less in the frog . . .whose genome has very little redundant DNA than in the axolotl, rich in DNA'.[35] Instead of changes in regulatory systems Grassé proposes the acquisition of new genes and functions.

> In order to create, evolution demands new materials, such as genes formed *de novo,* or untried patterns of over printed codons. It is not at all the same gene that, from one class of vertebrates to another, induces. . .ganoid, placoid, or cycloid scales in fishes, epidermal osseous scales in reptiles, feathers in birds, hair in mammals. Every novelty demands its own genes, which are themselves also novelties'.[36]

With regard to the evolution of new functional genes, Grassé requires not only that a new cistron be interpolated, either at a broken piece of DNA strand or at the end of a pre-existing strand, but also that a new enzyme involved in the expression of that cistron be evolved, implying a different new cistron coding for the enzyme. As to meaning: 'theoretically there is nothing to prevent them from organizing into codons which then order the formation of a protein used in the economy of the living organism'.[37] He does however admit that the question of probability of such events looms large. The process must be preceded by 'preliminary intracellular operations' of unspecified precision and,

> Autoadaptive Lamarckian variation is an adequate response by the organism to aggression from the environment. . .Evolution demands the acquisition over time, as organisms grow more complex, of novelties whose information is inserted into the DNA strands in the form of new genes.'[38]

Since Grassé demands that, 'Any system that purports to account for evolution must invoke a mechanism not mutational and aleatory', and presumably not based on natural selection as the main causative agent, it is not surprising that he chooses the Lamarckian mechanism. However, there are other options. The evolution of co-ordinative conditions discussed by Russell, Whyte and Ashby is aleatory, but the organism has the role of making best use of them, through organisational flexibility. Natural selection is redundant in such events. Aleatory epigenetic events, being more profound than events affecting the structural genes expressed in the mature organism, are much less likely, according to Grassé, to contribute to meaningful evolutionary change, but, as numerous epigeneticists have argued, the flexible behaviour and internal directive activity of the organism may have reduced the randomness. That the mechanisms of evolutionary epigenetics have not been fully discovered is no reason to pretend that the phenomena do not exist.

There is no simple black and white choice between Darwinism and Lamarckism. But the naïve belief that all experimental, theoretical and metaphysical and ideological eggs must be put in the same basket continues to produce explosions of hot air. The latest has been elicited by E.J. Steele's *Somatic Selection and Adaptive Evolution* (1979), which opens by expressing an intuitive dissatisfaction with theories that lack 'an element of "directional" progress in the complexity and sophistication of adapted living forms', and suggests a Lamarckist 'sense of direction' as a complement.[39] The central question that Steele addresses is the creativity of the immune system in its response to novel antigens. The current conventional *clonal selection theory* of immunology argues that the immune system is capable of producing a very wide spectrum of antibody structural sequences, one of which is likely to have a close structural fit with any known or unknown antigen. Once the appropriate antibody has been identified its carrier cell is then cloned to provide sufficient quantities of the defensive molecule to deal with the pathogenic molecule. The unanswered question is: how is the broad antibody spectrum produced?

Setting aside the older instructional model which proposed that the antigen acted as a template for antibody transcription, Steele observes that the clonal selection theory admits the possibility that all antibody types are the products of natural selection, or that some, e.g. those for common pathogens, are the products of natural selection, and the others arise by *somatic selection,* that is, they appear through somatic mutations in the individual during ontogeny, and are then cloned if they are immunologically useful. The argument up to this point does not involve the inheritance of acquired characteristics, only somatic events in the individual. The first possible mechanism for clonal selection does not place a great burden on the genome. It is estimated on the basis of the number of amino acyl residues in the antigen binding portion of immunoglobulins that 0.01 per cent of the total vertebrate genome is all that would be required to code for one million different antibody structures.[40] However, molecular incompatability estimations reduce this to approximately 100,000 possibilities, a fair assortment when one recalls that while the possible number of antigen structures is astronomical they are likely to belong to the 500-1,000 protein classes discussed by Zuckerkandl, and for the immune reaction to be successful the colinearity of antigen and antibody does not have to be exact. However, Steele meets the selectionist model, which poses that all antibody structures are or have been useful, with a utilitarian criticism: why would this great array of possible structures be retained? Some of them are useless under most circumstances, and many, though they might have been useful in the past, have not been recently used. Without the encouragement of natural selection they would wither on the vine. An orthogenetic mechanism of repeated duplication and random mutation would save the hypothesis, but Steele is drawn instead to the somatic selection hypothesis. The evidence for this is not in dispute; the controversy arises when he claims that certain of these somatically selected and cloned antibodies are somatogenically induced, i.e. that antibodies which have mutated and been selected outside the germ plasm, and would normally die off with the individual, are somehow fixed genetically and passed on to the next generation. There is a hint of physiogenesis in Steele's thesis that he calls 'positive selection feedback': change in the internal environment imposed by antigens is followed by the ontogenic response of cloning the most useful antibody, if one comes to be available.[41]

Steele also suggests that somatic selection is a universal principle, not simply confined to the lymphocytes that carry antibodies. The mechanism proposed for the somatogenic induction is viral transduction of the somatically selected and cloned mutant gene or its messenger RNA. Somatic cells with a rapid turnover or high proliferation rate, like blood cells and epithelial cells exposed to the external environment or in the gut, would be good candidates for the somatic mutation-selection mechanism since they would spread the somatic novelty more rapidly, and increase the probability of somatogenic induction. Viruses could infect the germinal products, ova or sperm, or simply be transmitted in spermatic fluid to infect the zygote. The virus could then synthesise a DNA copy by means of reverse transcriptase, and allow it to be integrated into the germ line DNA. Steele goes even further and argues that certain endogenous viruses have been naturally selected for their intimacy with their hosts and consequent utility as vectors and transducers of useful somatic mutations.[42] This is not by any means 'pure fiction' since the existence of viruses which can effect such transductions has been experimentally demonstrated, and investigations into the extent of these transductions have only scratched the surface of the phenomenon. Steele notes that the chemical-Lamarckist interpretations of Guyer and Smith's early experiments were consistent with his own ideas. However, he then generalises his hypothesis to the point of collapse by having it explain the evolution of the central nervous system as an accumulation of adaptive somatogenic events.[43] This might be permissible in explaining the refinement of sensory organs or the build-up of instinctive behaviour. The central nervous system provides adaptability, not adaptiveness, but Steele does not recognise this limitation of the Lamarckist doctrine. Organisation cannot be conferred by a random acquisition of structural genes.

Experiments conducted by Steele and his co-worker R.M Gorczynski (1980, 1981) purported to prove that a particular immune response could be transmitted through male mice to their offspring. These results have been challenged and have not been found repeatable by other investigators. L. Brent and his co-workers conclude that while,

> the immune response and its specific diminishment (tolerance) are admirably suited to make the basis of a test for Lamarckism. . .All examples of supposedly Lamarckian inheritance proposed so far have been mistaken or open to more plausible explanations, especially in terms of selection.[44]

While Brent can say on the basis of his newest evidence, that, 'in respect of the systems we have been investigating, Lamarckian inheritance does not obtain', the door remains ajar, as it always must, since that is the nature of scientific explanation.[44] Steele understands that the metaphysical battle lines match the individual's role in evolution against the gene pool's, and so, while he may be prepared for experimental disproof he should also be able to understand the subjective fury of some of his other critics. The publication of Gorczynski's and Steele's results was accompanied by a *Nature* editorial reminiscent of *The Times* in the Twilight of Empire.[45]

Despite Lamarckism's disreputable associations, successive generations of biologists have felt the need to embrace it, giving the lie to its epithet as a doctrine that grew old and died with its supporters. A concept that gives organisms a direct role in their own evolution appeals more strongly to many biologists than the aleatory process of neo-Darwinism, which depends on the random appearance of useful novelty, as by the throw of dice. However, the metaphysical battle lines do not coincide cleanly with the Lamarckist and Darwinist positions. The ranks of those who stress the evolutionary role of the individual organism include the whole motley crowd that *Nature* regards as the enemies of scientific enlightenment, biologists both transcendental and realistic who have been interested in epigenetics, internal coordination, physiological adaptability and behaviour.[46] Selectionism, whose demological propaganda deny the individual's role, has attracted and welcomed an equally motley crowd of neo-Democriteans, reductionists, selfish-geneticists, Panglossians and nihilists, as long as they have been prepared to accept natural selection as the one and only cause of evolution. The blurring of the battle lines has made it difficult for some biologists to decide which side they are on, and which they want to belong to, and in the anxiety to choose, some have taken to Lamarckism *faute de mieux,* or through induction by elimination, seeing it erroneously as the only alternative to selectionism.

This conflict is placed in the larger context of the social sciences and humanities by Jacques Barzun's *Darwin, Marx, Wagner* (1958). The antagonists are identified as over-simplified Rationalism and Voluntarism; reason in opposition to feelings, an unwinnable conflict since neither can exist without the co-existence of the other. Similar polarisations are found in Pirsig's classical motorcycle mechanic and the romantic musician, and in Roszak's scientism versus humanism, which seek a synthesis of the contradictions but fear that the grand illusion of pure Rationalism has prevailed and driven its opponents to untenable positions.[47] In *Personal Knowledge* (1958) Michael Polanyi argued that the mistaken belief in a pure, objective impersonal knowledge, or scientific detachment, was to blame for the 'massive modern absurdity' of naïve positivism that dominates twentieth-century thought, 'it exercises a destructive influence in biology, psychology and sociology, and falsified our whole outlook far beyond the domain of science'.[48] Laplacean mechanicism, which T.H. Huxley espoused, held that an objective universal mind with knowledge of the positions and movements of all bodies large and small, down to the atoms, could predict future events including human behaviour. This for Polanyi was a conjuring trick which equated knowledge of physical data with knowledge of human experience, and the Laplacean mind could really understand nothing because what it knew meant nothing. The consequence of the Laplacean delusion was that,

> a discovery will. . .no longer be valued by the satisfaction which it gives to the intellectual passions of scientists, but will be assessed according to its probable utility for strengthening public power and improving the standard of living. . .This is how a philosophic movement guided by aspirations of scientific severity has come to threaten the position of science itself. This self-contradiction stems from a misguided intellectual passion — a passion for achieving absolutely impersonal knowledge which being unable to recognise any persons presents us with a picture of the universe in which we ourselves are absent.[49]

The alternative is to recognise that knowledge is unavoidably personal. Words have 'personal coefficients' of confidence, scepticism, irony, sarcasm or misunderstanding, determined by the epistemological matrix of the individual who uses them, that is the complex of memory, culture, tradition, training and psychological conditioning that consciously and tacitly characterises the individual mind. Even quantitative expressions and mathematical symbols are chosen as a result of imprecise mental activities. Therefore, any system of philosophy that sets up strictness of meaning as an ideal is bound to be inconsistent. Lack of absolute objectivity in interpersonal communication is something we all have to live with, but as the new realists and Woodger were well aware, the problem of imprecision becomes serious at the interdoctrinal level, especially if the interlocutors believe that they are in possession of pure, quantifiable objectivity. A selectionist gives a certain meaning to 'natural selection' based on an epistemological matrix that I have been trying to clarify in the earlier chapters. Its meaning, i.e. its various definitions, its epistemological aura and psychological affectiveness, is comprehended by a large body of other selectionists who share the same matrix. But the complexity of meaning makes the concept extremely baffling to those who are shut out of that matrix whether voluntarily or for lack of the appropriate conditioning. The opinion that it is the consensus of informed biologists that natural selection is the cause of evolution tells us more about the selectionist matrix than it intends to convey. As Polanyi said,

> Our objectivism, which tolerates no open declaration of faith, has forced modern beliefs to take on implicit forms. . .And no one will deny that those who have mastered the idioms in which these beliefs are entailed do also reason most ingeniously within those idioms. [50]

It takes a crisis of conscience to break out of a strong belief web; its believers regard its all-sufficient interpretative powers as evidence of its truth, only appreciating its speciousness if they lose it. Polanyi was referring to Marxism and Freudianism, but the same arguments apply to selectionism. The stability of such belief systems was due to the piecemeal refutations of objections to them: 'The convincing power possessed by the interpretation of any particular new topic in terms of such a conceptual framework is based on past applications of the same framework to a great number of other topics not now under consideration'. [51] For example, if we were to ask how a phenomenon could possibly be regarded as a case of adaptation to environment the answer would be that many such examples worried Darwin himself and have ultimately been demonstrated as adaptations. There is therefore no reason not to believe that in time the unknowns in question will be discovered to fit the selectionist scheme. This may indeed be so, but this does not prove the universality of adaptation. Circularity, another aspect of such belief systems, was introduced by Darwin himself: for the purpose of proving natural selection evolution was taken as axiomatic, and for the purpose of proving evolution natural selection was taken as axiomatic. In reality artificial selection was the only reliable axiom. Polanyi also argued that such belief systems increase their stability by expanding to absorb inconvenient truths: 'All interpretative frameworks have structure which supplies a reserve of subsidiary explanations for difficult situations.' [52] For example, there is a different kind of natural selection to suit every occasion, and each apparent exception is labelled as a 'new' kind of selection added to the 'synthesis'. The last defensive characteristic of the belief-web is to deny rival conceptions the room to develop. A new conception can only be established by a series of relevant examples, but if these are dealt with one at a time by an *ad hoc* extension of the central system there is no opportunity for the concept to germinate and give rise to a sufficient number of growing points that could provide a plausible body of evidence with enough impact to alter the belief-web.

An analysis of dogmatic belief systems, or idols was conducted by Francis Bacon in *Novum Organum*. His theme was in the tradition of the Mediaeval nominalists, but he attributed his inspiration to Heracleitus. *Idols of the Tribe* were delusions based upon 'common sense' such as 'seeing is believing'. *The Idols of the Market* were founded on the consensus of public opinion, and included its tendency to reify the non-existent. *The Idols of the Den* included self-delusion in the absence of more objective critical opinion. *Idols of the Theatre* might arise from any philosophical system that became fashionable. The new realists expounded similar thoughts in their manifesto. Possibly the best known modern discussion of the Idols of the Market and Theatre is that of Irving Janis, who calls these phenomena *groupthink,* which obviously owes a debt to Orwell's *doublethink,* a system of political self-delusion described in *1984*.[53] The characteristics of Groupthink are the illusion of invulnerability; group rationalisation of uncomfortable data; stereotyped views of rival or external ideas as too ignorant or uninformed to be worthy of consideration; discouragement of internal dissent; self-censorship resulting in the shared illusion of unanimity; and the self-appointment of 'mindguards' who protect the group from the dangers of external disturbance and internal deviation. Many of the attributes of selectionism can be recognised here, though the group has a measure of internal dissent aimed against the narrower gradualistic orthodoxy, as well as an increasing tendency to look into some of the historical and philosophical alternatives. Kuhn has pointed out that in times of crisis normal scientists are more inclined to look beyond their narrow, puzzle-solving activities in an attempt to understand the epistemological foundations of their insecurity, so here again may be a sign of impending paradigm-shift.[54]

Group-spotting is itself an Idol of the Theatre, and it goes without saying that because there is a group it need not suffer from all the negative characteristics of Groupthink. Most scientists feel more comfortable as members of groups, and internal dissidents either meet in sub-groups to discuss their ideas out of sight of the mindguards, or, if they leave the group, they try to form liaisons with others of their kind. If they are evolutionists, they may head in the direction of the pre-established Lamarckist group. Kuhn condemns those who depart the paradigm before the proper time to exile from their science. The sentence can be commuted if a rival paradigm can be formed or found, and this may be a better explanation of the perennial durability of the shadow paradigm than faith in the second coming of the inheritance of acquired characteristics.

Narrow-minded positivism has helped to create an epistemological no-man's land where speculations of all kinds can flourish, as well as creating a general hunger for a lusher, more romantic kind of explanation. At the outset I suggested that current controversies might represent a period of crisis and re-equilibration such as Kuhn describes as the prerequisite for the formulation of a new paradigm. However, if Kuhn's comment on professional insecurity is examined properly in context the fit is not very close.

> Because it demands large-scale paradigm destruction and major shifts in the problems and techniques of normal science, the emergence of new theories is generally preceded by a period of pronounced professional insecurity. As one might expect, that insecurity is generated by the persistent failure of the puzzles of normal science to come out as they should. Failure of existing rules is the prelude to a search for new ones.[54]

In biology the puzzles of population genetics continue to come out as they should. The puzzles of the fossil record have always demanded some alternative explanation to gradualistic neo-Darwinism. The search for new rules, or alternative explanations, began as soon as *Origin of Species* was published, and its inadequacies had been quite clearly outlined by Kölliker, Mivart and the neo-Lamarckists before the turn of the century. Does Woodger's admonition that we should find our biological Galileo before we identify our Newton still apply to biology; is evolution theory still in the pre-paradigm phase? Since so many alternatives to neo-Darwinism have been available historically it is interesting to consider why there has been a relatively sudden reawakening of interest in some of them, especially since there have been no unexpected anomalies in the neo-Darwinist paradigm.

William Provine's explanation of the analogous shift in opinion in the 1930s is that it was not the novel and unequivocal findings of theoretical genetics, nor even the authority of their originators, that carried the day, so much as the popularisers who translated the theory into terms that the educated public could understand, and that other biologists could use as models of explanation for their students.[55] Arthur Koestler and Alister Hardy, who were popularisers of the 1960s, certainly made doubters of some of us, Koestler renewing the old concepts of 'reculer pour mieux sauter' in epigenetics, and the 'systems' organisation of organisms, Hardy reawakening interest in behaviour as an evolutionary mechanism and bringing the significance of genetic assimilation and Baldwin Effect to the fore. For most biologists the two sailed too close to Lamarckism for comfort; and their dabbling in extrasensory perception and Jungian psychology created the same odour that had clung to the spiritualism of A.R. Wallace and Henri Bergson. C.H. Waddington equivocated on the evolutionary significance of genetic assimilation, and was reluctant to go 'Beyond Reductionism' with Koestler at the Alpbach Symposium in 1969.[56] However, as an original contributor to epigenetics and theoretical biology, and as a populariser of these contributions, he must be regarded as a major stimulus of present demands for broader, or more holistic interpretations of evolution. Some of the current batch of PhDs who were born after the publication of *The Strategy of the Genes,* inform me that they are 'rediscovering' Waddington. The implications of allometry and heterochrony and its particular manifestation of neoteny have appeared in Koestler and Waddington, and might easily have been inferred from the readily accessible *Embryos and Ancestors* of Gavin De Beer, as well as from the frequently cited passages by D'Arcy Thompson. However, many modern biologists attribute these concepts most directly to S.J. Gould's *Ontogeny and Phylogeny.* Here again is an original contributor to the evolutionary literature, at the centre of the punctuated speciation controversy, and a populariser who explores the fringes of orthodoxy. On the face of it, Provine's hypothesis holds up under current conditions. However, the belief-web of neo-Darwinism is far from being breached, and even such 'shifts of emphasis' as Stanley's *Macroevolution* requests are far from acceptance.[57]

A real shift of emphasis would take us into Løvtrup's fourth, epigenetic theory of evolution, but the final question that I must address is whether or not this element alone is adequate for a comprehensive theory. In a way Løvtrup has created a trap for himself since a comprehensive theory is always what lies beyond the next rise, after the present, recognised summits have been surmounted.

17 PUTTING HUMPTY DUMPTY TOGETHER AGAIN

> Formal operations relying on *one* framework of interpretation cannot demonstrate a proposition to persons who rely on *another* framework. Its advocates may not even succeed in getting a hearing from these, since they must first teach them a new language, and no one can learn a new language unless he first trusts that it means something. A hostile audience may in fact deliberately refuse to entertain novel conceptions. . .precisely because its members fear that once they accepted this framework they will be led to conclusions which they — rightly or wrongly — abhor. Proponents of a new system can convince their audience only by first winning their intellectual sympathy for a doctrine they have not yet grasped.
>
> Michael Polanyi 1958[1]

The American New Realists attempted a synthesis of the conflict between analytical reductionism and holism by asserting that the former represented a valid epistemological or empirical technique while the latter was needed to integrate the products of that technique. However, the likely consequence of the aphorism that Humpty Dumpty does not have to be put together again, provided that it is understood that Humpty whole is something more than Humpty in pieces, is that Humpty will be left forever strewn around the landscape. I am going to propose in this final chapter that Humpty whole is a necessary element of evolution theory, and that emergence doctrine provides a means of integrating information on variation and its evolutionary consequences in a holistic manner that is in contrast to the *ad hoc* approach of selectionism. Since emergentism is not an entirely novel conception, and since 'emergence' continues to be used in a vague way to signify the bothersome evolution of phylogenetic archetypes, indicating an intuitive recognition of the problem, it may not be necessary to teach a new language, nor to bring on the full complement of intellectual siege-engines that Polanyi thinks is necessary for the breaching of belief systems.

Teilhard resurrected transcendental emergentism in *The Phenomenon of Man* (1959). *Personal Knowledge* also advocated

338

emergentism, Polanyi intending that his post-critical philosophy re-equip us with the faculties which centuries of critical thought have taught us to distrust, in the expectation that once we

> have been made to realise the crippling mutilations imposed by an objectivist framework — once the veil of ambiguities covering up these mutilations has been definitely dissolved — many fresh minds will turn to the task of reinterpreting the world as it is.[2]

Bringing his philosophy to bear upon life, he concluded that there were ascending levels of existence and behaviour; that systems must be understood as wholes as well as analytically; that the personal involvement of the investigator with the higher levels of existence was unavoidable, and that it was logically impossible to compare living systems, to which criteria of success and failure apply, with systems that had no such distinctions.

> Accordingly, it is as meaningless to represent life in terms of physics and chemistry as it would be to interpret a grandfather clock or a Shakespeare sonnet in terms of physics and chemistry, and it is likewise meaningless to represent mind in terms of a machine or of a neural model. Lower levels do not lack a bearing on higher levels; *they define the conditions of their success, and account for their failures, but they cannot account for their success, for they cannot even define it.*[3]

Polanyi then suggested that the opposition to Darwinism of the emergentism of Morgan and Alexander was evidence of its rectitude, since such 'only a prejudice backed by genius can have obscured such elementary facts'.[4] Emergent events could have been overlooked because they were masked by the ephemeral genetic variations that neo-Darwinists assumed to be the material of evolutionary progress. Progress could not however be regarded as merely adaptive except in the context of 'a continuous ascending evolutionary achievement'.[5] Although it was not his purpose to categorise emergences, nor to trace their evolutionary history, he mentioned the organisation of the prokaryote, the emergence of the eukaryote, sexual repro-duction, the emergence of centralised nervous systems, the appearance of consciousness, and the development of the noo-sphere, as emergences in the course of evolution. The whole of this progress was the product of an 'orderly innovating principle', the action of which was released by random physicochemical external events, with the corollary that a favourable environment was necessary for its sustenance.[6] The concept of the emergence of centres of self-control, subjective interest and judgement supported the philosophy of personal knowledge by reducing the meaningless-ness of the universe, and if the adduction of other than physical principles to evolution was vitalism, then it was also common sense. Darwinism's error was that it investigated the conditions of evolution and overlooked its action as progressive emergence. Polanyi also insisted on unpredictability as a fundamental character of emergence. Unpredictability is an ineffable quality with romantic appeal, but it should not be allowed to mystify. The difficulty in pinning it down is that all of the evolution that we are familiar with has already happened, and successful predictions are based upon familiar events in homologous or analogous systems. Ayala, following Harrison, has argued that the emergence problem is easily circumvented if, for example, instead of wondering at the unpredict-ability of the water molecule in terms of the qualities of hydrogen and oxygen we simply list the qualities of water as properties of hydrogen and oxygen in combination with one another.[7] This does not do away with the need for the concept of emergence, but it spreads the problem so thinly that it is less obtrusive.

K.R. Popper (1972a) has also addressed the problems of episte-mology that exercise Polanyi. Quantum mechanics had rescued humanistic physicists from 'the nightmare of the physical determinist', the conviction that, 'all our thoughts, feelings, and efforts can have no practical influence upon what happen in the physical world: they are, if not mere illusions, at best superfluous by-products ('epiphenomena') of physical events'.[8] Mechanistic evolu-tionism lends itself to similar sentiments, and Popper's alternative to the biological Laplacean delusion is a rational account of emergent evolution, 'a view of evolution as a growing hierarchical system of plastic controls', subsuming neo-Darwinism, with mutation treated as accidental trial-and-error gambits, and natural selection the eliminator of the erronious.[9] Popper's scheme proposes that organisms are fundamentally problem-solvers, proceeding by testing new reactions, structures, functions, behaviours and ultimately hypotheses. Here the individual organism is the spear-head of evolution, and the solution of one problem through the success of a new approach brings it to the next problem. The evolutionary solutions are essentially plastic controls, i.e. new functional-morphological elements that have a feedback loop to the lower levels in the organisatory hierarchy. Any new element might fail for intrinsic incongruity or because of environmental circumstances, and would be recognised as a successful emergent only with hindsight, just as problems are only recognised with clarity when they have been solved. For example, the solution of a problem of effective reproduction creates the problem of overcrowding. In unicellular organisms one solution to this new problem was to become multi-cellular, and share the available resources more effectively. This brings in the kind of problem-solving hinted at by Ashby, when he said that by a change of goal a problem could itself become a solution. In this case the 'problem' of the uni-cell being unable to go off and do its own thing becomes the 'solution' that potentiates metazoan and metaphytic evolution. Popper recognises this implication when he discusses the organism's 'choice of aims'.[10]

In the original formulation of this thesis Popper incorporated genetic assimilation as a necessary corollary, since not all solutions could have an immediate hereditary basis.

> The behaviour of the organism is thrown up experimentally as a probe by the physiological system and yet controls, largely, the fate of this system. Our conscious states are similarly related to our behaviour. They anticipate our behaviour, working out, by trial and error, its likely consequences; thus they not only control but they try out, *deliberate*.[11]

At times conscious choices even override the fundamental demands of survival. Popper remarks that the greatest hope of the hopeful monster resides in monstrous behaviour rather than epigenetic, anatomical deviation.[12] The 'ethological' monster might arise from a mutation in the nervous system or from novelty in the environment, either case producing unusual or innovative behaviour.

Another realistic treatment of emergence, or 'hierarchical jumps' is provided by John Platt (1970), who once more argues that a new hierarchical level can be attained because of new materials or information or organisms in the environment, which destabilise the orginal structure to the extent that it can only survive by reorganising in such a way as to incorporate the new elements. A common feature of sociological emergences is that they are preceded by cognitive dissonance, which means a condition provoked by an unresolved conflict or dissonance of two pieces of plausible information.[13] The initial response to this is often a nervous or anxious inaction; then the ultimate choice is rationalised as the best one. Kuhn observed that anomalous experimental results produce professional insecurity; a 'cold sweat' may be the physiological result. So, at the organismic level, physiological or behavioural responses to novelty may be unable to meet the exigency or preserve the pre-existing organisational integrity. Platt argues that the ensuing disequilibration affects the entire system: this would certainly be true of physicochemical physiogenesis. Next the adjustment, when it comes, is sudden; how sudden in biological terms is hard to assess, although physiogenic change can be instantaneous, due to the lack of a compensating mechanism, or by the fatigue of an inadequate one. Simplification, in Riedl's sense of decrease in the number of interconnecting decisions required to sustain a stable system, is also required to sustain stability. Related kinds of simplification are Pattee's (or Aristotle's) 'separation of offices', Ashby's semi-independence of subsystems in the hierarchy; and the most straightforward expression of simplification: regression (Lwoff). Finally, Platt reiterates the proviso that self-generated hierarchical changes are impossible because of homeostasis, and so the system must go outside itself to reach a new grade of organisation. Taking the opinions of the modern emergentists in account, along with the topics discussed in *The Homeostasis Paradox* it is possible to begin to outline the aromorphic process of emergence (Figure 17.1). The initiating causes in this scheme are direct changes in the physicochemical environment, and changes in the behaviour of the organism that bring it into contact with a changed environment. The immediate effect is functional-morphological disequilibration.

The theme of the system being disequilibrated from without, before reorganising within has been echoed often in the history of biology. Modern theoreticians like Bohm and Jantsch examine the simplest physical models of disequilibration, such as the semi-stable vortices in turbulent water flow, or the semi-stable flow patterns in a container of water that is being warmed.[14] This has led to the metaphysical conception of what Popper would call a 'tiny baby problem'.[15] Any tiny baby may be ugly, red, noisy, leaky and beautiful only to its progenitors. The metaphor was, however, prompted by the apocryphal tale of the unwed mother who replied to her censorious family, 'It may be a problem but it's such a tiny one'. Jantsch, for example, seems to imply that non-equilibrium thermodynamics provides a universal model that can be applied at all

Figure 17.1: The Elements of Aromorphosis Leading to Grade Evolution

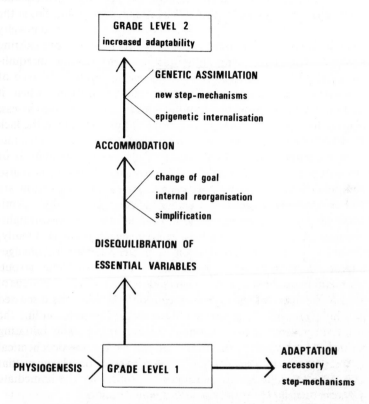

evolutionary grades from the prebiotic to the spiritual. Moreover the causation is autocatalytic, hence 'self-organising', or 'immanent', as some of the early emergentists suggested. I do not exclude this possibility: spontaneous random mutation, duplication and transposition are arguably immanent events, but all of the examples of physiological emergences that I have discussed resulted from the interaction of the organism with its environment, and physiogenesis is a distinctly exogenous or 'transeunt' cause. Jantsch's tiny baby matures into a preformationistic world view:

> The evolution of the universe is the history of an unfolding of differentiated order or complexity. Unfolding is not the same as building-up. The latter emphasizes structure and describes the

emergence of hierarchical levels by the joining of systems 'from the bottom up'. Unfolding, in contrast implies the interweaving of processes which lead simultaneously to phenomena of structuration at different hierarchical levels.[16]

Thus *evolutio,* in the old-fashioned preformationistic sense, is explicitly contrasted with an epigenetic, emergentistic interpretation. Although Jantsch eschews neo-Darwinistic reductionism, the tiny baby in autocatalytic non-equilibrium thermodynamics is a romantic reductionism, which in the manner of a baby cuckoo has heaved the organism, a static structural concept, out of the nest, just as effectively as the selectionists' selfish genes.

On the other hand, Jantsch's world view is imbued with a dynamic sense of process that is absent from most earlier writings, and the components of his evolutionary process are essentially those that I feel are strongly suggestive of epigenetic emergentism. Moreover, when he says, 'the levels of global stability or autopoietic [self-regulating and self-repairing] existence reached along this evolutionary path are not predetermined but result partly from the interaction between the "system and environment"', he breaks his metaphysical restraints, and includes the associative or 'horizontal' elements of evolutionary causality, which constitute the concluding element of my present discussion of emergent events.[17]

Interorganismic associations range from endosymbiotic mutualism, through viral transduction, to commensalism and simple co-evolution, such as the predator-prey relationship, or what C.L. Morgan called 'selection proper'.

Meyer-Abich (1964) proposed that *holobiosis,* a symbiotic relationship in which the several symbionts remain in intimate association in the cytoplasm or nuclei of the germ cells, was one of the fundamental causes of emergence. Under this heading would come the endosymbiotic theory of the origins of mitochondria and chloroplasts. The first known holobiont was the lichen, whose condition was discovered by Schwendener in 1860. This association of alga and fungus not only co-operates in the production of metabolites that cannot be synthesised by its individual algal and fungal relatives, but also forms unique reproductive bodies that guarantee the continued association of the two partners in subsequent generations. The sponges were also proposed as holobionts. Their four cell types fall into two different physiological classes, the flagellated choanocytes and the rest. Meyer-Abich was persuaded by the work of Farkas (1929) that the belief that all of the cell types came out of the same sponge germ cells was an over-simplification, since the amoeboid germ cells move through the sponge body, phagocytosing, but not digesting, choanocytes, which then give rise to the new cell line of choanocytes in the next generation. Pfeffer (1881) and others had argued that plant plastids were holobiotic, and Meyer-Abich speculated that not only were other cytoplasmic organelles of the same nature, but nuclei were also polyphyletic holobioses. In the same vein he postulated that higher plants had their origins as algal-fungal holobioses, the fungal elements being critical in the invasion of the land as rooting and absorbing organs. The higher metazoa might also have had early holobiotic origins: 'A higher rank of organismic organization during phylogenetic evolution is attainable only by producing new organism wholeness by holobioses'.[18] The principle advantage of these intimate associations was, he argued, novelty in energy-sharing. As I suggested earlier, a holobiosis in which the waste-product of one associate could be used by the other for the synthesis of a molecule essential for both would be of immediate utility. Meyer-Abich also implied that the differentiation and specialisation of tissues in metazoa was facilitated through the polyphyletic diversity of the original holobioses, a concept now supported by our knowledge of transposable regulatory genes.

Even as recently as 1964 Meyer-Abich's theories would have been regarded as flights of fancy by biologists who believed that organelles were somehow derived by complexification of the primi-tive cell. The popular autogenic hypothesis, or *direct-filiation theory* of mitochondrial evolution proposed the invagination of the prokaryote cell membrane, and only by some fancy topological foot-work did it account for the mitochondrial double membrane, leaving unexplained its apparent ability to reproduce itself independently of the nucleus. In the last fifteen years, however, the plausibility of the alternative holobiosis theory has gradually filtered into the consciousness of biologists, largely through the persuasive powers of Lynn Margulis's *Origin of Eukaryotic Cells* (1970), and Lewis Thomas's *The Lives of A Cell* (1974). In *Symbiosis in Cell Evolution* (1981) Margulis provides a historical introduction to holobiosis. Although she does not use Meyer-Abich's term it is useful for the most general category, and has both priority and clarity of definition. Margulis cites Mereschkovsky (1905 and 1909), Portier's 1918 *Les Symbiotes,* Wallin's 1927 *Symbionticism and the Origin of*

Species, and a 1952 review by Lederberg as the most significant expositions of the 'disreputable' theories of holobiosis, noting that their disrepute lay mostly in the absence of any scientific proof. From the point of view expressed in this chapter, Wallin's was the most interesting of the earlier contributions since he did not accept natural selection as the cause of evolutionary progress. Symbionticism, the establishment of intimate microsymbiotic complexes, was the fundamental factor involved in the origin of species, since by this means new genes could be acquired by the host organisms, and the physiological and morphogenic interactions between symbionts and hosts was already a matter of record. Nevertheless, Wallin did not believe that symbionticism together with natural selection could bring about progressive evolution, and like many of his contemporaries looked to a third, unknown factor as its cause.

What F.J.R. Taylor has called the *serial endosymbiotic theory* is outlined by Margulis thus:

> . . .mitochondria developed efficient oxygen-respiring capabilities when they were still free-living bacteria and. . .plastids derived from independent photosynthetic bacteria. Hence, the functions now performed by cell organelles are thought to have evolved long before the eukaryotic cell itself existed. Photosynthesis evolved in anaerobic bacteria in the absence of molecular oxygen very early in the history of life. The type of photosynthesis that gives off oxygen appeared later. Oxygen-respiring organisms evolved only after photosynthetically produced oxygen had accumulated in the environment. Many different metabolic mechanisms evolved to cope with and eventually utilize oxygen, a highly reactive and potentially poisonous gas. The nucleocytoplasmic host part of eukaryotic cells evolved separately from the organelles. It was neither photosynthetic nor adept at utilizing oxygen. However, ancestors of the nucleocytoplasm could withstand high temperatures and acidic conditions. Furthermore, the whiplike cilia absent in prokaryotes and nearly omnipresent in eukaryotes are thought to have derived from still another group of free-living bacteria. Eukaryotic cells are thus considered to have originated as communities of cooperating entities that had joined together in a definite order; with time, the members of the cooperative, already skilled in their specialties, became organelles.[19]

The evidence that at the DNA level mitochondria and plastids and nuclei are interdependent, the organelles being able to reproduce their own DNA independently, but having some of their functions coded by nuclear DNA, is paradoxical. From the endosymbiotic point of view it implies a somatogenic translocation of the symbionts' DNA. What passed from the cytoplasm to the nucleus-to-be was, however, preformed genetic material; not the inheritance of acquired characteristics, but *the acquisition of heritable characteristics*. From the direct-filiation point of view it implies that the end of independence of organelles from an original total dependence on the nucleus has somehow failed to be completed. This confused some biologists who might otherwise have accepted endosymbiosis on the grounds of strong circumstantial evidence. However, sequencing evidence of endosymbiosis provided by Schwartz and Dayhoff (1978) has converted many doubters. On the basis of sequencing of ferredoxin, ribosomal RNA and cytochrome C, chloroplasts of the Metaphyta are shown to have much closer affiliation with some unicellular algae and with the blue-green protists than with the protoplasm within which they reside. On similar evidence mitochondria are shown to have a close affiliation with some photosynthetic and aerobic bacteria. The case for holo-biosis has gone from strength to strength, and the position of direct-filiation has considerably weakened.

In 1975 Pirozynski and Malloch proposed a botanical evolutionary symbiotic hypothesis:

the colonization of land and indeed the very evolution of plants (and, indirectly, of animals and 'higher' fungi) was possible only through establishment (unique or repeated) of symbiotic association of a semi-aquatic ancestral alga and an aquatic fungus — an oomycete. In other words, terrestrial plants are the product of this ancient and continuing partnership.[20]

This is an explanation for the mysterious 'sudden' appearance in the late Silurian of advanced forms of terrestrial plants, just as Margulis's theory explained the 'explosive' evolution of the three higher kingdoms of the animals, plants and fungi in the Cambrian. The algal member of this Silurian partnership was a fresh or brackish-water form which probably already possessed the ability to produce drought-resistant spores. The fungus was an oomycete found in late Precambrian deposits. A number of known modern symbiotic relationships between filamentous fresh-water algae and water moulds, such as *Spirogyra* with *Pythium,* indicate the likelihood of the Silurian holobiosis. The fossil nematophytes of the Silurian contemporaries of more complex vascular plants had spores reminiscent of oomycetes and mycelia-like threads, and may have been relatively unsuccessful holobionts. Mycorrhizal symbionts can be identified in Silurian fossils of *Asteroxylon* and *Rhynia,* forming associations of the kind that are formed in many modern vascular plants as well as in the lower plants. According to Gerdemann (1968) over 90 per cent of plants which have been examined are mycorrhizal, and the growth of the metaphyte is difficult if not impossible in the absence of the fungi. Most of the familiar woodland 'mushrooms' are in fact the fruiting bodies of basidiomycetes that form mycorrhizal associations with the forest trees. In some cases, for example the truffles, there is a triangular relationship between the tree, the truffle, and the animal which excavates it. One mycophilic squirrel digs the truffle, carries it into the tree, eats the edible outer rind, and drops the remainder, which bursts on impact with a branch and disperses its spores to the wind.[21]

As Pirozynski and Malloch pointed out, neither member of the symbiotic association that first invaded the land had the equipment for independent survival, since the alga had no means of efficient nutrient extraction from soil, and the fungus could not photosynthesise. Some modern fungus-metaphyte symbioses demonstrate a co-operative mechanism for resistance to dehydration. For example, in dry conditions the fungal hyphae and part of the host tissue of the liverwort *Pellia* die and the cells become filled with oil, providing a desiccation-resistant layer. In ferns similar oil-filled cells protect the meristematic tissues. Furthermore, although the alga would not be able to extract nitrates and phosphates from soil efficiently the fungal symbiont could penetrate soil and use extracellular enzymes for the release of nutrients from both organic and inorganic insoluble sources. The classical holobionts, the lichens, excel in conditions of desiccation and nutrient-poor substrates. Pirozynski and Malloch stress the conclusion that,

> It appears that from earliest times, the alga-fungus partnership, though mandatory, was rather loose; the fungus did not become incorporated into spores or seeds, which thus remained the product of the autotrophic partner. Presumably, the fungus gained independence by becoming saprotrophic very early in the

evolution of land plants and became the standard component of the advancing edge of the flora, thus eliminating the necessity to develop a more foolproof mechanism to secure the unfailing presence of the fungal component.[22]

Although the association has not advanced to the status of full holobiosis, it is not necessary for my argument that it should have done so; the case that I wish to make is that these mutualistic associations meet the definition of emergence in the sense that the two participants together have new, holistic qualities not possessed by the associates in their previous independent existence, and their novel qualities potentiated what has been called 'explosive' diversification and evolution. Kunicki-Goldfinger (1980) pursues a similar argument in some detail. Natural selection may properly be allowed to explain only the refinement of the association, and so the serial endosymbiotic theory does not, as Margulis claims, depend heavily on neo-Darwinist thought. It is the direct-filiation theory that is the natural epistemological ally of selectionism, and its adherents have regarded holobiosis as a disreputable theory because they tacitly know that if they accept it they will be led to conclusions that they would abhor. Historically, emergentists have been motivated by disatisfaction with selection theory, and in Margulis we find a closet emergentist trying to cram the product of a holistic intuition into an incongruous reductionist mould.

Another significant associative element in emergence is viral transduction, which Steele suggested as a mechanism whereby immunity could be transferred from one organism to another. Again, this would represent the acquisition of heritable characteristics, an established phenomenon which has become almost commonplace in the context of genetically tailoring bacteria with desirable enzyme-synthetic qualities. This may be of general evolutionary significance, along with Steele's assertion that some viruses could have been selected for their ability to benignly co-exist with, and stimulate evolutionary change in their host species through the orderly transduction of genetic material: vehicles that would extend the leaps of the jumping genes. Could the birth defects caused by some viral pathogens be the distortion of a normal evolutionary epigenetic process? To paraphrase W.C. Wells, is what genetic engineers do by science done with equal efficiency, though more slowly by nature? Anderson (1970) has observed that viral transduction across wide phyletic gaps such as between invertebrates and vertebrates is possible, and has suggested that this could be a mechanism of convergent evolution.

At a more profane level of discourse are the well-known and undisputed symbioses that are the foundations of several ecosystems, such as the algal-coral relationship that builds the great reef complexes in barren tropical seas; the chemo-autotrophic sulphide-oxidising bacteria that are endosymbiotic in gutless bivalves, providing the host with energy derived from the hydrogen sulphide of the anoxic environment.[23] Similar bacteria are the primary producers, both endogenous and exogenous, for the communities of deep-sea animals that have colonised the volcanic thermal vents of the Pacific.[24] Cellulose-digesting and other symbiotic bacteria and protozoa potentiated the evolution of the insects, and the mammalian herbivores and all of their ecological dependents. Thus, interorganismic associations are important from the level of basic metabolism to the level of societies.

With this final major associative element in place it is now possible to systematise the processes and mechanisms of emergent evolution.

Emergence has been used in a number of different senses by biologists, and some definition is desirable before attempting to place the concept in the context of evolution theory. The common usages are these, and I have added no. 7 in the light of the above discussion:

Emergence: 1. To come out as, in the hatching of an egg, or a fish coming out of water on to land.
2. An evolutionary novelty of any kind gradual or saltatory.
3. An advance in functional-morphological grade.
4. An auto-adaptive innovation, rather than an adaptation to environment.
5. An autonomous or autocatalytic novel product, (transcendental or material) of a living system (rather than one evoked by external conditions).
6. A novelty induced by the environment.
7. An association of the two or more organisms that has novel or enhanced qualities not possessed by the symbionts in their earlier independent existence.

Traditionally, emergence doctrine stressed autonomy and unpredictability in evolutionary innovation, but this imposed severe restrictions on a concept with greater holistic potential. Evolution cannot be understood by considering the whole organism and ignoring the whole physical and biotic environment that the organism inhabits. Physiogenesis is usually a necessary factor in physiological grade evolution, and the co-existence of potential associates is the prerequisite of emergence by association. Because of the variety of usages of emergence and the resultant contradictions, Lovejoy (1927) attempted to formalise the concept by saying that emergence could be said to have occurred if in comparing two time periods any of the following could be observed:

1. A new improved method of doing the same general thing.
2. A new quality for a pre-existing organism.
3. New types having some new qualities and lacking some old qualities.
4. New events irreducibly different from old ones.
5. A proportionate numerical increase.[25]

The more of these categories that could be discerned the greater the emergence. Essentially, Lovejoy was categorising the different kinds of biological change from simple adaptive changes to archetypal functional-morphological ones, bringing us back almost full circle to the 'revolt of the clay against the potter', and the nineteenth-century biologists who insisted that in order to understand evolution the different qualities of variation must be understood. It is still a tenable position, since selection theory remains content to deal only with the consequences of variation. Also, the evidence from the fossil record that indicates that speciation may include rapid initial change followed by stasis suggests that something beyond natural selection is going on, but sheds no light on the significant qualities of biological change. The comparative physiologist C.L. Prosser has mentioned that evolutionary emergence is given meaning by the fact that qualitative differences exist between integrated and disintegrated systems; the fact that multiple forms of the same gene or its product can give rise to a variety of functions through a variety of mutations; the importance of the whole organism in the context of behaviour and response to environment, and finally the possibility that biological concepts such as homeostasis, species and population dynamics may have a holistic and heuristic significance analagous to philosophical generalisations such as beauty and value.[26] There are

obviously different degrees of emergence in terms of their consequences for evolution, or in orthodox language, in terms of their
fitness, although advance in grade may not always be immediate
manifest as reproductive success.

Table 17.1: Some Major Biological Emergences from the Origin of Life to the Present

Linguistic functions
Logical functions (conscious and tacit)
Consciousness
Memory (long term)
Higher neural-sensory integrations (e.g. hand-eye, audiovisual)
Thermoregulatory integrations
Viviparity
Emergence of animals on to land
Emergence of plants on to land
Entry of primitive fish into fresh water
Neotenic changes in pre-vertebrate evolution
Segmentation
Acquisition of coelom
Acquisition of gut
Nerve myelination
Neurotransmitters evolution
Striated muscle-fibre evolution
Gastrulation
Multi-cellularity
Eukaryote evolution by endosymbiosis
Photosynthetic oxygenation of biosphere
True reproduction
Association of abiogenic protobionts

Lovejoy's scheme, encompassing everything from minor point
mutations and population shifts to archetypal saltations, is just too
comprehensive, and it is desirable to focus more narrowly upon the
events which are associated with hierarchical emergences. These
include the following: behavioural actions, environmental effects,
DNA acquisitions, and mutualistic associations that, (1) cause
pervasive changes; (2) cause sudden changes; (3) provide instant
benefit; (4) feed back to alter the environment. The events associated
with the disequilibration of these changes are internal reorgani-
sations that reduce energy costs and maximise flexibility (Ashby,
Whyte, Reidl) and the various kinds of simplification discussed
above (Ashby, Reidl, Patee, Hardy, Koestler, Lwoff). In Table 17.1
are listed some of the major biological emergences going back in

time from recent human history to the origin of life, events which satisfy at least three of the above criteria.

Table 17.2: The Causes of Emergent Evolution

	I Extrinsic Causes
Physical:	mutagenic: heat and radiation
	physiogenic: temperature, light
	epigenetic: temperature, light
Chemical:	mutagenic: free radicals, metals and metallic compounds etc.
	physiogenic: salinity, water (or lack of it), oxygen, carbon dioxide
Biological:	Loose co-evolutionary association: predators and prey, pollinators, fructivores, commensals, biocoenoses
	parasitic association
	societies: human, other mammals, birds insect colonies (protistan, lower invertebrates)
	multicellular associations
	endosymbionts
	viral DNA transducers

	II Intrinsic Causes
Organismic:	behaviour
	intelligence
	hormonal changes
	autocatalytic functional-morphological changes
Cellular:	recombinations
	structural gene mutations
	regulator mutations
	gene duplications
	gene transpositions
	chromosomal mutations
	polyploidy

Some of these events, it could be argued, might have been partly the result of a gradual accumulation of elements by natural selection, but for the most part they are macro-evolutionary changes that make little sense in terms of micro-evolutionary mechanisms. I have omitted numerous other possible candidates for emergence that have been important in grade evolution, such as the evolution of mammary glands, eyes, ears and eggs; the build up of the fundamental metabolic pathways of aerobic and sulphide-oxidising respiration; the emergence of the mitotic apparatus and sexual reproduction, the refinement of DNA transcription and translation, and the genetic code. Most of these have been subjects in the gradualist versus saltationist debates; but although they probably have some emergent, orthogenic, or 'internal selection' components they are at least partly accountable in neo-Darwinistic terms. Stebbins (1974) can think of seventy such controversial events of grade evolution; so few that selectionists have been justified in ignoring them.

It is impossible to identify any of the emergent events with single effective causes, since they mostly have associative, behavioural, functional-morphological, epigenetic, molecular biological and physicochemical components. Table 17.2 attempts to sort out the various causes in two groupings, the extrinsic (= exogenous, environmental, transeunt) and the intrinsic (= endogenous, immanent). (NB: a physiogenic influence is defined as an extrinsic physicochemical condition that causes an organismic physico-chemical change.)

Firstly it is clear that the individual intrinsic, effective causes at the cellular level may provide no more than the small changes that are the raw material for natural selection. Selectionists have in fact devoted most of their attention to structural gene mutations, and although all of the other cellular events are known to exist their implications have been largely ignored. Ohno was one of the first evolutionists to realise the implications of a molecular 'repetitive addition' mechanism, namely repetitive DNA or gene duplication. In Chapter 10, I proposed that this would provide a molecular mechanism for orthogenesis. Dover (1982) suggests a hypothesis of *molecular drive* which subsumes some of these ideas. This hypothesis is based upon the known occurrence of non-random repeated sequences of DNA which gives evolution a direction that is not subject to natural selection:

> The widespread fixation of variants by molecular drive is different [from natural selection and genetic drift] in that it is an outcome of a variety of sequence exchanges within and between chromosomes that give rise to persistent non-mendelian patterns of inheritance. Significantly, there are circumstances in which the activities of the genomic mechanisms, in spreading sequence information between chromosomes, would lead to the progressive increase of a variant through a family more or less simultaneously in each individual of a sexual population. This concerted pattern of fixation by molecular drive may provide an explanation for the origins of species discontinuities and biological novelty.[27]

These processes may also explain orthogenesis, and the phenomenon of differing degrees of 'adaptation' in similar species in similar circumstances that intrigued J.C. Willis. 'Inherent tendencies', and 'orderly innovating principles' are beginning to acquire materialistic substance. Molecular drive, gene transposition, and chromosomal mutations and position effects are now the subjects of intensive research. A brief review is provided by D.R. Brooks (1983).

Extrinsic and intrinsic categories are a rather arbitrary division. For example, as Schmalhausen argued, extrinsic epigenetic causes may be internalised; extrinsic physiogenic causes may be genetically assimilated. Once the preliminary emergent events such as the first association of previously independent symbionts has been established extrinsic causes become intrinsic; so also with the first multi-cellular associations. The intrinsic causes in certain organisms have effects that become the extrinsic causes for others: bacterial mutations may be transposed to other bacteria; viral mutations may be transposed to their hosts. Behaviour produces novel environmental encounters. The physiology of masses of organisms can bring about fundamental physicochemical changes in the environment, such as the depletion or provision of nutrients. The first photosynthetic production of oxygen caused a mega-evolutionary chain reaction, providing an ozone layer that reduced ultraviolet penetration into the biosphere, permitting the proliferation of life at the water-air interface, and putting a stop to the abiotic synthesis of organic compounds. The presence of the initially toxic oxygen resulted in the genetic assimilation of the aerobic condition, the development of aerobic respiration out of the detoxifying mechanism, the potential for the symbiosis of aerobic and anaerobic types and for ultimate aerial oxygen physiogenesis, and respiration. This in turn created the possibility of a metabolic rate that would ultimately support a conscious nervous system. Moreover, the organismic intrinsic causes are the effects of the cellular intrinsic causes; and some of the associative extrinsic causes are in turn the effect of organismic events such as the evolution of behaviour, locomotion, and sensory systems. While it is desirable to interpret the organism both as a self-sufficient whole that persists in its own being, and as a whole within the larger whole of the environment, some analysis is also useful, but the Aristotelian or Baconian approach of making lists and tables is totally inadequate, because of mutual feedback between causes and effects.

Figure 17.2: The Evolutionary Gyre

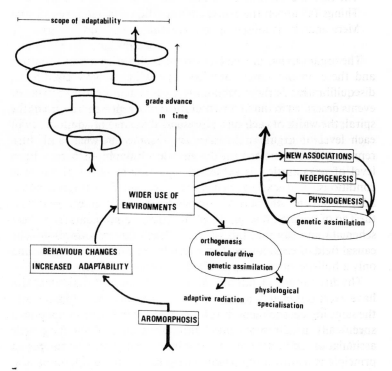

The ideograph in the top left illustrates the form of grade evolution. After each vertical aromorphic step organisms have an expanded potential for environmental exploitation, adaptive radiation and physiological specialisation. The horizontal element is represented (bottom right) as a cycle to emphasise the repetitive nature of the process. The cycle need not be completed before the next aromorphosis and a series of incomplete cycles would leave the impression of a large gap in the fossil record.

Figure 17.2 is an ideographic depiction of the cycle of interconnected elements of organismic activity following emergence to a new grade level of evolution. The cycle need not be completed before the next emergent step is taken, but parts of the cycle are repeated time and again at each grade level. With the addition of the dimension of time and the scope of adaptability the pattern becomes an irregular gyre, the symbol so popular with the transcendentalists. The first four lines of W.B. Yeats's 'The Second Coming' (1920) provide an apt metaphor for aromorphosis, and the whole poem a salute to the hopeful monster.

Turning and turning in the widening gyre,
The falcon cannot hear the falconer;
Things fall apart: the centre cannot hold;
Mere anarchy is loosed upon the world.

The organism has moved beyond the old constraints and controls and these are no longer able to cope with the new conditions; disequilibration is the attribute of the emergent new. The cycle of events described in the first part of the figure represents a turn of the spiral; the width of each turn represents the range of adaptability of each level of organisation, and the distance between each turn represents both time and the degree of evolutionary progress. Each level of organisation is subjected to a similar range of extrinsic conditions, and each turn repeats the pattern of exploration, adaptation, and interplay of cause and effect as the previous. This is the Tao of biology. It is impossible to isolate components from these schemata and say that this or that is the cause of evolution. The causal field of each component intermingles with many others, and only a holistic view can approach comprehension.

The difficulty of identifying distinct causes and effects should also have exercised selectionists, since differential reproduction creates the same logical problems. Natural selection has not been mentioned specifically in these schemes, although it is implicit in genetic assimilation and adaptive and specialised changes. The utilitarian principle is assumed to contribute to the spread of any biological novelty through a population. But correlating categories of emergent change with fitness values is a futile assignment unless it is possible to stretch the concept of inclusive fitness through time as well as through space. While the metaphor of selection pressure has the heuristic value of focusing attention upon areas of variation likely to be useful in particular circumstances, nature is often remiss in providing the predicted adaptations and has often advanced by secret paths: 'Expect surprises' is not an attractive slogan for positivists.[28] In short, although natural selection, like death and taxes, is always with us it no more explains evolution than death explains life nor taxes explain the production of wealth.

The biological changes that selectionists concentrate upon seem to be the least significant ones, and to make everything else subordinate to these, while serving the tacit demands of reductionism, is to persist in looking through the wrong end of the evolutionistic telescope, as well as being fundamentally anti-intellectual. Although a substantial number of biologists from Kölliker and Mivart to the present have understood this, the question remains why more have not. Ernst Mayr argues that even now Darwinism has scarcely managed to surmount all of the obstacles that lay in its path. Darwin's evolutionism had to survive in the struggle with the creationists and Cuvierians; his explanatory hypothesis had to contend with scientific critics and rival hypotheses such as mutationism, neo-Mendelism, neo-vitalism and Lamarckism. Darwin encountered professional jealousy and attracted a coterie of single-visioned disciples, but the centre held in the face of shifts of emphasis brought about by Mendelian, classical and population genetics. And the belief that unreconstructed neo-Darwinism is the all-sufficient explanation of evolution illustrates a survival of the polemically fittest.

A formal or even cohesive holistic theory of evolution has to face a similar array of obstacles, including the historical association of emergentism and holism with the 'disreputable' theories of Lamarckism, transcendentalism and dialectical materialism. Both Marx and Engels extended the Marxist view of human history as a discontinuous, revolutionary process to natural history and evolution. That Engels's thoughts on this subject came to the attention of a general public too late to have a fundamental epistemological impact was partly due to a historical accident, the death of Marx, which obliged him to turn his attention to political realities.[29] As Needham, himself a dialectical materialist, pointed out, Soviet opinion on the history of biology came as a surprise to English audiences. Nevertheless, emergentistic concepts had been developing in Germany, Britain and the United States outside Marxist ideology, though with common roots in Hegelian dialectic. I have also come to my present conclusions by an independent route. My choice of title and the explicit theme of dialectical progress was chosen only after the first draft had been completed, and I was able to step outside the set of concepts that had subjectively intrigued me. Engels imputed that this was the long way round; but he also admitted that such a route might be just as effective in the long term; Marx had had to follow it himself, deriving a system of thought from the evidence, instead of forcing the evidence into a doctrinaire mould. Skolimowski (1974) has noted,

> a variety of authors who have investigated the properties of complex hierarchical systems. . .conspicuously avoid mention-

ing dialectical materialism, and the canons of dialectical development, although quite often their analysis is carried along
lines very similar to those of Marxist dialecticians. Is this because
we have carried an understanding of the dialectical process of
nature far beyond the Marxist tenets, or is it because we still
remain in the ideological straitjacket which has imposed on us
certain conceptual and linguistic taboos? If the latter is the case,
and I suspect this is so, then this should make us aware how
vulnerable we are to the subtle pressure of ideology under which
we as scientists and philosophers happen to be born and live.[30]

There is a natural reluctance, however, to exchange one restraining
garment for another that, for example, under Lysenko's sartorial
influence, cuts off the circulation of ideas altogether. Dialectical
materialism has been used as a smear and a rallying slogan in recent
debates over sociobiology, punctuated equilibrium and cladism, and
it is deplorable that a particular epistomological view should auto-
matically be politicised or seen to be so.

A holistic or emergentistic approach courts accusations of
insulting the mythic hero of evolutionism, Charles Darwin. I use
'mythic' in the anthropological sense of imbuing the hero with
immaculate qualities of virtue and perfection that the original did
not possess, not in the sense that the original's abilities were fictionalised or false. Emergentism also confronts the general reluctance,
even among biologists, to think about evolution in general terms,
free from the reification of natural selection as the agent of
evolution, an axiom in which there has been so much investment of
time, emotion and research funds. Also in the ranks of opposing
forces are blind reliance on Occam's Razor, the metaphysics of
reductionism and positivism, and the Laplacean delusion. These
associations and obstacles make the task of persuading a hostile
audience much more difficult than the more simple task of revealing
a scientific factoid that makes the existing theories 'wrong'. The
problem is largely that selection *is* scientific explanation, but it
explains something other than what it purports to explain.
Emergentism and holism are *not* yet scientific explanation, but are
inductive guides that illuminate fundamental aspects of evolution
that have barely been considered. Polanyi suggests the rhetorical
approach of justifying refusal to enter into an opponent's way of
argument by making it appear altogether unreasonable. The danger
of this is that it escalates into an attempt to discredit the opposition

with accusations of gullibility, stupidity or dishonesty. There is ample historical illustration of this on both sides.

The politics of disillusionment with any belief system are fraught with dangers. The first is the temptation to fill the vacuum that has been formed with the system that has just been rejected since it is the one that is most easily available. A reconversion to the true faith is just as vehement as a simple conversion. Or, an alternative belief-web may be sought; Lamarckism has provided this refuge historically, and its appeal has lain as much in this as in its intellectual qualities. One of the results of disillusionment has been an incredible scholarly schism between transcendentalism and positivism. Each side has cultivated its own incestuous interests and felt no urge to pay attention to the arguments echoing faintly across the void. Moreover, scholarly specialisation leaves so little time for generalisation that it generates a peculiar intellectual gullibility, wherein issues are judged by proxy.

I have not tried to present any single new idea that would shatter orthodox complacency. Instead, having been tacitly persuaded of the power of a holistic treatment of evolution, I have chosen a historical epistemological approach to meet the objectives of gaining intellectual sympathy for certain doctrines and introducing the unfamiliar vocabulary that is required to understand them. By discussing the transcendental doctrines at some length I hope to have demonstrated that they need not be abhorred, that they have not only been necessary to exhaust certain possibilities that naturally occur to the human mind, that even if deprived of their revelatory dimension they retain heuristic qualities, and that the polarisation of metaphysics and biology is consequently absurd. As C.L. Prosser has suggested, 'the principles of emergence, if fully understood, could provide the bridge between the objective and subjective worlds'.[31]

The delusion of the finished synthesis places restrictions on freedom of thought of which its believers are unaware. Selectionists point to the internal debates as evidence of free discussion, but the freedom is bounded by the dead hand of Darwin. These debates need a real dialectical freedom to realise their full potential, which the broader context of evolutionary emergentism can provide. The belief that the synthesis is finished has caused dismay for the end of intellectual progress in some biologists, and smugness in those who believe themselves responsible for the finishing touches. To propose that intellectual progress in biology is possible for anyone who is prepared to shake the scales from their eyes may seem like dangerous mystical adventurism to the members of the neo-Darwinist establishment, but in the larger realm of the human intellect it should go without saying.

BIBLIOGRAPHICAL REMARKS

There are two categories of debt that I owe to the literature. The first is the catalytic writing whose textual use in the present work may seem insignificant, but has nevertheless been valuable. Kerkut (1960) stimulated many zoologists to consider their own evolutionary opinions more carefully. Waddington (1957), Hardy (1965) and Samuel (1972) demonstrated the breadth of historical and theoretical thought required for proper comprehension. Koestler (1967, 1971), Barzun (1958) and Himmelfarb (1959) challenged evolutionary orthodoxy, and Polanyi (1962, 1966), Croizat (1962) and Ohno (1970) were particularly stimulating to me.

The second debt that I especially wish to acknowledge is owing to secondary sources that I have exploited heavily. These are Løvtrup (1974-1983) Provine (1971 and 1978), Haraway (1976 and 1978), Gould (1977), Burkhardt (1977) and Gruber (1960). Others, such as Hitching (1982), Mayr (1982) and Keller (1983) came too late to have a formative influence on the first draft, but were useful in plugging some obvious gaps.

The acquisition of much of the historical reference material was due to the assistance of Betty Gibbs and the University of Victoria inter-library loan service.

In the reference list I have put the date of first publication after the title of some works, where appropriate. Where two editions or issues are referenced the second is the one for which page references are given in the notes.

NOTES

Introduction

1. Spilsbury 1974 p.46. **2.** The opening sentence is plagiarised from the introduction to Haldane 1932. **3.** Muller 1959. **4.** Polanyi 1966 p.47. **5.** Reid 1976. **6.** Macbeth 1971. **7.** Weismann 1883. **8.** Cf. Bock and Wahlert 1965. **9.** Darwin 1859. **10.** Gould 1977 p.177, citing Weismann. **11.** Wilson E.B. 1928 p.3, 11. **12.** Driesch 1908. **13.** Himmelfarb 1959 applies the expression 'subsuming the antithesis under the thesis'; Feyerabend 1978 uses 'self-immunising'. **14.** See Chapters 16 and 17.

Chapter 1

1. Marchant 1916 p.131, citing a letter from Wallace to Darwin, 1864. **2.** Needham 1943 p.32. **3.** Osborn 1929 p.143, citing Leibnitz's *Protogae* XXVI. **4.** Osborn 1929 p.188. citing Linnaeus's *Systema naturae*, 1762. **5.** Butler, 1879 p.90,91, citing Buffon's *Histoire naturelle*, vol. 4. **6.** Biographical notes on Lamarck are drawn from Brukhardt 1977. **7.** Osborn 1929 p.308. **8.** Owen 1894 p.248 ff cites a letter from Owen to Chambers 1844. **9.** Published as a monograph in 1818. **10.** Poulton 1908, citing Prichard 1826. **11.** Osborn 1929 p.308. **12.** Ibid. p.296. **13.** Ibid. p.311; Marchant 1916 pp. 140-3, cites a letter from Wallace to Darwin in 1866 urging the substitution of 'survival of the fittest' for 'natural selection', to get over the metaphorical problem. **14.** Darwin F. 1892 p.13 gives Darwin's own account. **15.** Himmelfarb 1959 p.155 discusses the contradiction. **16.** Hull 1973; see Introduction. **17.** Hull 1973 pp.6,7; also Himmelfarb 1959. **18.** Samuel 1972 p.115, citing Chambers 1844. **19.** Millhauser 1959 p.4. **20.** Ibid. **21.** Letter, Darwin F. 1958 p.196. **22.** Mayr 1964 p.xxi. **23.** Kohlbrugge 1915. **24.** The context of the idea of progress in relation to evolution is further discussed by Roppen 1956. **25.** Darwin 1859 p.489. **26.** Darwin F. 1892 p.226 citing a letter from T.H. Huxley to C. Darwin, Nov. 23, 1859. **27.** Ibid. p.226. **28.** Mayr 1970 p.5. **29.** See Schrödinger 1944 p.56 ff; also Cairns *et al*. 1966. **30.** Avery *et al*. 1944. Hershey and Chase 1952. **31.** Schrödinger 1944. **32.** Watson 1968. **33.** Chargaff 1950. **34.** Muller 1930.

Chapter 2

1. Spilsbury 1974 p.21. **2.** E.g. Moorehead and Kaplan 1967, Salisbury 1969. **3.** Popper 1959. **4.** Dobzhansky 1970 p.i. **5.** Lucretius 1969 (1st Century BC) p.191. **6.** Ibid. p.189. **7.** Ibid. p.189. **8.** Lyell 1830-3. **9.** Darwin 1872 p.60. **10.** Darwin 1859 p.61. **11.** Ibid. p.84. **12.** Spencer 1898 vol. 1, p.335. **13.** Argyll, Duke of 1886 p.335. **14.** Kimura 1961. **15.** Simpson and Beck 1965; Hardy 1965 p.135. **16.** Poulton 1908 p.xxii. **17.** Mayr 1970 p.119. **18.** Romanes 1892-1916 vol. 1, p.275. **19.** Lloyd 1914 p.44. **20.** This definition is traditionally attributed to Dobzhansky. **21.** Goudge 1961. **22.** Carroll 1871; see also Gardner 1965 p.306 **23.** Haldane J.B.S. 1932 p.4. **24.** Russell 1962 p.95. **25.** Cannon 1959. **26.** Thompson 1917 p.672. **27.** Carter 1954 p.191. **28.** Darwin 1859 p.184. **29.** Hudson 1870 p.160. **30.** Darwin 1859 p.186. **31.** Ibid. p.184. **32.** Huxley J.S. 1942 p.421. **33.** Bateson W. 1894 p.11. **34.** Willis 1940 p.21. **35.**

Cuénot 1914 pp. 67, 68 **36.** Carter 1954 p.275. **37.** Willis 1940 p.15. **38.** Darwin 1872. **39.** Ibid. p.107. **40.** Lloyd 1914 p.32. **41.** Whyte 1965 p.27. **42.** Grant 1977 p.262. **43.** Riedl 1978 p.262. **44.** Carter 1954 p.269. **45.** Grant 1977 p.264, citing Hardy 1954. Hardy attributed the idea to Garstang. **46.** Ohno 1970 preface. **47.** Bolk 1915. **48.** Koestler 1967 p.196. **49.** Stanley 1979 p.22. **50.** See Chapter 13 and Goldschmidt 1940. **51.** Eldredge and Gould 1972. **52.** Darwin 1859 p.298. **53.** Stanley 1979 p.142 **54.** Ibid. p.151. **55.** Thorpe 1974 p.111. **56.** Ayala 1974 citing Nagel 1961. **57.** Ayala 1974. **58.** Watson 1965 p.67. **59.** Grassé 1977 p.5; Spilsbury 1974 p.21. **60.** Thompson 1917 p.672. **61.** Feyerabend 1978 p.78. **62.** Huxley T.H. 1880. **63.** Lewontin 1978 p.215 discusses the evolutionary red queen hypothesis. **64.** Huxley T.H. 1908 pp. 349-50. **65.** Feyerabend 1978 p.78.

Chapter 3

1. Dixon 1885 p.20. **2.** Kölliker 1864; Huxley T.H. 1864. **3.** Huxley 1864, citing Kölliker 1864, in Huxley T.H. 1908 p.183. **4.** Ibid. p.178. **5.** Ibid. p.179. **6.** Ibid. p.186. **7.** Ibid. pp.187-8. **8.** Poulton 1908 p.219; first mentioned in Poulton's 1905 Huxley Lecture at the University of Birmingham. **9.** Biographical notes on Mivart are drawn from Gruber 1960. **10.** Mivart 1897. **11.** Mivart 1872 in Mivart 1892 pp.60-1. **12.** Gruber 1960 p.87 citing a letter from Huxley to Hooker, Sept. 11 1871. **13.** Gruber 1960 p.89, citing a letter from Darwin to Huxley, Sept. 1871. **14.** Gruber 1960 p.87, citing a letter from Hooker to Huxley, Sept. 1871. **15.** Wright 1871 p.84. **16.** Gruber 1960 p.110, citing a letter from Hooker to Huxley, Dec. 1871. **17.** Mivart 1899. **18.** Gruber 1960 p.73. **19.** Mivart 1871 p.4. **20.** Ibid. pp.11-12. **21.** Ibid. p.21. **22.** Ibid. pp. 36,37. **23.** Ibid. pp.37,38. **24.** Lissman 1958 and 1963. **25.** Mivart 1871 p.63. **26.** Ibid. p.64. **27.** Ibid. p.68, citing T.H. Huxley's Hunterian Lectures of 1866. **28.** Mivart 1871 p.78. **29.** Ibid. p.99, citing Carpenter 1839 p.978. **30.** Owen 1868 vol. 3 p.785. **31.** Mivart 1871 p.113. **32.** Ibid. p.113. **33.** Jenkin 1867. **34.** Vorzimmer 1970 p.116 ff. **35.** Darwin 1872 p.133. **36.** Darwin 1869 p.45. **37.** Thomson 1868. **38.** Huxley T.H. 1893b p.333. **39.** Mivart 1871 pp. 155-6. **40.** Lyell 1830-3 vol. 1 p.180. **41.** Mivart 1871 p.229. **42.** Ibid. p.233. **43.** Ibid. p.238 citing Murphy 1869; also Murphy 1879 p.1. **44.** Murphy 1879 noted that this was in agreement with Hermann 1863 p.252. **45.** Murphy 1869 p.68. **46.** Bennett 1871 p.273. **47.** Bennett 1870 p.33. This article provoked an exchange with Wallace and others in the later Nov. and Dec. issues of *Nature* in 1870. **48.** Mivart 1892 vol. 2 p.417. **49.** Ibid. p.386. **50.** Ibid. p.414. **51.** Spencer 1898 vol. 1 pp. 357-8, 372. **52.** Dixon 1885 p.9. **53.** Ibid p.20. **54.** Romanes 1885 p.26 and Romanes 1886 p.314. **55.** Romanes 1886 p.315. **56.** Dixon 1885 p.49. **57.** Crow and Kimura 1970; also King and Jukes 1969.

Chapter 4

1. Osborn 1895 p.438. **2.** Mivart 1871 p.11. **3.** Galton 1872. **4.** Biographical notes on Spencer are drawn from Eliot 1917. **5.** Eliot 1917 p.14. **6.** Ibid. pp. 23,24. **7.** Spencer 1898 vol. 1 p.326. **8.** Ibid. p.340. **9.** Ibid. p.511. **10.** Ibid. p.510. **11.** Cope 1896 p.9. **12.** Cope 1887 p.78. **13.** Morgan C.L. 1891 pp. 79,80. **14.** Ibid. p.97. **15.** Ibid. pp.119,120. **16.** Ibid. p.174. **17.** Ibid. p.227, citing Galton 1889 p.32. **18.** Ibid. p.226. **19.** Wallace 1870 p.334. **20.** Rensch 1959 p.334. **21.** Ibid. p.335. **22.** Wallace thought that primitive man could not use all that brain power, and if that was the case what had been the selective advantage? Wallace 1889 p.467. **23.** Osborn 1895 p.418. **24.** Ibid. p.424. **25.** Ibid. p.427. **26.** Brooks 1883 p.82. **27.** Bateson B. 1928 p.42,43, citing a letter from W. Bateson to his sister Anna. **28.** W. Bateson in Bateson B. 1929 p.29. **29.** Bateson B. 1929 p.144. **30.** Bateson B. 1928 p.38, citing a letter from W. Bateson to his sister

Margaret 1887. **31.** Bateson W. 1894 p.vi. **32.** Bateson W. 1894. **33.** Bateson W. 1894 pp. 11,12. **34.** Ibid. p.71. **35.** Himmelfarb 1959 p.93. **36.** Bateson W. 1894 p.16. **37.** Løtrup *et al.* 1974. **38.** Bateson W. 1894 p.568. **39.** The expressions 'genotype' and 'phenotype' were not used by Mendel but coined by Johanssen in 1903. Mendel called the genes 'elementen'. **40.** De Vries 1889. **41.** Provine 1971 p.66. **42.** Johanssen 1903. **43.** De Vries 1909-10 p.28. **44.** Ibid. p.652.

Chapter 5

1. Driesch 1908 p.284. **2.** Lloyd 1914 p.10. **3.** Lem 1976 pp. 227-8. **4.** Spaulding 1912 pp.244-6. **5.** Roux 1888. **6.** Driesch 1892. **7.** Morgan, T.H. 1895. **8.** Driesch and Morgan T.H. 1895. **9.** Himmelfarb 1959 p.155. **10.** Vorzimmer 1970 p.116. **11.** Kirk 1975 p.414. **12.** Driesch 1908 vol. 2 p.247. **13.** Ibid. p.257. **14.** Ibid. p.246. **15.** Ibid. vol. 1 p.77. **16.** Ibid. p.107. **17.** Ibid. vol. 2 p.263. **18.** Ibid. p.264. **19.** Waddington 1957. **20.** Driesch 1908 vol. 1 p.295. **21.** Ibid. vol. 2 p.333. **22.** Ibid. vol. 1 p.262. **23.** Ibid. p.286. **24.** Ibid. p.292. **25.** Oppenheimer 1967 p.78. **26.** Haraway 1978 p.195. **27.** Biographical notes on Bergson are drawn from Ruhe and Paul 1914. **28.** Bergson 1911 p.102. **29.** Ibid. p.85. **30.** Lovejoy 1942 pp.11-12. **31.** Bergson 1911 pp.127-9. **32.** Ibid p.237. **33.** Ibid. p.238. **34.** Ibid. p.255. **35.** Teilhard 1961 p.102. **36.** Ibid. p.107. **37.** Ibid. p.150. **38.** Ibid. p.257. **39.** Ibid. p.230. **40.** Ibid. p.262. **41.** Ibid. p.263. **42.** Ibid. p.285. **43.** Ibid. p.29; 'pious bunk' was the epithet used by Medawar in Moorehead and Kaplan 1967 p.xi.

Chapter 6

1. Huxley A. 1970 (1944) p.viii. **2.** These experiments are mentioned in a number of the essays in the anthology by Solomon 1966. **3.** Huxley A. 1970 p.viii. **4.** Haraway 1976 p.34; also Beckner 1967; Ritter 1919; Ritter and Bailey 1928. **5.** Bernard 1878 Lecture 2 part 3, in Bernard 1974 p.84. **6.** Biographical notes on Bernard are drawn from Robin 1979, especially from the contributions by Robin and Banard. **7.** Medawar 1969. **8.** Bernard 1927 p.89. **9.** Rostan 1831, cited by Hall 1969 vol. 2 pp.251-2. **10.** Bernard 1907 p.119. **11.** Ibid. p.112. **12.** Olmsted 1938. **13.** Bernard 1878 Lecture 8 in Bernard 1974 p.241. **14.** Virtanen 1967 p.20, citing a letter from Bernard to Mme Raffalovitch. **15.** Bernard 1967 p.15. **16.** Barcroft 1934; originally stated in a lecture at Harvard in 1929. **17.** Biographical notes on Lewes are drawn from Hirschberg 1970. **18.** Lewes, 1874-5 vol. 1 pp.110-11. **19.** Ibid. p.114. **20.** Ibid. pp.116,122. **21.** Ibid. p.124. **22.** Ibid. vol 2. p.489. **23.** Ibid. p.490. **24.** Ibid. vol. 1 pp.116-17. **25.** Driesch 1908 vol. 1 pp.109-10. **26.** Holt *et al.* 1912 p.2. **27.** Ibid. p.25. **28.** Ibid. p.18. **29.** Ibid. p.6. **30.** Ibid. p.40. **31.** Ibid. pp.14-19. **32.** Spaulding 1912 p.161. **33.** Haldane J.S. 1913 p.27. **34.** Ibid. p.61. **35.** Ibid. p.80. **36.** Ibid. p.88. **37.** Haldane J.S. 1932 p.48. **38.** Loeb 1916 pp.vi, 8. **39.** Ibid. p.4-5, paraphrasing Uexküll 1913. **40.** An outline of the *umwelt* concept is given by T. Uexküll 1978. **41.** Uexküll 1926 pp.336-7. **42.** Cannon 1929. **43.** Cannon 1932 p.24. **44.** Wilson E.B. 1928 p.1, citing Virchow 1858 p.12. **45.** Montgomery 1880 p.319. **46.** Ibid. p.485. **47.** Sharp 1926 p.73. **48.** Ritter 1919 vol. 1 p.220. **49.** Ritter and Bailey 1928, citing Whitehead 1925 p.23. **50.** Whitehead 1925 p.111. **51.** Ritter and Bailey 1928 p.351. **52.** Clark R. 1968, citing a letter from J.B.S. Haldane to Robert Graves.

Chapter 7

1. Alexander 1920 pp.46-7. **2.** Sellars 1922. **3.** Engels 1975 (1878) p.67. **4.** Mill 1973 (1843) pp.371-2. **5.** Lewes 1868 p.372. **6.** Lewes 1874-5 vol. 1 p.189. **7.** Drummond 1894 pp. 1,9. **8.** Ibid. p.9. **9.** Ibid. p.17. **10.** Drummond 1883 p.405. **11.** Ibid pp.405-6. **12.** Drummond 1894 p.426. **13.** Ibid. p.426. **14.** Ibid. p.208. **15.** Biographical notes on Drummond are drawn from Smith 1899. **16.** This is discussed by Moore 1979. **17.** Nietzsche 1887 pp.77-8. **18.** Roszak 1975 p.118. **19.** Blavatsky 1888 vol. 2 p.260. **20.** Srivastava 1968 p.87, citing Maitra 1946 p.58. **21.** Biographical notes on Alexander are drawn from Laird 1939. **22.** Laird 1939 p.58. **23.** Ibid. p.60, citing a paper read by Alexander to the British Academy, January 1914. **24.** Hardy 1965 pp.10-12. **25.** Laird 1939 p.61. **26.** Ibid. p.62. **27.** Ibid. p.62 ff. **28.** Ibid. p.65. **29.** Alexander 1920 vol. 2 pp.45-6. **30.** Haldane J.S. 1913 p.36. **31.** Alexander 1920 vol. 2 p.63. This statement is complicated by Alexander's use of 'quality' in the conventional sense as well as for a metaphysical category. **32.** Ibid. vol. 2 pp.65-6. **33.** Spaulding 1918 p.257. **34.** Alexander 1920 vol. 2 p.365. **35.** Brettschneider 1964 pp.57-8. **36.** Koestler 1967 p.273, citing MacLean 1958 p.613. **37.** Brettschneider 1964 p.74. **38.** Biographical notes on C.L. Morgan are drawn from Clarke E. 1974 and from the autobiographical introduction to Morgan C.L. 1931. **39.** Morgan C.L. 1895 p.59. **40.** Morgan C.L. 1931 p.vii. **41.** Ibid. p.302. **42.** Morgan C.L. 1915 p.255. **43.** Morgan C.L. 1931 p.5. **44.** Ibid. p.9. **45.** Ibid. p.13. **46.** Ibid pp.1920. **47.** Ibid. pp.104-5. **48.** Ibid. p.107. **49.** Ibid. p.285.

Chapter 8

1. Smuts 1926 p.97. **2.** Ibid. preface. **3.** Biographical notes on Smuts are drawn from the biography by his son J. SMmuts 1952. **4.** Smuts 1926 p.11. **5.** Ibid. p.91. **6.** Ibid. p.113. **7.** Ibid. p.103. **8.** Ibid. p.142. **9.** Ibid. p.144. **10.** Ibid. p.167. **11.** Ibid. p.205. **12.** Ibid. p.206. **13.** Ibid. pp.209-10. **14.** Smuts 1952 pp.287,292. **15.** Smuts 1926 p.332. **16.** Smuts 1952 pp.331-5, citing J.C. Smuts's address to the British Association: *Science from the South African Point of View*, July 6 1925, Capetown. **17.** Croizat 1958. **18.** Smuts 1952 pp.307-12, citing J.C. Smuts's Rhodes Memorial Lecture at Oxford, 1929. **19.** Phillips 1970 p.34. **20.** Smuts 1952 pp.315-22. **21.** Ibid. p.357-62, citing J.C. Smuts's speech to the Royal Institute of International Affairs in 1934. **22.** Ibid. pp.350-7, citing J.C. Smuts's rectorial address at the University of St Andrews, 1934. **23.** Cannon W.B. 1932 p.xv, citing E.H. Starling's Harveian Oration to the Royal College of Surgeons, 1918. **24.** Henslow 1908. **25.** Cannon W.B. 1932 p.301. **26.** Ibid. p.302. **27.** Ibid. p.302. **28.** Ritter and Bailey 1928 p.334. **29.** Biographical notes on Wheeler are drawn from Evans and Evans 1970. **30.** Wheeler 1939 pp.145-6. **31.** Ibid. p.150. **32.** Ibid. pp.150-1. **33.** Ibid. p.159. **34.** Ibid. p.197. **35.** Ibid. p.197. **36.** Ibid. p.198. **37.** Ibid. p.204. **38.** Lull 1929, as suggested by G.G. Simpson in the foreword to Grant 1977 p.vi. **39.** Bacon 1620 and 1627. **40.** Jennings 1908. **41.** Jennings 1910. **42.** Provine 1971 p.106 ff. gives a full account. **43.** Jennings 1917. **44.** Jennings 1927 p.19. **45.** Ibid. p.20. **46.** Ibid. p.22. **47.** Wilson E.O. 1975 p.7. **48.** Jennings 1927 p.23. **49.** Lovejoy 1927. **50.** Ibid. p.169. **51.** Ibid. p.172. **52.** Lovejoy 1924 p.177. **53.** Ibid. p.178. **54.** Lovejoy 1927 p.173. **55.** Morgan, C.L. 1933 p.80. **56.** Ibid. p.122. **57.** Haldane J.S. 1929 p.261. **58.** Haldane J.B.S. 1932 p.156. **59.** Morgan C.L. 1933 p.123. **60.** Ibid. p.165, citing Tennant 1932, lecture 6 p.102. **61.** Tennant 1932 p.56. **62.** Morgan C.L. 1933 p.165. **63.** Ibid. pp.38-9, referring to Tyndall's Presidential Address to the British Association at Belfast.

Chapter 9

1. Bateson W. 1909 p.216 in Bateson B. 1928. 2. A noteworthy exception is Haraway 1976 p.1 ff. who gives a sympathetic discussion of the Kuhnian controversy and finds that the paradigm is a useful if not formally perfect concept. 3. Haraway 1976 p.188, citing Mastermann 1970; Kuhn 1970 p.175. 4. Kuhn 1970 p.24. 5. Packard A.S. 1901 p.398. 6. Osborn 1929 p.237 citing Lamarck's 1802 *Hydrogéologie* p.67. 7. Ibid. pp.238-9, translating Lamarck 1809. 8. Ibid. p.240. 9. Lamarck 1960 (1809) pp.221-2, my translation. 10. Ibid. p.250, see also Burkhardt 1977 p.170, citing Lamarck 1815-22 vol. 1 p.185. 11. Burkhardt 1977 pp.195-6. 12. Darwin F. 1958 (1892) p.219, citing a letter from Lyell to Darwin, Oct. 3 1859. 13. Darwin F. 1958 (1892) pp.219-20, citing a letter from Darwin to Lyell, 1859. 14. Cannon G.H. 1959 p.22. 15. Huxley T.H. 1859. 16. Bateson 1909. 17. Huxley T.H. 1894. 18. Burkhardt 1977 p.41. 19. Packard A.S. 1901 pp.384-5; Hutton 1899. 20. Spencer 1898 p.511 ff. 21. Biographical notes on Butler are drawn from Holt 1964. 22. Butler 1879 p.51. 23. Ibid. p.109. 24. Ibid. p.137. 25. Ibid. pp.175-6. 26. Ibid. p.231, paraphrasing Darwin 1859 p.206. 27. An autobiographical account of these shifts in opinion was given by Butler 1880. 28. Butler 1910 p.7. 29. Ibid. p.45. 30. Butler 1910 p.45, citing a letter from Darwin to Butler. 31. Butler 1880. 32. Hartog 1910 pp.xiii-xiv. 33. Butler 1886 pp.49,58, referring to Romanes 1883. 34. Ibid. p.97, referring to Carroll 1876. 35. Gardner 1962 p.38. 36. Ibid. p.38. 37. Haldane J.B.S. 1968 p.118. 38. Butler 1886 p.278, citing Lankester 1884. 39. Shaw 1921 p.xl. 40. Routtenberg A. 1984 'Neuronal Plasticity: Changes in Brain Cells with Experience'. University of Victoria Seminar, 16 Feb. 1984. See also Pribram 1971. 41. Nordenskiöld 1928 p.570; Russell 1916. 42. Hartog 1910 p.xxx. 43. Cope 1896 Chapter 11. 44. Osborn 1929. 45. Biographical notes on Cope are drawn from Maline 1978. 46. Cope 1887 p.196. 47. Ibid. p.207. 48. Ibid. p.210. 49. Hering was mentioned by Cope in his 1882 essay on archaesthetism as if he had only just come across the German author. In *On Catagenesis* 1884 (see Cope 1887). Cope noted that he had not met with any 'scientific statement' of the subject, indicating that he had read Butler but did not care to admit it. 50. Dall 1889 p.447. 51. Cope 1896 pp.520-1, citing Hyatt 1889. 52. Hutton 1889 p.135. 53. Ibid. p.138. 54. Morgan T.H. 1903 p.250 ff. 55. Ibid. p.250 ff. 56. Ibid. p.256. 57. Biographical notes on Packard are drawn from Norland 1974. 58. Packard A.S. 1901 p.390. 59. Ibid. p.396 footnote. 60. Ibid. p.406. 61. Poulton 1908 pp.95-119. 62. Ibid. p.101. 63. Poulton 1908, citing Courthope 1870. 64. Poulton 1908 p.117. 65. Biographical notes on Kammerer are drawn from Koestler 1973. 66. Kammerer 1907. 67. Koestler 1973 p.41. 68. Kammerer 1909. 69. Kammerer 1923. 70. Koestler 1973 pp.45,169. 71. Ibid. p.46. 72. Ibid. pp.62-3, citing a letter from W. Bateson to his wife. 73. Bateson W. 1913 p.199. 74. MacBride 1919. 75. Bateson W. 1919. 76. Koestler 1973 p.71. 77. MacBride 1923. 78. Koestler 1973 p.85. 79. Goldschmidt 1956. 80. Hardy 1965 p.159, referring to MacBride 1924. 81. MacBride 1924 p.107. 82. Ibid. p.224. 83. Ibid. p.234 ff. 84. Ibid p.245. 85. Guyer and Smith, 1918 and 1920. 86. Eldridge 1926 p.333.

Chapter 10

1. Eimer 1898 p.2. 2. Osborn 1895 p.436. 3. Darwin 1859 p.197. 4.Waagen 1869. 5. Nägeli 1884, my description of the principle is based largely on Morgan T.H. 1903 pp.324-39. 6. Radl 1930. 7. Morgan T.H. 1903 p.339. 8. Osborn 1895 pp.433-4. 9. According to Rensch 1959 it was Cope 1884 and Doederlein 1887 who first argued that direct lines of evolution could pass beyond the point of utility. 10. Eimer 1890 p.49. 11. Ibid. p.21. 12. Ibid. p.29. 13. Ibid. p.408. 14. Berg 1922 gave a brief historical account of orthogenesis crediting Danilevsky 1885, Koken 1893, Scott 1894, and

368 *Notes*

Cope 1896 with ideas similar to Eimer's. See Berg 1922 for references. **15.** Eimer 1898
p.22. **16.** Osborn 1895 p.435. **17.** Biographical notes on Berg are drawn from
Dobzhansky's introduction to the 1969 MIT edition of *Nomogenesis.* **18.**
Dobzhansky 1969 p.xvii. **19.** Berg 1969 (1922) p.8. **20.** Ibid. p.15. **21.** Ibid. pp.137-8.
22. Ibid. p.139. **23.** Plate 1903, 1913. **24.** Darwin 1859 pp.193-4. **25.** Handlirsch
1902-1908; Punnett 1915. **26.** Berg 1969 p.403. **27.** Ibid. pp.406-7. **28.** Dobzhansky
1969 p.x-xi. **29.** Biographical notes on Vavilov are drawn from Adams 1978. **30.**
Osborn 1902 p.270. **31.** Gould 1981 p.16. **32.** Ibid. p.18. **33.** Thompson 1926. **34.**
Thompson 1917 p.13. **35.** Ibid. p.79, Thompson observes that a similar idea is found
in Bacon. **36.** Haller 1766 vol. 8 p.114. **37.** Darwin 1859 p.444. **38.** Thompson 1917
p.270. **39.** Ibid. p.281. **40.** Ibid. p.1095. **41.** Ibid. p.1064. **42.** Willis 1922 p.v. **43.** Ibid.
p.3, citing Lyell 1856 (1830-3) p.702. **44.** Owen 1847. **45.** Willis 1940 p.51. **46.** Catton
1966 p.170 attributes this to Nathan Bedford Forrest. **47.** Croizat 1958. **48.** Croizat
1962 pp.466-7. **49.** Ibid. p.467 ff. **50.** Simpson 1971 pp.118-19. **51.** Ibid. p.133. **52.**
Ibid. p.138. **53.** Ibid. p.127. **54.** Grassé 1977 p.50. See also Chapter 2 of this book. **55.**
Ibid. p.61. **56.** Ibid. p.70. **57.** Conn 1900 p.92. **58.** Gould 1974. **59.** Gould 1977 p.90.
60. Gould 1981 p.20. **61.** Ohno 1970 pp.60-1, citing Brown and David 1968. **62.** Ohno
1970 p.61. **63.** Croizat 1962 p.482.

Chapter 11

1. Barzun 1958 p.120, citing Kellog 1907. **2.** Darwin F. 1958 (1892) p.51. **3.** Weismann
1893b. **4.** Butler 1910 (1880) p.181. **5.** Wagner 1873 p.4. **6.** Gulick 1887 p.198. **7.**
Romanes 1886 p.315. **8.** Gulick 1887 p.198. **9.** Medawar 1969 p.4, citing Galton 1872.
10. Weldon 1893 p.329. **11.** Weldon 1894 p.24. **12.** Provine 1971 p.63. **13.** Ibid. p.64.
14. Ibid. p.69, citing Bateson's Presidential Address to the Zoological Section of the
British Association, Cambridge, 1904. Provine also notes that Yule had suggested a
solution to the conflict in 1902. **15.** Ibid. p.51, citing a letter from Pearson to Galton,
Feb. 12 1897. **16.** Poulton 1908 pp.95-6. First given as an address to the Boston
Natural History Society in 1894. **17.** Ibid. p.97. **18.** Ibid. p.110. **19.** Ibid. pp.xiii-xiv.
20. Ibid. p.xvii. **21.** Ibid. p.xvii. **22.** Boulenger 1907. **23.** Poulton 1908 p.xxxix. **24.**
Ibid. p.xxxviii. **25.** Ibid. p.xxxiii. **26.** Castle 1905 p.523. **27.** Castle 1911. **28.** Nillson-
Ehle 1909, translated by Provine 1971 p.117. **29.** Biographical notes on T.H. Morgan
are drawn from Sturtevant 1965. **30.** Morgan T.H. 1903 p.165. **31.** Ibid. p.370. **32.**
Cited by Evans and Evans 1970. **33.** Sturtevant 1965 p.46 ff. provides some personal
reminiscences. **34.** Morgan T.H. 1925 p.7. **35.** Ibid. p.128. **36.** Ibid. p.128. **37.** Mayr
1970 p.4. **38.** Morgan T.H. 1925 p.135. **39.** Ibid. p.139. **40.** Ibid. p.141. This is the
same argument as was used by the proponents of organic selection. **41.** Samuel 1972
p.216, citing Morgan T.H. 1928. **42.** Samuel p.217. **43.** Ibid. p.218. **44.** Morgan T.H.
1925 p.141. **45.** Barzun 1958 p.120. **46.** Hardy 1908, Weinberg 1908. **47.** Pearson 1910
p.381. **48.** Provine 1971 p.135. **49.** Ibid. p.144, citing Fisher 1918. **50.** Fisher and Ford
1926; Fisher 1927. **51.** Wright 1917. **52.** 1930; Mayr 1955. **53.** Wright 1930 p.355. **54.**
Smith 1968 p.x. **55.** Ibid. p.ix. **56.** Clark R. 1968 provides the following biographical
notes on J.B.S. Haldane. **57.** Haldane J.B.S. 1932 p.2. **58.** Ibid. p.32. **59.** Ibid. p.110.
60. Ibid. p.156. **61.** Ibid. pp.157-8. **62.** p.162. **63.** Provine 1971 p.176. **64.**
Waddington 1957 p.61. **65.** Mayr 1959 p.2. **66.** Waddington 1953 p.186. **67.** Provine
1978 p.180. **68.** Ibid. p.182. **69.** Ibid. p.182. **70.** Mayr 1959 p.6. **71.** Ibid. p.13.

Chapter 12

1. Woodger 1929 p.483. **2.** Waddington 1969 vol. 2 p.122. **3.** Sellars 1922 pp.334-5. **4.**
Woodger 1929 p.xv. **5.** Ibid. p.23. **6.** Ibid. pp.29,40,54. **7.** Ibid. pp.39-40. **8.** Ibid.

p.57. **9.** Ibid. p.157 footnote. **10.** Ibid. p.209. **11.** Ibid. pp.16-17, citing C.D. Broad 1925 pp.5-6. Broad was referring to behaviourism in particular; he said that the originators of silly theories were not silly people. 'Only very acute and learned men could have thought of anything so odd or defended anything so preposterous against the continual protests of common-sense'. **12.** Woodger 1929 p.7. **13.** Ibid. p.266. **14.** Woodger 1930a p.4. **15.** Woodger 1929, citing the German theoretician Schaxel's concept of organism. **16.** Woodger 1930b and Woodger 1931 p.206. **17.** Haraway 1976 p.95, citing unpublished lecture notes of Harrison's Silliman Lectures of 1949. See also Haraway 1976 p.93. **18.** Molière 1663. **19.** Haraway 1976 p.95. **20.** Ibid. p.102, citing Hopkins 1906. **21.** Needham 1936 p.8. **22.** Needham 1943 p.244; originally given as the Herbert Spencer Lecture, Oxford, 1937. **23.** Ibid. p.244. This is a portmanteau quotation from Lenin 1931 and Bernal 1934. **24.** Needham 1943 pp.207-32. **25.** Ibid. p.212. **26.** Ibid. p.39. **27.** Eddington 1935 p.56. **28.** Bertalanffy 1952 p.127. These views had been expounded in 1934, before the publications of Schrödinger and Eddington. **29.** Needham 1943 p.227. He attributed the recognition of this distinction to J.S. Haldane 1936 p.25. **30.** Haraway 1976 p.131 ff. Other founding members of the club were L.L. Whyte, A.D. Ritchie, J.D. Bernal and M. Black. **31.** Bertalanffy 1933. **32.** Winterstein 1928. **33.** Bertalanffy 1933 p.17. **34.** Ibid. p.49. **35.** Ibid. p.88. **36.** Ibid. p.90. **37.** Ibid. pp.119-20, citing Weiss 1920. **38.** Goldschmidt 1960 p.319. **39.** Bertalanffy 19 p.183. **40.** Blake 1927 (1790) p.46. **41.** Bertalanffy 1952 p.45. **42.** Ibid. p.46. **43.** Ibid. p.52. **44.** Ibid. p.79. **45.** Ibid. pp.84-5. **46.** Ibid. p.87. **47.** Ibid. p.92. **48.** Ibid p.96. **49.** Ibid. p.108. **50.** Bertalanffy 1968 p.2. **51.** De Bono 1971, 1973. **52.** Phillips 1976. **53.** Bacon 1902 (1620) pp.21-2. **54.** Cairns-Smith 1971 provides some assistance in simplifying the problem. **55.** Pattee 1972 p.37. **56.** Weiss, Needham and Harrison are the central subjects in Haraway 1976. **57.** Haraway 1976 p.153, citing Weiss 1936 pp.512-13. **58.** Morgan C.L. 1896 p.312 ff. **59.** Osborn 1896. **60.** Ibid. **61.** Ibid. **62.** Poulton 1897. **63.** Baldwin 1917 p.341, citing Osborn 1896. **64.** Ibid. pp.23-4. **65.** Ibid. p.37. **66.** Ibid. p.48. **67.** Baldwin 1896 p.94. **68.** Ibid. p.95. **69.** Ibid. p.96. **70.** 1979. **71.** Baldwin 1896 p.97. **72.** Simpson 1953. **73.** Waddington 1953. **74.** Waddington 1975 pp.59-98. **75.** Ibid. p.71. **76.** Koestler and Smythies 1969 p.387 (Waddington's contribution to the post-seminar discussion).

Chapter 13

1. Goldschmidt 1940 p.271. **2.** Mayr 1960 p.351. **3.** Hall 1969 vol. 1 p.242. **4.** Singer 1959. **5.** Spemann 1938, citing Loeb 1894. **6.** Spemann 1938 provides a general account. **7.** Wilson E.B. 1928 p.1108. **8.** Spemann 1938 p.25 ff. **9.** Gould 1977 p.49, citing Serres 1830. **10.** Biographical notes on Goldschmidt are drawn from Goldschmidt 1960. **11.** Goldschmidt 1960 p.29. **12.** Gaissinovitch 1974. **13.** Goldschmidt 1940 p.206. **14.** Ibid, pp.250-1. **15.** Ibid. p.271. **16.** Harms 1934. **17.** Goldschmidt 1950 p.279, citing Stockard 1931. **18.** Ibid. p.316, citing J.S. Huxley 1932. Severstov was credited with similar ideas. **19.** Goldschmidt 1940 p.390. **20.** Ibid. p.391, citing Bonavia 1895. **21.** Goldschmidt 1960 p.324. Goldschmidt noted that Wright had been exceptionally objective in his criticism. **22.** Severtsov 1927 p.161, from a translation in Russell 1962 p.133-4. **23.** Dobzhansky 1969 provides definitions of Severtsov's terminology. **24.** Beurlen 1930. **25.** Schindewolf 1937 pp.208-9; from a translation by Russell 1962 p.195. **26.** Clark A.H. 1930 p.195. **27.** Russell 1962 p.132. **28.** Ibid. p.132. **29.** Ibid. p.137. **30.** Dobzhansky 1949 provides some biographical notes on Schmalhausen. **31.** Schmalhausen 1949 p.2. **32.** Ibid p.8. **33.** Dobzhansky 1949 p.ix. **34.** Waddington 1975 pp.1-11. **35.** This was at the time of writing his *Organisers and Genes* 1940. **36.** Waddington 1957 p.5. **37.** Ibid. p.9. **38.** Ibid. p.32. **38.** Homeorhesis literally means 'the same [direction of] flow'. **39.** Ibid. p.158. **40.** Ibid. p.157. **41.** Matsuda 1982. **42.** Dalq 1951. **43.** Waddington 1957 p.80. **44.** Ibid.

p.306. **45.** Britten and Davidson 1971. **46.** Uexküll 1926 p.219. **47.** Ibid. p.266. **48.** Løvtrup 1975 pp.508,511. **49.** Løvtrup 1976 p.280. **50.** Løvtrup 1978 p.348. **51.** Løvtrup 1974. p.165 ff. **52.** McClintock 1951,1952; Jacob and Monod 1961. **53.** Løvtrup 1974 p.170. **54.** Ibid. p.400. **55.** Ibid. p.401. **56.** Ibid. p.404. **57.** Løvtrup 1977 p.28. **58.** Løvtrup 1974 p.416. **59.** Ibid. p.417; see also Cole 1967, Villée 1942 and Grüneberg 1943. **60.** Ibid. p.435. **61.** Britten and Davidson 1971 p.118; see also Flamm, Walker and McCallum 1969. **62.** Britten and Davidson 1969,1971. **63.** Britten and Davidson 1971 p.126. **64.** Ohno 1970 pp.77-80,145. **65.** Riedl 1978 p.xv. **66.** Ibid. p.1. **67.** Ibid. p.26. **68.** Ibid. p.99-100. **69.** Ibid. p.76. **70.** Ibid. p.87. **71.** Ibid. p.92. **72.** Ibid. p.104. **73.** Murray *et al.* 1981, Palka *et al.* 1983. **74.** Riedl 1978 p.166. **75.** Ibid. p.195. **76.** Keller 1983 provides a good summary. **77.** McClintock 1951, p.42. **78.** I concur with the interpretation given by Keller 1983. **79.** Fink *et al.* 1981, Shapiro 1980. **80.** McClintock 1978,1980.

Chapter 14

1. Woodger 1929 p.417. **2.** Russell 1946 p.148. **3.** Child, 1924 p.43. **4.** Ibid. p.237. **5.** Russell 1946 p.7. **6.** Cannon 1932 p.20. Russell 1946 p.191 pointed out that the importance of self-maintenance, 'persistence in being', was clearly indicated by Spinoza in 1677. **7.** Russell 1946 p.22. **8.** Ibid. p.110 off. **9.** Beckner 1968 pp.143,147 attributed the distinction to Braithwaite 1947, and R.B. Perry the new realist was one of the early analysts of purpose in behaviour. **10.** Russell 1962. Russell died in 1954 with the ms. almost complete. **11.** Waddington 1957 p.146. **12.** Whyte 1965 gives the definitions in the preface. **13.** Whyte 1965 p.xviii. **14.** Ibid. p.6. **15.** Ibid. p.28. **16.** Ibid. pp.75-6. **17.** Locker 1973 pp.663-91. **18.** Needham 1931. **19.** Baldwin 1937 p.58. **20.** Lwoff 1944 p.11. **21.** Kollar and Fisher 1980; see also Grant and Wiseman 1982, Kollar and Fisher 1982. **22.** Keilin 1959. **23.** Zuckerkandl 1975 p.34. **24.** Ibid. p.31. **25.** Ibid. p.37. **26.** Ibid. p.42. **27.** Horowitz 1945. **28.** Hochachka 1973.

Chapter 15

1. Schmidt-Nielson 1959. **2.** Belding 1967 p.497. **3.** Schmalhausen 1949 pp.184,188. **4.** Ibid. p.190. **5.** Cary *et al.* 1971; Bennett and Ruben 1979. **6.** *Nature* 1984, 308, p.685 ff: editorial and various authors. **7.** Ashby 1952 p.57 ff. **8.** Ibid. p.79. **9.** Ibid. p.144. **10.** Ibid. p.156. **11.** Ibid. p.142. **12.** Bateson G. 1972 p.375 ff. calls W. Bateson's version of this 'Bateson's Rule'. **13.** Ashby 1952 p.233. **14.** Hoar 1983 pp.310-11 gives a brief account. **15.** Ibid. p.311-13. **16.** Jerison 1973 p.309, citing Clark, W. L.G. 1962 p.227. **17.** Jerison 1973 p.21. **18.** Ibid. p.21. **19.** Ohno 1970 p.145. **20.** *Nature* 1982 vol. 308 p.685 ff: editorial and various authors. **21.** Jerison 1973 p.336. **22.** Ibid. p.386. **23.** Rosen 1973 p.115. **24.** Jantsch 1980 p.49,58. **25.** Ibid. p.28, describing the views of I. Prirogine. **26.** Ibid. p.146. **27.** Although I have not referred specifically to the opinions of Gregory Bateson 1972, a number of his essays provide parallel arguments.

Chapter 16

1. Kuhn 1970 pp.67-8. **2.** Løvtrup 1982. **3.** See Chapter 12 for a discussion of evolution in relation to the second law of thermodynamics. **4.** Eldredge and Gould 1972 p.84. **5.** Løvtrup 1981. **6.** Himmelfarb 1959 p.93. **7.** Hitching 1982 p.186. **8.** *Nature* 1981 vol. 289 p.735: editorial: 'Darwin's Death in South Kensington'.

9. Halstead 1980 p.208. **10.** McDougall 1927. This period of neo-Lamarckism is discussed by Hardy 1965 pp.157-8. **11.** Crewe 1936. **12.** Agar *et al.* 1954. **13.** Lysenko 1928. Zirkle 1949 gives a brief account of Lysenko's rise. **14.** Dobzhansky 1949 p.82. **15.** Lysenko 1948. **16.** Ibid. p.7, citing a letter from Engels to Lavroy, Nov. 1875. **17.** Ibid. p.12,15. **18.** Darlington 1949. **19.** Jones 1943 p.14. **20.** Ibid. p.16. **21.** Jones 1953 p.77. **22.** Ibid. p.120. **23.** Cannon H.G. 1959 p.43. **24.** Ibid. p.91. **25.** Ibid. pp.99-100. **26.** Ibid. p.102. **27.** Ibid. p.109. **28.** Ibid. p.129, citing Vandel 1958 p.199. **29.** Hardy 1965 p.216. **30.** Ibid. p.257. **31.** Wintrebert 1962 p.5. **32.** Grassé 1977 p.23. **33.** Ibid. p.24. **34.** Ibid. p.48,31 respectively. **35.** Ibid. p.191. **36.** Ibid. pp.203-4. **37.** Ibid. p.229. **38.** Ibid. p.245. **39.** Steele 1979 p.1. **40.** Ibid. p.23; Dyson 1974 pp.535-6. **41.** Steele 1979 p.35. **42.** Ibid. p.37. **43.** Ibid. p.49. **44.** Brent *et al.* 1982 p.244. **45.** *Nature* 1981 vol. 289 pp.631-2. editorial: 'Too Soon for the Rehabilitation of Lamark'. **46.** *Nature* 1981 vol. 293 p.245. editorial: 'A Book for Burning?' **47.** Roszak 1972; Pirsig 1974. **48.** Polanyi 1962 p.37. **49.** Ibid. pp.141-2. **50.** Ibid. p.288. **51.** Ibid. p.289. **52.** Ibid. p.291. **53.** Janis 1982. **54.** Kuhn 1970 p.88. **55.** Provine 1978. **56.** Koestler and Smythes 1969; see also Haraway 1976 p.203. **57.** Stanley 1979.

Chapter 17

1. Polanyi 1962 pp.151-2. **2.** Ibid. p.291. **3.** Ibid. p.382. **4.** Ibid. p.382. **5.** Ibid. p.386. **6.** Ibid. p.386. **7.** Ayala 1972 p.6. **8.** Popper 1972c p.217. **9.** Ibid. p.244. **10.** Ibid. p.251. **11.** Ibid. p.251. **12.** Ibid. p.282. **13.** Festinger 1962. **14.** Brooks 1983, Jantsch 1980, Platt 1970, Wiley and Brooks 1982, Løvtrup 1983, Wiley and Brooks 1983. **15.** Popper 1972 p.233. **16.** Jantsch 1980 p.75. **17.** Ibid. p.49. **18.** Meyer-Abich 1964 p.142. **19.** Margulis 1981 p.3, Taylor 1974. **20.** Pirozynski and Malloch 1975 pp.154-5. **21.** J.M. Trappe, University of Victoria seminar March 1983; see also Maser *et al.* 1978. **22.** Pirozynski and Malloch 1975 p.162. **23.** Reid and Bernard 1980, Felback 1983 and Cavanaugh 1983. **24.** Lovejoy 1927 p.173. **25.** Prosser 1965 pp.377,387. **26.** Dover 1982 p.111. **27.** Jantsch 1980 suggests that this summarises the predictability of evolution. Whyte (1965) p.xiii 'Expect surprises!' should be the watchword of all scientists who try to look beyond the fashions of the day.' **28.** Mayr 1972. **29.** Engels 1954 (1925). **30.** Skolimowski 1974 p.213; see also Olsen 1968 and 1973 for pertinent comments. **31.** Prosser 1965 p.364.

REFERENCES

Adams M.B. (1978) 'Vavilov, Nikolay Invanovich, in C.C. Gillispie (ed.), *Dictionary of Scientific Biography 15*, 505-13, Scribners, New York
Agar W.E., Drummond, F.H., Tiegs O.W. and Gunson M.M. (1954) 'Fourth (final) Report on a Test of McDougall's Lamarckian Experiment', *J. Exptl. Biol 31*, 307-21
Alexander S. (1899) *Moral Order and Progress*, Kegan Paul, Trübner, London
———(1920) *Space, Time and Deity*, Macmillan, London
Anderson N.C. (1970) 'Evolutionary Significance of Viral Infection', *Nature 227*, 1346-7
Argyll, J.G. Campbell, Duke of (1886) 'Organic Evolution', *Nature, 34*, 355-56
Ashby W.R. (1945) 'The Effect of Controls on Stability', *Nature, 155*, 242
(1952) *Design for a Brain*, Wiley, New York
Avery O.T., MacLeod C.M. and McCarty C. (1944) 'Studies on the Chemical Nature of the Substance Inducing Transformation of Pneumococcal Types', *J. Exptl. Med. 79*, 137-58
Ayala F.J. (1972) 'The Autonomy of Biology as a Natural Science', in A.D. Breck and W. Yourgrau (eds), *Biology, History and Natural Philosophy*, pp. 1-16, Plenum Press, New York
———(1974) 'The Concept of Biological Progress', in F.J. Ayala and T. Dobzhansky (eds), *Studies in the Philosophy of Biology*, pp. 339-55. Macmillan London

Bacon F. (1902) *Novum Organum* (1620) Collier, New York
———(1974) *New Atlantis* (1627) Clarendon, Oxford
Baden-Powell R.F. (1908) *Scouting For Boys*, Pearson, London
Bakker R. (1975) 'Dinosaur Renaissance', *Sci. Am., 232*, 58-78
Baldwin E. (1937) *An Introduction to Comparative Biochemistry*, Cambridge University Press, Cambridge
Baldwin J.M. (1896) 'A New Factor in Evolution', *Am. Nat., 30*, 354-536
———(1897) 'Organic Selection', *Nature, 55*, 558
———(1902) *Development and Evolution*, Macmillan, New York
———(1917) *Development and Evolution*, 2nd edn, Macmillan, New York
Balfour A.J. B. (1911) 'Creative Evolution and Philosophic Doubt', *Hibbert Journal*, October
Barash D. (1979) *Sociobiology: The Whisperings Within*, Souvenir, London
Barcroft J. (1934) *Features in the Architecture of Physiological Function*, Macmillan, New York
Barzun J. (1958) *Darwin, Marx, Wagner: Critique of a Heritage*. 2nd edn. Doubleday, Garden City, NY
Bateson B. (1928) *William Bateson F.R.S.*, Cambridge University Press, Cambridge
Bateson G. (1972) *Steps to An Ecology of Mind*, Chandler, New York, Ballantine edn, 1972
Bateson W. (1894) *Materials for the Study of Variation Treated with Especial Regard to Discontinuity in the Origin of Species*, Macmillan, London
———(1905) 'Evolution for Amateurs', *The Speaker*, June 24, pp. 449-55
———1909) 'Heredity and Variation in Modern Lights', in A.C. Seward (ed.), *Darwin and Modern Science*, Cambridge University Press, Cambridge
———(1913) *The Problems of Genetics*, Yale University Press, New Haven

————(1919) 'Dr Kammerer's Testimony on the Inheritance of Acquired Characteristics, *Nature, 103,* 344-5

————(1922) 'Evolutionary Faith and Modern Doubts', *Science, 55,* 55-61

Beament J.W. (1961) 'The Role of Physiology in Adaptation and Competition Between Animals', *Symp. Soc. Exptl. Biol. 15,* 62-71

Beckner M. (1967) 'Organismic biology', in P. Edwards (ed.), *Encylopaedia of Philosophy, 5*

————*(1968) The Biological Way of Thought,* University of California Press, Berkeley

Belding H.S. (1967) 'Resistance to Heat in Man and Other Homeothermic Animals', in A.H. Rose (ed.), *Thermobiology,* Academic Press, London

Bennett A.F. and Ruben J.A. (1979) 'Endothermy and Activity in Vertebrates', *Science, 206,* 649-54

Bennett A.W. (1870) 'The theory of Natural Selection from a Mathematical Point of View.' *Nature, 3,* 30-3

————(1871) 'Review of *The Genesis of Species', Nature, 3,* 270-3

Berg L.S. (1969) *Nomogenesis or Evolution Determined by Law,* (1922), MIT Press, Cambridge, Mass

Bergson H. (1908) *L'Évolution créatrice,* Alcan, Paris

    (1911) *Creative Evolution,* Holt, New York

Bernal J.D. (1934) Aspects of Dialectical Materialism, Watts, London

Bernard C. (1865) *Introduction à l'étude de la médecine experimentale,* Bailliere, Paris

————(1878) *Leçons sur les phénomènes de la vie commune aux animaux et aux végétaux,* Baillière, Paris

————(1927) *An Introduction to the Study of Experimental Medicine,* H.C. Green (trans.), Macmillan, New York

————(1967) *The Cahier Rouge,* R. Guillemin (trans.), Schenkman, Cambridge, Mass

————(1974) *Lectures on the Phenomena of Life Common to Plants and Animals,* H.E. Hoff, R. Guillemin and L. Guillemin (trans.), Thomas, Springfield, Ill.

Bertalanffy L. von (1933) *Modern Theories of Development; An Introduction to Theoretical Biology,* J.H. Woodger (trans.), Oxford University Press, Oxford

————(1949) *Das Biologische Weltbild,* Franske, Bern

————(1952) *Problems of Life,* Wiley, New York

————(1968) *General System Theory,* Braziller, New York

Beurlen K. (1930) 'Vergleichende Stammesgeschichte, Grundlagen, Methoden, Probleme, unter besonderer Berucksiehtigung der höheren Krebse', *Fortschr, Geol., 7,* 317-586

Blake W. (1927) 'The Marriage of Heaven and Hell', (1790), in *Blake's Poems and Prophecies,* Everyman's Library Edition, Dent, London. p.46

Blavatsky H.P. (1888) *The Secret Doctrine,* Theosophical Society, London

Blyth E. (1835) 'An Attempt to Classify the "Varieties" of Animals', *Magazine of Nat. Hist., 8,* 40-53

————(1837) 'On the Psychological Distinctions between Man and All the Other Animals.' *Magazine of Nat. Hist., N.S., 1,* 1-9, 77-85, 131-41

Bock W.J. and Wahlert G, von (1965) 'Adaptation and the Form-function Complex', *Evolution, 19,* 269-99

Boesigner E. (1974) 'Evolutionary Theories after Lamarck', in F.J. Ayala and T. Dobzhansky (eds), *Studies in the Philosophy of Biology,* Macmillan, London, pp.21-43

Bolk L. (1915) Überlagerun, Verschiebung, und Neigung des Foramen magnum am Schädel der Primaten, 3', *Morph. Anthrop., 7,* 611-92

Bonavia E. (1895) *Studies in the Evolution of Animals,* Constable, London

Boveri T. (1901) 'Über die Polarität des Seeigeleies'; *Verhandl. Phys.-Med. Ges. Wurzburg, N.F., 34,* 145-76.

Brackman A.C. (1980) *A Delicate Arrangement,* Times Books, New York

Braithwaite R.B. (1947) 'Teleological Explanations', *Proc Aristot. Soc., N.S. 47*

Brent L., Chandler P., Fierz W., Medawar P.B., Rayfield L.S. and Simpson E. (1982) 'Further Studies on Supposed Lamarckian Inheritance of Immunological Tolerance', *Nature, 295,* 242-4

Brettschneider B.D. (1964) *The Philosophy of Samuel Alexander, Idealism in 'Space, Time and Deity',* Humanities Press, New York

Britten R.J. and Davidson E.H. (1969) 'Gene Regulation for Higher Cells: a Theory', *Science, 165,* 349-57

――――and Davidson E.H. (1971) 'Repetitive and Non-repetitive DNA Sequences and a Speculation on the Origins of Evolutionary Novelty', *Quart. Rev. Biol., 46,* 111-38

Broad C.D. (1925) *The Mind and its Place in Nature,* Routledge & Kegan Paul, London

Brooks D.R. (1983) 'What's Going On in Evolution? A Brief Guide to Some New Ideas in Evolutionary Theory', *Can. J. Zool., 61,* 2637-45.

Brooks W.K. (1883) *The Law of Heredity, a Study of the Cause of Variation, and the Origin of living Organisms,* Murphy, Baltimore

Brown D.D. and David I.B. (1968) 'Specific Gene Amplification in Oocytes', *Science, 160,* 272-80

Buffon G.L. (1749-1804) *Histoire naturelle, générale et particulière,* Imprimerie Royale, Plassan, Paris

Burkhardt R.W. (1977) *The Spirit of System,* Harvard University Press, Cambridge, Mass

Butler S. (1872) *Erewhon,* Trübner, London

――――(1877) *Life and Habit,* Trübner, London

――――(1879) *Evolution Old and New,* Hardwicke & Bogue, London

――――(1880) 'Letter'. *The Athenaeum,* Jan. 31

――――(1886) *Luck or Cunning,* Fifield, London

――――(1910) *Unconscious Memory,* 2nd edn (1880), Cape, London

Cairns J., Stent G.S. and Watson J.D. (eds) (1966) *Phage and the Origins of Molecular Biology,* Cold Spring Harbor Laboratory of Quantitative Biology, NY

Cairns-Smith A.G. (1971) *The Life Puzzle,* Oliver & Boyd, Edinburgh

Campbell J.W. (1973) 'Nitrogen Excretion', in C.L. Prosser (ed.), *Comparative Animal Physiology,* Saunders, Philadelphia, pp.279-316

Cannon H.G. (1959) *Lamarck and Modern Genetics,* University of Manchester Press, Manchester

Cannon W.B. (1929) 'Organization for Physiological Homeostasis', *Physiol. Rev., 9,* 1109-15

――――(1932) *The Wisdom of the Body,* Norton, New York

Carey F.G., Teal J.M., Kanwisher J.W., Lawson K.D. and Beckett J.S. (1971) 'Warm-blooded Fishes', *Am. Zool., 11,* 137-45

Carpenter W.B. (1854) *Principles of Comparative Physiology,* 4th edn (1839), Churchill, London

Carroll L. (1871) *Through the Looking Glass and What Alice Found There,* Macmillan, London

――――(1876) *The Hunting of the Snark,* Macmillan, London

Carter G.S. (1954) *Animal Evolution, a Study of Recent Views of its Causes,* rev edn, Sidgwick & Jackson, London

Castle W.E. (1905) 'The Mutation Theory of Organic Evolution from the Standpoint of Animal Breeding', *Science, N.S. 21,* 521-525

————(1909) 'Is Selection or Mutation the More Important Agency in Evolution?' *Sci. Monthly,* 91-8

————(1911) *Heredity in Relation to Evolution and Animal Breeding,* Appleton, New York

Catton B. (1966) *The Penguin Book of the American Civil War,* Penguin, Harmondsworth, Middlesex

Cavanaugh C.M. (1983) 'Symbiotic Chemoautotrphic Bacteria in Marine Invertebrates from Sulphide-rich Habitats', *Nature, 302,* 58-61

Chambers R. (1969) *Vestiges of the Natural History of Creation,* (1844), Humanities Press, New York

Chargaff I. (1950) 'Chemical Specificity of Nucleic Acids and Mechanism of Their Enzymatic Degradation'. *Experientia, 6,* 201-9

Child C.M. (1924) *Physiological Foundations of Behaviour,* Holt, New York

Clark A.H. (1930) *The New Evolution: Zoögenesis,* Williams & Wilkins, Baltimore

Clark R. (1968) *J.B.S. The Life and Work of J.B.S. Haldane,* Hodder & Stoughton, London

Clark W. L.G. (1962) *The Antecendents of Man,* 2nd edn, Quadrangle, Chicago

Clark E. (1974) 'Morgan, Conwy Lloyd' in C.C. Gillispie (ed.), *Dictionary of Scientific Biography, 9,* 512-13, Scribners, New York

Cole R.K. (1967) 'Ametapodia, a Dominant Mutation in the Fowl', *J. Heredity, 58,* 141-6

Conn H.W. (1900) *The Method of Evolution,* Putnam, New York

Cope E.D. (1866) 'On the Cyprinidae of Pennsylvania.' *Trans. Am. Phil. Soc., 13,* 351-99

————(1868) 'The Origin of Genera', *Proc. Philadelphia Acad. Sci.,* October, in Cope E.D. (1887) pp. 41-123

————(1871) 'The Method of Creation of Organic Types', *Proc. Amer. Phil. Soc.,* December, in Cope E.D. (1887) pp.172-214

————(1884) 'Progressive and Regressive Evolution Among Vertebrates', in Cope E.D. (1896)

————(1887) *The Origin of the Fittest,* Open Court, Chicago

————(1896) *The Primary Factors of Organic Evolution,* Open Court, Chicago

Courthope W.J. (1870) *The Paradise of Birds,* Blackwood, Edinburgh

Crewe F.A.E. (1936) 'A Repetition of McDougall's Lamarckian Experiment', *J. Genetics, 33,* 61-84

Croizat L. (1958) *Panbiogeography,* Croizat, Caracas

————(1958) 'An Essay on the Biogeographic Thinking of J.C. Willis', *Archivo Botanico e Biogeografico Italiano, 34,* 90-116

————(1962) *Space, Time, Form: the Biological Synthesis,* Croizat, Caracas

Crow J.F. and Kimura, M. (1970) *An Introduction to Population Genetics Theory,* Harper & Row, New York

Cuénot L. (1914) 'Theorie de la préadaptation', *Scientia, 16,* 60-73

Cunningham J.T. (1921) *Hormones and Heredity,* Constable, London

Cuvier G. (1836) 'Biographical Memoirs of M. de Lamarck by the Baron Cuvier', Anon (trans.), *Edinburgh New Philosophical Journal, 20,* 1

Dalq A.M. (1951) 'Form and Modern Embryology', in L.L. Whyte (ed.), *Aspects of Form,* Indiana University Press, Bloomington, pp.91-120

Dall W.H. (1889) 'Hinge of Pelecypoda and its Development', *Am. J. Science and Arts, 38,* 445-91

Darlington C.D. (1949) 'The Retreat from Science in Soviet Russia', in C. Zirkle (ed.), *Death of a Science in Russia,* University of Pennsylvania Press, Philadelphia, pp. 157-68

376   *References*

Darwin C. (1859) *On the Origin of Species by Means of Natural Selection,* Murray, London

―――(1868) *The Variation of Animals and Plants Under Domestication,* 2 vols, Murray, London

―――(1869) *On the Origin of Species by Means of Natural Selection,* 5th edn, Murray, London

―――(1871) *The Descent of Man, and Selection in Relation to Sex,* 2 vols, Murray, London

―――(1872) *On the Origin of Species by Means of Natural Selection,* 6th edn, Murray, London

―――(1879) *Life of Erasmus Darwin,* Murray, London

and Wallace A.R. (1858) 'On the Tendency of Species to Form Varieties; and on the Perpetuation of Varieties and Species by Means of Natural Selection', *J. Linn. Soc. (Zool.), 3,* 45-62

Darwin E. (1794) *Zoonomia,* Johnson, London

Darwin F. (ed.) (1958) *The Autobiography of Charles Darwin and Selected Letters,* (1892) Dover, New York

Davenport C.B. (1903) 'The Animal Ecology of the Cold Spring Sandspit, with Remarks on the Theory of Adaptation', *Decennial Publications of the University of Chicago, 10*

Dawkins R. (1976) *The Selfish Gene,* Oxford University Press, London

De Beer G. (1930) *Embryology and Evolution,* Oxford University Press, Oxford

―――(1940) *Embryos and Ancestors,* Oxford University Press, Oxford

De Bono E. (1971) *The Use of Lateral Thinking,* Penguin, Harmondsworth, Middlesex

―――(1973) *PO: Beyond Yes and No,* Penguin, Harmondsworth, Middlesex

De Vries H. (1889) *Intracelluläre Pangenesis,* Fischer, Jena

―――(1901-3) *Die Mutationstheorie,* 2 vols, Veit, Leipzig

―――(1909-10) *Mutation Theory,* 2 vols, J.B. Farmer and A.D. Darbishire (trans), Open Court, Chicago

Dixon C. (1885) *Evolution Without Natural Selection,* Porter, London

Dobzhansky T. (1937) *Genetics and the Origin of Species,* Columbia University Press, New York

―――(1947) 'N.I. Vavilov, a Martyr of Genetics', *J. Hered., 38,* 227-32

―――(1949) 'Introduction' to I.I. Schmalhausen (1949)

―――(1969) 'Introduction' to L.S. Berg (1969)

―――(1970) *Genetics of the Evolutionary Process,* Columbia University Press, New York

Doederlein L. (1887) 'Phylogenetische Betrachtungen', *Biol. Zentralbl., 7,* 394-402

Dover G.A. (1982) 'Molecular Drive: a Cohesive Mode of Species Evolution', *Nature, 299,* 111-17

Driesch H. (1892) 'Entwickelungsmechanisches', *Anatomischer Anzeiger,* Jena, *7*

―――*(1908) The Science and Philosophy of the Organism,* Black, London

―――and Morgan T.H. (1895) 'Zur analysen der ersten Entwickelungsstadien des Ctenophoreneies', *Archiv für Entwicklungsmechanick,* Berlin, 2

Drummond H (1883) *Natural Law in the Spiritual World,* Hodder & Stoughton, London

―――(1894) *The Ascent of Man,* Hodder & Stoughton, London

Dyson R.D. (1974) *Cell Biology: A Molecular Approach,* Allyn & Bacon, Boston

Eddington A.S. (1935) *New Pathways in Science,* Cambridge University Press, Cambridge

Eimer T. (1890) *Organic Evolution as the Result of the Inheritance Of Acquired Characteristics,* J.T. Cunningham (trans.), Macmillan, London

————(1898) *On Orthogenesis and the Impotence of Natural Selection in Species Formation,* T.J. McCormack (trans.), Open Court, Chicago

Eldridge S. (1926) *The Organization of Life,* Crowell, New York

Eldredge N. and Gould S.J. (1972) 'Punctuated Equilibria: an Alternative to Phyletic Gradualism', in Schopf T.J.M. (ed.), *Models in Paleobiology,* Freeman, Cooper, San Francisco

Eliot H. (1917) *Herbert Spencer,* Holt, New York

Engels F. (1954) *The Dialectics of Nature,* Foreign Languages Publishing, Moscow. First published in Russian and German in 1925

————(1975) *Socialism: Utopian and Scientific,* Foreign Languages Press, Peking. Abstracted from F. Engels (1878) *Herr Dühring's Revolution in Science,* Leipzig

Evans M.A. and Evans H.E. (1970) *William Morton Wheeler, Biologist,* Harvard University Press, Cambridge, Mass.

Farkas B. (1929) 'Beitrage zur Kenntis des feineren Baues und der Entwicklung der Spongien', *Int. Congr. Zool., 10.* Budapest, 1927

Felbeck H. (1983) 'Sulfide Oxidation and Carbon Fixation by the Gutless Clam *Solemya reidi:* an Animal-bacteria Symbiosis,', *J. Comp. Physiol., 152,* 3-11

Festinger L. (1962) 'Cognitive Dissonance', *Sci. Am., 207,* 93-101

Feyerabend P.K. (1978) *Against Method: Outline of an Anarchistic Theory of Knowledge,* NLB Verso, London

Fink G., Farabaugh F, Roeder G. and Chaleff D. (1981) 'Transposable Elements in Yeast', Cold Spring Harbor Symposia on Quantitative Biology *45,* 575-80

Fisher R.A. (1918) 'The Correlation Between Relatives on the Supposition of Mendelian Inheritance', *Trans. Roy. Soc. Edinburgh, 52,* 399-433

————(1927) 'On Some Objections to Mimicry Theory: Statistical and Genetic', *Trans. Entomol. Soc. Lond. 75,* 269-78

————(1930) *The Genetical Theory of Natural Selection,* Oxford University Press, Oxford

————:and Ford E.B. (1926) 'Variability of Species', *Nature, 118,* 515-16.

Flamm W.G., Walker M.B. and McCallum M. (1969) 'Some Properties of the Single Strands Isolated from the DNA of the Nuclear Satellite of the Mouse (*Mus musculus*)', *J. Mol. Biol., 40,* 320-443

Florkin M. (1944) *Bioche nical Evolution,* S. Margulis (ed.), Academic Press, New York

Futuyama D.J. (1983) Science on Trial, Pantheon, New York

Gaissinovitch A.E. (1974) 'Wolff, Caspar G', in C.C. Gillispie (ed.), *Dictionary of Scientific Biography, 15,* 524-6, Scribners, New York

Galton F. (1869) *Hereditary Genius, and Inquiry into its Laws,* Macmillan, London

————(1872) 'On Blood Relationships', *Proc. Roy. Soc., 20,* 394-402

————(1872) 'Statistical Inquiries into the Efficacy of Prayer', *Fortnightly Review,* August 1

————(1889) *Natural Inheritance,* Macmillan, London

Gardner M. (1962) *The Annotated Snark,* Simon & Schuster, New York

————(1965) *The Annotated Alice,* Penguin, Harmondsworth, Middlesex

Gerdemann J.W. (1968) 'Vesicular-arbuscular Mycorrhiza and Plant Growth', *Ann. Rev. Phytopath., 6,* 397-416

Goethe J.W. von (1816) 'Metamorphosis of Animals', in *Oevres d'histoire naturelle,* (1837) C.F. Martins (trans.), Paris

Goldschmidt R.B. (1927) *Physiologische Theorie der Vererbung,* Springer, Berlin

————(1940) *The Material Basis of Evolution,* Yale University Press, New Haven

————(1951) 'Chromosomes and Genes', *Cold Spring Harbor Symposia on Quantitative Biology, 16,* 1

————(1956) *Portraits from Memory: Recollections of a Zoologist*, University of Washington Press, Seattle

————(1960) *In and Out of the Ivory Tower*, University of Washington Press, Seattle

Gosse P.H. (1857) *Omphalos, An Attempt to Untie the Geological Knot*, Van Vorst, London

Gorczynski R.M. and Steele E.J. (1980) 'Inheritance of Acquired Immunological Tolerance to Foreign Histocompatability Antigens in Mice', *Proc. Nat. Acad Sci. USA*, *77*, 2871-75

————and Steele E.J. (1981) 'Simultaneous Yet Independent Inheritance of Somatically Acquired Tolerance to Two Distinct H-2 Antigenic Haplotype Determinants in Mice', *Nature*, *289*, 678-81

Goudge T.A. (1961) *The Ascent of Life: A Philosophical Study of the Theory of Evolution*, University of Toronto Press, Toronto

Gould S.J. (1974) 'The Evolutionary Significance of "Bizarre" Structures: Antler Size and Skull Size in the "Irish Elk"', *Megaloceros giganteus'*, *Evolution*, *28*, 191-220

————(1977) *Ontogeny and Phylogeny*, Harvard University Belknap Press, Cambridge, Mass.

————(1981) 'A Most Chilling Statement', *Natural History*, *90*, 13-20

Grant B. and Wiseman L.L. (1982) 'Fossil Genes: Scarce as Hen's Teeth?' *Science*, *215*, 698-9

Grant V. (1977) *Organismic Evolution*, Freeman, San Francisco

Grassé P.-P. (1973) *L'Évolution du vivant*, Albin Michel, Paris

————(1977) *Evolution of Living Organisms*, Academic Press, New York

Gruber J.W. (1960) *A Conscience in Conflict*, Temple University, Columbia University Press, New York

Grüneberg K. (1943) *The Genetics of the Mouse*, Cambridge University Press, Cambridge

Gulick J.T. (1887) 'Divergent Evolution, Through Cumulative Segregation', *J. Linn, Soc. Zool.*, *20*, 189-274, 312-80

Guppy, H.B. (1906) *Observations of a Naturalist in the Pacific between 1891 and 1899*, 2 vols, Macmillan, London

Guyer M.I. (1922) 'Serological Reactions as a Probable Cause of Variations', *Am. Nat.*, *56*, 80-96

————and Smith E.A. (1918) 'Studies on Cytolysins. 1. Some Prenatal Effects of Lens Antibodies', *J. Exp. Zool.*, *26*, 65-82

————and Smith E.A. (1920) 'Transmission of Eye Defects Induced in Rabbits by Means of Lens-sensitized Fowl Serum', *Proc Nat. Acad. Sci.*, *6*, 134-9

————and Smith E.A. (1920) 'Studies on Cytolysins II. Transmission of Induced Eye Defects', *J. Exp. Zool.*, *30*, 171-216

Haake W. (1895) *Gestalten und Vererbung*, Weigel, Leipzig

Haldane J.S. (1913) *Mechanism Life and Personality*, Murray, London

————(1929) *The Sciences and Philosophy*, Hodder & Stoughton, London

————(1932) *Materialism*, Hodder & Stoughton, London

————(1936) *The Philosophy of a Biologist*, Oxford University Press, Oxford

Haldane J.B.S. (1924) 'A Mathematical Theory of Natural and Artificial Selection', *Trans. Cam. Phil. Soc.*, *23*, 26-58. Continues in 9 parts through 1932

————(1932) *The Causes of Evolution*, Longmans, London

————(1968) 'Aunt Jobisca, the Bellman and the Hermit', in *Science and Life*, pp. 115-23, The Rationalist Publishing Co., London

Hall T.S. (1969) *Ideas of Matter and Life*, 2 vols, University of Chicago Press, Chicago

Haller A. von (1766) *Elementa Physiologiae Corporis Humanis*, Bousquet, Lausanne

Halstead B. (1980) 'Letter: Museum of Errors'. *Nature, 288,* 208

Handlirsch A. (1902-8) *Die Fossilen Insecten und die Phylogenie der Rezenten Formen,* Engelmann, Leipzig

Haraway D.J. (1976) *Crystals, Fabrics and Fields: Metaphors of Organicism in Twentieth Century Developmental Biology,* Yale University Press, New Haven
————(1978) 'Reinterpretation or Rehabilitation: an Exercise in Contemporary Marxist History of Science', *Studies in History of Biology, 2,* 193-209

Hardy G.H. (1908) 'Mendelian Proportions in a Mixed Population', *Sciences,* N.S. 28, 49-50

Hardy A.C. (1954) 'Escape from Specialization', in J.S. Huxley, A.C. Hardy and E.B. Ford (eds), *Evolution as a Process,* Allen & Unwin, London
————(1965) *The Living Stream,* Collins, London

Harms J.W. (1934) *Wandlungen des Artgefüges,* Heine, Tübingen

Hartog M. (1910) 'Introduction' to S. Butler *Unconscious Memory,* Cape, London

Harvey W. (1651) *Exercitationes de Generatione Animalium,* London

Hegel G. (1977) *The Phenomenology of Spirit* (1807), A.V. Miller (trans.), University of Chicago Press

Henderson L.J. (1917) *The Order of Nature,* Harvard University Press, Cambridge, Mass

Henslow G. (1908) *The Heredity of Acquired Characters in Plants,* Murray, London

Hering E. (1895) *'Memory as a Universal Function of Organised Matters',* (1870) in E. Hering, *On Memory and the Specific Energies of the Nervous System,* Open Court, Chicago

Hermann, L. (1863) *Grundriss der Physiologie des Menschen,* Hirschwald, Berlin

Hershey A.D. and Chase M. (1952) 'Independent Functions of Viral Protein and Nucleic Acid in Growth of Bacteriophage', *J.Gen Physiol., 36,* 39-56

Himmelfarb G. (1959) *Darwin and The Darwinian Revolution,* Doubleday, Garden City, NY

Hirschberg E.W. (1970) *George Henry Lewes,* Twayne, New York

His W. (1874) *Uber unsere Körperform,* Vogel, Leipzig

Hitching F. (1982) *The Neck of the Giraffe,* Pan Original, London; Mentor Books, New York

Hoar W.S. (1983) *General and Comparative Physiology,* 3rd edn, Prentice-Hall, Englewood, Cliffs, NJ

Hockachka P.W. (1973) 'Comparative Intermediary Metabolism', in C.L. Prosser (ed.), *Comparative Animal Physiology 3rd edn,* pp. 212-78, Saunders, Philadelphia

Holt E.B., Marvin W.T., Montague W.P., Perry R.B., Pitkin W.B. and Spaulding E.G. (1912) *The New Realism. Cooperative Studies in Philosophy,* Macmillan, New York

Holt L. (1964) *Samuel Butler,* Twayne, New York

Holtzer H. and Abbot J. (1968) 'Oscillations of the Chondrogenic Phenotypes *in vitro*', in H. Ursprung (ed.), *The Stability of the Differentiated State,* Springer Verlag, Berlin, pp. 1-16

Hopkins F.G. (1906) 'The Analyst and the Medical Man', *The Analyst, 31,* 385-94

Horowitz N.H. (1945) 'Evolution of Biochemical Syntheses', *Proc. Nat. Acad Sci., 31,* 153-7

Hudson W.H. (1870) 'Letter from Mr W.H. Hudson', *Proc Zool. Soc. Lond.,* March 24, pp. 158-60

Hull D.L. (1973) *Darwin and His Critics,* Harvard University Press, Cambridge, Mass

Hutton, F.W. (1899) *Darwinism and Lamarckism,* Duckworth, London

Huxley A. (1944) *The Perennial Philosophy,* Harper & Row, New York, Harper Colophon (1970)

Huxley J.S. (1932) *Problems of Relative Growth,* Methuen, London
———(1942) *Evolution: the Modern Synthesis,* Allen & Unwin, London
———(1974) *Evolution: the Modern Synthesis,* 3rd edn, Allen & Unwin, London
Huxley T.H. (1859) 'Darwin on the Origin of Species', *The Times,* London, Dec. 6, p. 7
———(1864) 'Criticisms on "The Origin of Species"', *Nat. Hist. Rev.* in T.H. Huxley (1908) *Lectures and Essays,* Cassell, London, pp. 177-93
———(1871) 'Mr Darwin's Critics', *Contemporary Review, 18,* 443-76
———(1908) 'On the Advisableness of Improving Natural Knowledge', First given as a lay sermon in 1866. *Lectures and Essays* pp. 337-50, Cassell, London
———(1880) 'The Coming of Age of the "Origin of Species"', Evening lecture to the Royal Institute, in T.H. Huxley (1893) *Darwiniana,* Macmillan, London
———(1893a) *Collected Essays,* Macmillan, London
———(1893b) *Discourses Biological and Geological,* Macmillan, London; Appleton, New York, (1913)
———(1894) 'Owen's Position in the History of Anatomical Science', in R. Owen
———(1894) *The Life of Richard Owen,* Murray, London
Hyatt A. (1889) 'The Genesis of the Arietidae', *Smithsonian Contributions to Knowledge and Memoirs of the Museum of Comparative Zoology, 16*

Jacob F. and Monod J. (1961) 'Genetic Regulatory Mechanisms in the Synthesis of Proteins', *J. Mol. Biol., 3,* 318-56
Janis I.L. (1982) *Groupthink: Psychological Studies of Policy Decisions and Fiascoes,* Houghton Mifflin, Boston
Jantsch E. (1980) *The Self-Organizing Universe,* Pergamon Press, New York
Jenkin F. (1867) 'The Origin of Species', *North British Review, 46,* 149-71
Jennings H.S. (1906) *Behaviour of the Lower Organisms,* Macmillan, New York
———(1908) 'Heredity, Variation and Evolution in Protozoa, 2. Heredity and Variation of Size and Form in *Paramecium,* With Studies of Growth, Environmental Action and Selection', *Proc. Am. Phil Soc., 47,* 393-546
———(1910) 'Experimental Studies on the Effectiveness of Selection', *Am. Nat., 44,* 136-45
———(1917) 'Heredity Characters in Relation to Evolution, II', *Nature, 100,* 213-16
———(1927) 'Diverse Doctrines of Evolution, Their Relation to the Practice of Science and of Life', *Science, 65,* 19-25
Jerison H.J. (1973) *Evolution of the Brain and Intelligence,* Academic Press, New York
Johansen K. (1968) 'Air-breathing Fishes', *Sci. Am., 213,* 102-11.
Johanssen W. (1903) *Ueber Erblichkeit in Populationen und in Reiner Linien,* Fischer, Jena
Jones F.W. (1943) *Habit and Heritage,* Routledge, London
———(1953) *Trends of Life,* Arnold, London

Kammerer P. (1907) 'Vererbung der erworbenen Eigenschaft habituellen Spätgebarens bei Salamandra maculosa', *Zb. f. Physiol.,* Leipzig, *21,* 99-102
———(1909) 'Vererbung erzwungener Fortpflanzungsanpassungen. 111 Mitteilung: Die Nachkommen der nicht brutfahigen Alytes obstetricans', *Arch., 28,* 447-546
———(1919) 'Vererbung erzwungener Formveranderungen. 1. Mitteilung: Bruntschweile der Alytes — Mannchen aus Wassereiern (Zugleich: Vererbung erzwungener Fortpflanzungsanpassungen. V. Mitteilung) *Arch., 45,* 323-70
———(1923) 'Breeding Experiments on the Inheritance of Acquired Characteristics', *Nature, 111,* 637-40
Keilin J. (1959) 'The Biological Significance of Uric Acid and Guanine Secretion', *Biol. Rev., 34,* 265-96

Keller E.F. (1983) *A Feeling for the Organism,* Freeman, San Fancisco
Kellog V.L. (1907) *Darwinism Today,* Holt, New York
Kerkut G.A. (1960) *Implications of Evolution,* Pergamon, Oxford
King J.L. and Jukes T.H. (1969) 'Non-Darwinian Evolution', *Science, 164,* 788-9
Kimura M. (1961) 'Natural Selection as the Process of Accumulating Genetic Information in Adaptive Evolution', *Genet. Res. Camb., 2,* 127-40
Kirk D. (ed.) (1975) *Biology Today,* Random House, New York
Koestler A. (1967) *The Ghost in the Machine,* Hutchinson, London
———(1971) *The Case of the Midwife Toad,* Random House, New York Vintage Edition (1973)
———and Smythies J.R. (1969) *Beyond Reductionism,* Hutchinson, London
Kohlbrugge J.H.F. (1915) 'War Darwin ein originelles Genie? *Biol. Zentralbl., 35,* 93-111
Kollar E.J. and Fisher C. (1980) 'Tooth Induction in Chick Epithelium: Expression of Quiescent Genes for Enamel Synthesis', *Science, 207,* 993-5
———and Fisher C. (1982) Response to Grant and Wiseman. *Science, 215,* 699
Kölliker R.A. von (1864) 'Ueber die Darwinsche Schöpfungstheorie ein Vortrag', Leipzig, *Zeit. für Wiss. Zool.,* Bd *14*
Kuhn T.S. (1970) *The Structure of Scientific Revolutions,* (1962) 2nd edn, University of Chicago Press, Chicago
Kunicki-Goldfinger W.J.H. (1980) *Evolution and symbiosis,* pp. 969-84 in W. Schweminter and M.E.A. Schenk (eds), *Endocytobiology* de Gruyter, Berlin

Laird J. (1939) 'Introduction' to Alexander S. *Philosophical and Literary Pieces,* Macmillan, London
Lamarck J.-B. (1877) *Flore française,* 2 vols, Paris
———(1802) 'Recherches sur l'organisation des corps vivants', Le Museum National d'Histoire Naturelle, Paris
———(1802) *Hydrogéologie,* Agasse, Maillard, Paris
———(1809) *Philosophie zoologique,* Paris. Reissued in *Naturalis Classica* series, J. Cramer and H.K. Swann (eds), Wheldon and Wesley, Codicote, Herts
———(1815-22) *L'Histoire naturelle des animaux sans vertèbres,* 7 vols, Paris
Lankester R. (1884) 'Letter', *The Athenaeum,* March 29
Lawrence W. (1832) *Lectures on Comparative Anatomy, Physiology, Zoology And the Natural History of Man,* delivered to the Royal College of Surgeons, 1816-18, Carlile, London
Lederburg J. (1952) 'Cell Genetics and Heredity Symbiosis', *Physiol. Ref V., 32,* 403-30
Lem S. (1976) *The Cyberiad,* Seabury, New York; Avon Books, New York
Lenin V.I. (1931) *The Teaching of Karl Marx,* Lawrence & Wishart, London
Lewes G.H. (1858) *Seaside Studies,* Blackwood, Edinburgh
———(1859-60) *Physiology of Common Life,* Blackwood, Edinburgh
———(1868) 'Mr Darwin's Hypothesis', *Fortnightly Review,* N.S. *3,* 353-73
———(1874, 1875) *Problems of Life and Mind,* 2 vols, Trübner, London
Lewontin R.C. 1978 'Adaptation', *Sci. Amer, 239,* 212-30
Linné C. von (1762) *Systema Naturae,* Stockholm
Lissman H.W. (1958) 'On the Function and Evolution of Electric Organs in Fish', *J. Exptl. Biol., 35,* 156-91
———(1963) 'Electric Location by Fishes', *Sci. Amer., 208,* 50-9
Lloyd R. (1914) *What is Adaptation?* Longmans & Green, London
Locker A. (ed.) (1973) *Biogenesis, Evolution, Homeostasis,* Springer-Verlag, New York
Loeb J. (1894) 'Ueber eine einfache Methode, zwei oder mehr zusammengewachsene Embryonen aus einen Ei hervorzubringen', *Pflügers Arch., 55,* 525-30

——(1916) *The Organism as a Whole: from the Physico-chemical Viewpoint,* Putnam, New York

Lovejoy A.O. (1924) 'The Discontinuities of Evolution', University of California Publications in Philosophy 5, 173-220

——(1927) 'The Meaning of Emergence and its Modes', *J. Philos. Studies, 2,* 167-89

——(1942) *The Great Chain of Being,* Harvard University Press, Cambridge, Mass

Lovelock J.E. (1979) *Gaia: A New Look at Life on Earth,* Oxford University Press, Oxford

Løvtrup S. (1974) *Epigenetics,* Wiley, London

——(1975) 'Letters', *Systematic Zoology, 24,* 507-11

——(1976) 'On the Falsifiability of neo-Darwinism', *Evolution Theory, 1,* 267-83

——(1977) *The Phylogeny of Vertebrata,* Wiley, London

——(1978) 'On von Baerian and Haeckelian Recapitulation', *Systematic Zoology, 27,* 348-52

——(1981) 'Macroevolution and Punctuated Equilibria', *Systematic Zoology, 30,* 498-500

——(1982) 'The Four Theories of Evolution', *Rivista di biologia, 75,* 53-9, 231-398

——(1983) 'Victims of Ambition: Comments on the Wiley and Brooks Approach to Evolution', *Systematic Zoology, 32,* 90-6

——Rahemtulla F. and Hoglund N.-G. (1974) 'Fisher's Axiom and the Body Size of Animals', *Zool. Scr., 3,* 53-8

Lucretius (1st century BC) *De Rerum Natura, On the Nature of Things* M.F. Smith (trans.), Sphere Books, London

Lull R.S. (1929) *Organic Evolution,* (rev. edn), Macmillan, New York

Lwoff A. (1944) *L'Évolution physiologique,* Hennann, Paris

Lysenko T.D. (1928) *Bulletin, 3,* Azerbayan Plant Breeding Station, USSR

——(1948) *The Science of Biology Today,* International Publishers, New York

Lyell C. (1830-3) *Principles of Geology,* 3 vols, Murray, London

Macallum A.B. (1926) 'The Palaeochemistry of the body Fluids and Tissues', *Physiol.Rev., 6,* 316

Macbeth N. (1971) *Darwin Retried,* Dell, New York

MacBride E.W. (1919) 'Letter', *Nature, 103,* 225-6

——(1923) 'Letter', *Nature, 112,* 98

——(1924) *An Introduction to the Study of Heredity,* Williams & Norgate, London

McClintock B. (1951) 'Chromosome Organization and Genic Expression', *Cold Spring Harbor Symposia on Quantitative Biology, 16,* 40

——(1956) 'Controlling Elements and the Gene', *Cold Spring Harbor Symposia on Quantitative Biology, 21,* 215

——(1978) 'Mechanisms that Rapidly Reorganize the Genome', in G.P. Reder (ed.), *Stadler Symposium, 10* Columbia, pp. 25-48

——(1980) 'Modified Gene Expressions Induced by Transposable Elements', in W.A. Scott *et al.* (eds), *Mobilization and Reassembly of Genetic Information,* pp. 11-19, Academic Press, New York

McDougall W. (1927) 'An Experiment for the Testing of the Hypothesis of Lamarck', *Brit. J. Psych., 17,* 268-304

MacLean P. (1958) 'Contrasting Functions of Limbic and Neocortical Systems of the Brain and Their Relevance to Psycho-physiological Aspects of Medicine', *Am. J. Med., 25,* 611-26

Maitra S.K. (1946) *Sri Aurobindo Annual,* August

Maline, J.M. (1978) 'Cope, Edward Drinker', in C.C. Gillispie (ed.), *The Dictionary of Scientific Biography, Suppl. 1.,* 91-93, Scribners, New York

Malpighi M. (1689) *De formatio pulli in ovo*

Malthus T.R. (1798) *An Essay on the Principle of Population,* Johnson, London
Mangum C.P. and Towle D. (1977) 'Physiological Adaptation to Unstable Environments', *Amer. Scientist, 65,* 67-75
Marchant J. (1916) *Alfred Russell Wallace, Letters and Reminiscences,* Harper, New York
Margulis L. (1970) *Origin of Eukaryotic Cells,* Yale University Press, New Haven
———(1981) *Symbiosis in Cell Evolution, Life and its Environment On the Early Earth,* Freemand, San Francisco
Maser C., Trappe J.M. and Nussbaum R.A. (1978) 'Fungal-small mammal Interrelationships with Emphasis on Oregon Coniferous Forests', *Ecology, 59,* 799-809
Mastermann M. (1970) 'The Nature of a Paradigm', in *Criticism and the Growth of Knowledge,* Cambridge University Press, Cambridge
Matsuda R. (1982) 'The Evolutionary Process in Talitrid Amphipods and Salamanders in Changing Environments with a Discussion of "Genetic Assimilation" and some other Evolutionary Comments', *Can. J. Zool., 60,* 733-49
Matthew P. (1831) *On Naval Timber and Arboriculture,* Longman, London
Mayr E. (1955) 'Integration of Genotypes: Synthesis', Cold Spring Harbor Symposia, *Quantitative Biology, 24,* 327-33
———(1959) 'Where Are We?' Cold Spring Harbor Symposia, *Quantitative Biology, 24,* 1-14
———(1960) 'The Emergence of Evolutionary Novelties', in S. Tax (ed.), *Evolution After Darwin,* 3 vols, *1,* 349-80, University of Chicago Press, Chicago
———(1964) 'Introduction' to *On the Origin of Species,* facsimile edn, Harvard University Press, Cambridge, Mass.
———(1970) *Populations, Species and Evolution,* Belknap Press of Harvard University, Cambridge, Mass.
———(1972) 'The Nature of the Darwinian Revolution', *Science, 1976,* 981-7
———(1982) *The Growth of Biological Thought,* Belknap Press of Harvard University, Cambridge, Mass.
Medawar P.B. (1969) *Induction and Intuition in Scientific Thought,* Methuen, London
Mereschkovsky K.C. (1905) 'La plante considérée comme un complex symbiotique', *Bul, Soc. Nat. Sci. Ouest, 6,* 17-98
———(1909) *Theory of Two Plasms as the Basis of Symbiogenesis, New Studies about the Origin of Organisms,* Kazan, Russia
Meyer-Abich A. (1964) *The Historico-Philosophical Background of Modern Evolution Biology,* Brill, Leiden
Mill J.S. (1973) *A System of Logic, Ratiocinative and Inductive,* (1843) vols 7,8, in *The Collected Works of John Stuart Mill,* University of Toronto Press
Millhauser M. (1959) *Just Before Darwin: Robert Chambers and Vestiges,* Wesleyan University Press, Middletown, Conn.
Mivart St G. J. (1871) *On the Genesis of Species,* Macmillan, London
———(1871) 'Darwin's *Descent of Man', Quarterly Review, 131,* 47-90
———(1872) 'Evolution and its Consequences. A Reply to Professor Huxley', *Contemporary Review, 19,* 168-97
———(1876) *Contemporary Evolution,* King, London
———(1889) 'Professor Weismann's Hypothesis', *Dublin Review, 22,* 269
———(1892) 'Eimer on Growth and Inheritance', in *Essays and Criticisms,* Osgood, London
———(1892) 'Happiness in Hell'. *Nineteenth Century, 32,* 899-919
———(1897) 'Reminiscences of T.H. Huxley', *Nineteenth Century, 42,* 985-98
———(1899) 'The Dreyfus Affair and the Roman Catholic Church', *The Times,* London, Oct. 17

Mocek R. (1974) *Wilhelm Roux — Hans Driesch. Zur Geschichte der Entwickelungs-physiologie der Tiere ('Entwickelungsmechanik'),* Fischer, Jena
Molière J.B.P. (1663) *L'École des femmes,* Lolly, Paris
Montgomery E. (1880) 'The Unity of the Organic Individual', *Mind, 5,* 318-36, 465-89
Moore J.R. (1979) *The Post-Darwinian Controversies,* Cambridge University Press, Cambridge
Moorehead P.S. and Kaplan M.M. (eds) (1967) *Mathematical Challenges to the Neo-Darwinian Interpretation of Evolution,* Wistar Institute Press
Morgan C.L. (1891) *Animal Life and Intelligence,* Arnold, London
————(1895) *An Introduction to Comparative Psychology,* Scott, London
————(1896) *Habit and Instinct,* Arnold, London
————(1915) 'Mind and Body in Their Relation to Each Other and to External Things', *Scientia, 18,* 244-56
————(1923) *Emergent Evolution,* Williams & Norgate, London
————(1931) *Emergent Evolution,* 3rd edn, Holt, New York
————(1933) *The Emergence of Novelty,* Williams & Norgate, London
Morgan T.H. (1895) 'Half Embryos and Whole Embryos from One of the First Two Blastomeres', *Anat, Anz., 10*
————(1903) *Evolution and Adaptation,* Macmillan, New York
————(1916) *A Critique of the Theory of Evolution,* Princeton University Press
————(1925) *Evolution and Genetics,* Princeton University Press
————(1928) *The Theory of the Gene,* Yale University Press, New Haven
Muller H.J. (1930) 'Types of Visible Variations Induced by X-rays in *Drosophila', J. Genet., 22,* 299
————(1959) 'One Hundred Years Without Darwinism are Enough', *School Sci. Math., 304*
Murphy J.J. (1869) *Habit and Intelligence,* (2nd edn 1879) Macmillan, London
Murray M.A. Schubinger M., and Palka J. (1981) 'Preferred Axon Paths in the Wing of *Drosphila', Soc Neurosci. Abstracts, 7,* 347

Nagel E. (1961) *The Structure of Science,* Harcourt, Brace & World, New York
Nägeli C. von (1884) *Mechanisch-physiologische Theorie der Abstammungslehre,* München, Leipzig
Needham J. (1931) *Chem·cal embryology,* Cambridge University Press, Cambridge
————(1936) *Order and Life,* Yale University Press, New Haven
————(1943) *Time the Refreshing River,* Allen & Unwin, London
————(1943) 'Evolution and Thermodynamics', in Needham J. *Time the Refreshing River,* pp. 207-32
————(1943) 'Integrative Levels; a Revaluation of the Idea of Progress', in Needham J. *Time the Refreshing River,* pp. 233-72
Nelkin D. (1982) *The Creation Controversy,* Norton, New York
Nevers P. and Saedler H. (1977) 'Transposable Genetic Elements as Agents of Gene Instability and Chromosomal Rearrangement', *Nature, 268,* 109-11
Nietzsche F. (1969) *On the Genealogy of Morals,* (1887) W. Kaufmann and R.J. Hollingsdale (trans), Random House Vintage edn, New York
Nillson-Ehle H. (1909) 'Kreuzungsuntersuchen an Hafer und Weizen'. *Lunds Universitets Arsskrift,* N.S. 2. vol. *5,* no 2
Nordenskiöld E. (1928) *The History of Biology,* L.B. Eyre (trans.), Knopf, New York
Norland C.E. (1974) 'Packard, Alpheus Spring.' in C.C. Gillispie (ed.), *Dictionary of Scientific Biography,* vol. *10,* 272-4, Scribners, New York

Ohno S. (1970) *Evolution by Gene Duplication,* Springer-Verlag, New York
Olmsted J.M.D. (1938) *Claude Bernard, Physiologist,* Harper, New York

Olson E.C. (1968) 'Book Review. Dialectics in Evolutionary Studies', *Evolution, 22,* 426-36

————(1973) 'Science and Philosophy in the Soviet Union', *Evolution, 26,* 675-6

Omdea P. (1975) 'Evolution of the Genome Considered in the Light of Information Theory', *Bull. Zool., 42,* 351-79

Oppenheimer J. (1967) *Essays in the History of Embryology and Biology,* MIT Press, Cambridge, Mass.

Orwell G. (1949) *Nineteen Eighty Four,* Secker & Warburg, London

Osborn H.F. (1895) 'The Heredity Mechanism and the Search for the Unknown Factors of Evolution', *Am. Nat., 29,* 418-39

————(1896) 'A Mode of Inheritance Requiring Neither Natural Selection Nor Inheritance of Acquired Characters', *Trans. New York Acad. Sci., 15,* 141-8

————(1902) 'Homoplasy as a Law of Latent or Potential Homology', *Am. Nat., 36,* 259-71

————(1929) *From the Greeks to Darwin,* (1894) 2nd edn, Scribners, New York; Arno Press, New York, reissue (1975)

Owen R. (1847) 'Report on the Archetype and Homologies of the Vertebrate Skeleton', *Rep. Brit. Assoc. Adv. Sci.,* 169-340

————(1868) *The Anatomy of the Vertebrates,* 3 vols, Longmans & Green, London

Owen R. (1894) *The Life of Richard Owen,* Murray, London

Packard A.S. (1901) *Lamarck, the Founder of Evolution,* Longmans, New York

Packard A. (1966) 'Operational Convergence between Cephalopods and Fish: an Exercise in Functional Anatomy', *Arch. Zool. Ital., 51,* 523-42

Palka J., Schubinger M. and Ellison R.L. (1983) 'The Polarity of Axon Growth in the Wings of *Drosophila melanogaster*', *Devel. Biol., 98,* 481

Pattee H.H. (1972) 'The Evolution of Self-simplifying Systems', in E. Laszlo (ed.), *The Relevance of General Systems Theory,* pp. 31-42, Braziller, New York

Pearson K. (1910) 'Darwinism, Biometry and Some Recent Biology', *Biometrika, 7,* 368-85

Pfeffer, W. (1881) *Pflanzenphysiologie,* Bd. *1,* 2

Phillips D.C. (1976) *Holistic Thought in Social Science,* Stanford University Press, Stanford

Phillips J.F.W. (1970) 'Smuts, Ecology and Holism', in Friedlander Z. (ed.), *Jan Smuts Remembered,* Allan Wingate, London

Pirozynski K.A. and Malloch D.W. (1975) 'The Origin of Land Plants: a Matter of Mycotrophism', *Biosystems, 6,* 153-64

Pirsig R. (1974) *Zen and the Art of Motorcycle Maintenance,* Morrow, New York

Plate L. (1903) *Uber die Bedeutung der Darwinischen Selektionsprinzip und Probleme der Artbildung,* Engelmann, Leipzig

————(1913) *Selektionsprinzip und Probleme der Artbildung, Ein Handbuch des Darwinismus,* Engelmann, Leipzig

Platt J. (1970) 'Hierarchial Growth', *Bull. Atomic Scientists.,* Chicago, Nov, pp.2-4, 46-8

Polanyi M. (1958) *Personal Knowledge,* Routledge & Kegan Paul, London

————(1962) *Personal Knowledge,* Revised edn, Routledge & Kegan Paul, London; Harper Torchbook edn, New York, (1964)

————(1966) *The Tacit Dimension,* Doubleday, Garden City, NY

Popper K.R. (1959) *The Logic of Scientific Discovery,* Basic Books, New York

————(1972a) *Conjectures and Refutations: the Growth of Scientific Knowledge,* 4th edn, Routledge & Kegan Paul, London

————(1972b) 'Of Clouds and Clocks', in Popper K.R. *Objective Knowledge, An Evolutionary Approach,* Oxford University Press, London

————(1972c) 'Evolution and the Tree of Knowledge', in Popper, K.R. *Objective Knowledge. An Evolutionary Approach*, Oxford University Press, London
Portier P. (1918) *Les Symbiotes*, Masson, Paris
Poulton E.B. (1897) 'Report of the Proceedings of the AAAS. Oct 15', *Science*
————(1908) *Essays on Evolution 1889-1907*, Oxford University Press, Oxford
Pribram K. (1971) *Languages of the Brain*, Prentice-Hall, Englewood Cliffs, NJ
Prichard J.C. (1826) *Researches into the Physical History of Mankind*, 2nd edn, Arch, London
Prosser C.L. (1965) 'Levels of Biological Organization and Their Physiological Significance', in J.A. Moore (ed.), *Ideas in Modern Biology* pp. 357-90, Natural History Press, Garden City, NY
————(ed.) (1973) *Comparative Animal Physiology*, 3rd edn, Saunders, Philadelphia
Provine W.B. (1971) *The Origin of Theoretical Population Genetics*, University of Chicago Press, Chicago
————(1978) 'The Role of Mathematical Population Geneticists in the Evolutionary Synthesis of the 1930s and 1940', *Studies in History of Biology, 2*, 167-92
Punnett R.C. (1915) *Mimicry in Butterflies*, Cambridge University Press, Cambridge

Radl E. (1930) *The History of Biological Ideas*, Oxford University Press, Oxford
Reid R.G.B. (1976) 'Letter', *Systematic Zoology, 25*, 296-7
————and Bernard F.R. (1980) 'Gutless Bivalves', *Science, 208*, 609-10
Rensch B. (1959) *Evolution Above the Species Level*, Methuen, London
Riedl R. (1978) *Order in Living Systems*, R.P.S. Jefferies (trans.), Methuen, London
Ritter W.E. (1919) *The Unity of the Organism*, 2 vols, Gorham Press, Boston
————and Bailey E.W. (1928) 'The Organismal Conception, Its Place in Science and Bearing on Philosophy', *University of California Publications in Zoology, 31*, 307-58
Robin E.D. (ed.) (1979) *Claud Bernard and the Internal Environment. A Memorial Symposium*, Dekker, New York
Romanes G.J. (1885) 'Review of Charles Dixon's "Evolution without Natural Selection"', *Nature*, 26-7
————(1886) 'Physiological Selection. An Additional Suggestion on the Origin of Species', *Nature, 34*, 314-16
————(1892-1916) *Darwin and After Darwin*, Open Court, Chicago
Roppen G. (1956) *Evolution and Poetic Belief*, Oslo University Press, Blackwell, Oxford
Rosen R. (1973) 'On the Generation of Metabolic Novelties in Evolution', in Locker A. (ed.), *Biogenesis Evolution Homeostasis*, Springer-Verlag, New York, pp. 113-24
Roszak T. (1972) *Where the Wasteland Ends*, Doubleday, Garden City, NY
————(1975) *The Unfinished Animal*, Harper & Row, New York
Rostan L. (1831) *Exposition des principes de l'organisme*, Labe, Paris
Roux W. (1888) 'Beitr. z. Entwl Mech. des Embryo. V. Über die künstliche Hervorbringung "halber" Embryonen durch Zerstörung einer der beiden ersten Furchungszellen, sowie über die Nachentwicklung der fehlenden Körperhalfte'. *Virchows Archiv., 14*, 419-521
Ruhe A. and Paul N.M. (1914) *Henri Bergson*, Macmillan, London
Russell E.S. (1916) *Form and Function*, Murray, London
————(1946) *The Directiveness of Organic Activities*, Cambridge University Press, Cambridge
————(1962) *The Diversity of Animals*, Brill, Leiden

Salisbury F.B. (1969) 'Natural Selection and the Complexity of the Gene', *Nature, 224*, 342

Samuel E. (1972) *Order: in Life*, Prentice-Hall, Engelwood Cliffs, NJ

Schindewolf H. (1937) 'Beobachtung und Gedanken zur Deszendenzliehre', *Acta Biother.*, *3*, Leiden, 195-212

Schmalhausen I.I. (1949) *Factors of Evolution. The Theory of Stabilizing Selection*, Blakiston, Philadelphia

Schmidt-Nielsen K. (1959) 'The Physiology of the Camel', *Sci. Am.*, *201*, 140-51

Schrödinger I. (1944) *What is Life?* Cambridge University Press, Cambridge

Schwartz R.M. and Dayhoff M.O. (1978) 'Origins of Prokaryotes, Eukaryotes, Mitochondria, and Chloroplasts', *Science, 199,* 395-403

Schwendener S. (1860) *Untersuchung über den Flechtenhallus*, Engelmann, Leipzig

Sellars R.W. (1922) *Evolutionary Naturalism*, Open Court, Chicago

Serres E.R.A. (1830) 'Anatomie transcendante — Quatrième memoire: Loi de symétrie et de conjugaison du système sanguin', *Ann Sci. Nat.*, *21*, 5-49

Severtsov A.N. (1927) 'Über die Beziehungen zwischen der Ortogenes und der Phylogenese der Tiere', *Jena Z. Naturw.*, *63*, 51-180

Shapiro J.A. (1980) 'Changes in Gene Order and Gene Expression', *Proc. First Intern. Symp. on Research Frontiers in Aging and Cancer,* Washington DC, 1980

Sharp L.W. (1926) *An Introduction to Cytology*, 2nd edn, McGraw Hill, New York

Shaw G.B. (1921) *Back to Methuselah*, Constable, London

Sheldrake R. (1981) *A New Science of Life*, Blond & Briggs, London

Sherrington C.S. (1906) *The Integrative Nature of the Nervous System*, Yale University Press, New Haven

Simpson G.G. (1953) 'The Baldwin Effect', *Evolution. 7,* 110-17

———(1967) *The Meaning of Evolution*, rev, edn, Yale University Press, New Haven; Bantam, New York, 1971 edn

———(1977) 'Foreword', to Grant V. *Organismic Evolution*, Freeman, San Francisco

and Beck W.S. (1965) *Life. An Introduction to Biology*, 2nd edn, Harcourt, Brace & World, New York

Singer C.J. (1959) *A History of Biology*, Abelard-Schuman, London

Skolimowski H. (1974) 'Problems of Rationality in Biology', in F.J. Ayala and T. Dobzhansky (eds), *Studies in the Philosophy of Biology*, pp. 205-24, Macmillan, London

Smith G.A. (1899) *The Life of Henry Drummond*, Hodder & Stoughton, London

Smith J.M. (1968) 'Introduction' to J.B.S. Haldane *Science and Life*, Rationalist Publishing, Islington

Smuts J.C. (1926) *Holism and Evolution*, Macmillan, London

———(1973) *Walt Whitman: a Study in the Evolution of Personality*, Wayne State University Press, Detroit

Smuts J. (1952) *Jan Christian Smuts*, Cassell, London

Solomon D. (ed.) (1964) *L.S.D. The Consciousness Expanding Drug*, Putnam, New York

Sommerhoff, G. (1950) *Analytical Biology*, Oxford University Press, Oxford

Spaulding E.G. (1912) 'A Defense of Analysis', in Holt *et al.* (1912) pp. 155-247

———(1918) *The New Rationalism,* Holt, New York

Spemann H. (1914) 'Über verzögerte Kernversorgung von Keimteilen', *Verh. d. D. Zool. Ges. Freiburg*, 216-21

———(1938) *Embryonic Development and Induction*, Yale University Press, New Haven; Hafner, New York, reprint (1962)

Spencer H. (1852) 'The Development Hypothesis'. *The Leader*

———(1857) 'Progress, Its Law and Cause', *Westminster Review*

———(1862) *First Principles,* Mannering, London

———(1864-6) *Principles of Biology*, Williams & Norgate, London

———(1886) 'The Factors of Organic Evolution', *Nineteenth Century, 107,* 570-89, 749-70

————(1893) 'The Inadequacy of Natural Selection', *Contemporary Review, 63,* 153-66

————(1898) *Principles of Biology,* 2nd edn, Williams & Norgate, London

Spilsbury R. (1974) *Providence Lost: a Critique of Darwinism,* Oxford University Press, Oxford

Srivastava R.S. (1968) *Sri Aurobindo and the Theories of Evolution,* Chonkhamba Sanskrit Office

Stanley S.M. (1979) *Macroevolution. Pattern and Process,* Freeman, San Francisco

Stebbins, C.L. (1974) 'Adaptive Shifts and Evolutionary Novelty: a Compositional Approach', in F.J. Ayala and T. Dobzhansy (eds), *Studies in the Philosophy of Biology,* Macmillan, London

Steele E.J. (1977) *Somatic Selection and Adaptive Evolution,* Williams & Wallace, Toronto

Stent G. (1969) *The Coming of the Golden Age,* Natural History Press, Garden City, NY

Stockard C.R. (1931) *The Physical Basis of Personality,* Norton, New York

Sturtevant A.H. (1965) *A History of Genetics,* Harper & Row, New York

Sutton W.S. (1903) 'The Chromosomes in Heredity', *Biol. Bull., 4,* 231-52

Taylor G.R. (1982) *The Great Evolution Mystery,* Harper & Row, New York

Taylor F.J.R. (1974) 'Implications and Extensions of the Serial Endosymbiosis Theory of the Origin of Eukaryotes', *Taxon, 23,* 229-58

Teilhard, de Chardin P. (1955) *Le Phénomène humain,* Éditions de Seuil, Paris

————(1959) *The Phenomenon of Man.* B. Wall (trans.), Harper & Row, London; Harper Torchbook Edition (1961)

Tennant F.R. (1932) *Philosophy of the Sciences,* Cambridge University Press, Cambridge

Thomas L. (1974) *The Lives of a Cell,* Viking, New York

Thompson D'Arcy W. (1917) *On Growth and Form,* Cambridge University Press

————(1926) 'Introduction', to L.S. Berg *Nomogenesis,* (first English edn)

Thomson W. (1868) 'On Geological Time', *Trans. Geol. Soc. Glasgow, 3*

Thorpe W.H. (1974) 'Reductionism in Biology', in F.J. Ayala and T. Dobzhansky (eds), *Studies in the Philosophy of Biology,* pp. 109-36, Macmillan, London

Uexküll J.J. von (1913) *Bausteine zu einer biologischen Weltanschauung,* München

————(1926) *Theoretical Biology,* D.L. MacKinnon (trans.), Harcourt Brace, New York

Uexküll T. (1978) 'Uexküll, Jakob Johann von', in C.C. Gillispie (ed.), *Dictionary of Scientific Biography,* Suppl. *1,* pp. 503-5

Valentine J.W. and Campbell C.A. (1975) 'Gene Regulation and the Fossil Record', *Am. Sci., 63,* 673-80

Vandel A. (1958) *L'Homme et l'évolution,* Gallimard, Paris

Villée C.A. (1942) 'The Phenomenon of Homeosis', *Am. Nat., 76,* 494-506

Virchow R. (1858) *Die Cellularpathologie in ihrer Begründung auf physiologische und pathologische Gewebelehre,* Berlin

Virtanen R. (1967) 'Claude Bernard and the History of Ideas', in *Claude Bernard and Experimental Medicine,* F. Grand and M.B. Visscher (eds), pp. 9-23, Schenkman, Cambridge, Mass.

Vorzimmer P.J. (1970) *Charles Darwin: the Years of Controversy,* Temple University Press, Philadelphia

Waagen W. (1869) 'Die Formenreiche des Ammonites subradiatus', *Geognostisch Palaeontologische Beitrage, 2,* 179-256

Waddington C.H. (1940) *Organisers and Genes*, Cambridge University Press, Cambridge
———(1952) 'Genetic Assimilation', *Nature, 169,* 278
———(1953) 'Epigenetics and Evolution', *Symposia of the Society for Experimental Biology, 7*
———(1957) *The Strategy of the Genes*, Allen & Unwin, London
———(1969) 'Paradigm for an Evolutionary Procession', in C.H. Waddington (ed.), *Towards a Theoretical Biology*, vol. 2, 106-24
———(1975) *The Evolution of an Evolutionist,* Edinburgh University Press, Edinburgh
Wagner M. (1873) *The Darwinian theory and the Law of the Migration Of Organisms*, Stanford, London
Wallace A.R. (1855) 'On the Law Which has Regulated the Introduction of New Species', *Annals and Magazine of Natural History, 16,* 184-96
———(1858) 'On the Tendency of Varieties to Depart Indefinitely from the Original Type', *J. Proc. Linn, Soc. (zool.) 3,* 53-62
———(1889) *Darwinism*, Macmillan, London
Wallin J.E. (1927) *Symbionticism and the Origin of Species*, Williams & Wilkins, Baltimore
Watson J.D. (1965) *Molecular Biology of the Gene,* Benjamin, New York
———(1968) *The Double Helix*, Atheneum, New York
———and Crick F.H.C. (1953) 'Molecular Structure of Nucleic Acids: a Structure for Deoxyribose Nucleic Acid', *Nature, 171,* 737-8
Weinberg W. (1908) 'Ueber den Nachweis der Vererbung beim Menschen', *Jahreshefte des Vereins für Väterlandische Naturkunde in Weitemburg, 64,* 368-82
Weismann A. (1883) *Uber die Vererbung*, Fischer, Jena
———(1885) *Die Kontinuität des Keimplasmas als Grundlage einer Theorie der Vererbung*, Fischer, Jena
———(1893a) *The Theory of the Germ Plasm*, W.N. Parker and H. Rönnfeld (trans), Scott, London
———(1893b) 'The All-sufficiency of Natural Selection', *Contemporary Review, 64,* 596-612
Weiss P.A. (1920) 'Morphodynamische Feldtheorie und Genetik', *Zeitschr. f. indükt Supplements. Band II Abstammungs — u. Vererbungslehre*
———(1936) 'Selectivity Controlling the Central-peripheral Relations in the Nervous System', *Biol. Rev., 11,* 494-531
Weldon W.F.R. (1893) 'On Certain Correlated Variations in *Carcinus moenas*,' *Proc. Roy. Soc., 54*
———(1894) 'The Study of Animal Variation', *Nature, 50,* 25-6
Wells W.C. (1818) *An Account of a Female of the White Race of Mankind Part of Whose Skin Resembles that of a Negro, With Some Observations of The Cause of the Differences in Colour and Form Between the White and Negro Races of Man*, Constable, London
Wheeler W.M. (1929) 'Present Tendencies in Biological Theory', *Scientific Monthly*, February, in W.M. Wheeler (1939) *Essays in Philosophical Biology*, G.H. Parker (ed.), Russell & Russell, New York, pp. 185-210
———(1929) 'Emergent Evolution and the Development of Societies', in W.M. Wheeler (1939) *Essays in Philosophical Biology*, Russell & Russell, New York, pp. 141-69
Whitehead A.N. (1925) *Science and the Modern World*, Macmillan, New York
———(1920) *The Concept of Nature*, Cambridge University Press, Cambridge
Whyte L.L. (1949) *The Unitary Principle in Physics and Biology,* Cressett, London
———(1965) *Internal Factors in Evolution*, Braziller, New York, Social Science

Paperbacks Edition 1968; Associated Book Publishers, London
Wiley E.O. and Brooks D.R. (1982) 'Victims of History — a Nonequilibrium Approach to Evolution', *Systematic Zoology, 31,* 1-24
————(1983) 'Nonequilibrium Thermodynamics and Evolution: a Response to Løvtrup', *Systematic Zoology, 32,* 209-19
Willis J.C. (1922) *Age and Area,* Cambridge University Press, Cambridge
————(1940) *The Course of Evolution,* Cambridge University Press, Cambridge
Wilson E.B. (1928) *The Cell in Development and Heredity,* 3rd edn with corrections, Macmillan, New York
Wilson E.O. (1975) *Sociobiology: The New Synthesis,* Harvard University Belknap Press, Cambridge, Mass.
Winterstein H. (1928) 'Kausalität und Vitalismus vom Standpunkt der der Denkökonomie', 2nd edn, *Abhandlungen zur Theorie d. org. Entwickelung, IV*
Wintrebert P. (1962) *Le vivant, créateur de son évolution,* Marsen, Paris
Wolff C.G. (1759) *Theoria Generationis,* Halle
Woodger J.H. (1929) *Biological Principles,* Routledge & Kegan Paul, London
————(1930a) 'The "Concept of Organism" and the Relation Between Embryology and Genetics, I', *Quart. Rev. Biol., 5,* 1-22
————(1930b) 'The "Concept of Organism" and the Relation Between Embryology and Genetics, II', *Quart. Rev. Biol., 5,* 438-63
————(1931) 'The "Concept of Organism" in the Relation Between Embryology and Genetics, III', *Quart. Rev. Biol., 6,* 178-207
Wright C. (1871) 'On the Genesis of Species', *N. Am. Rev., 113,* 63-103
Wright S. (1917) 'On the Nature of Size Factors', *Genetics, 3,* 367-74
————(1930) 'Review of R.A. Fisher's "The Genetical Theory of Natural Selection"', *J. Heredity, 21,* 249-356
————(1931) 'Evolution in Mendelian Populations', *Genetics, 16,* 97-159
————(1932) 'The Roles of Mutation, Inbreeding and Crossbreeding, and Selection in Evolution', *Proc. 6th Int. Cong. Ithaca., 1,* 356-66

Yeats W.B. (1920) 'The Second Coming' in *W.B. Yeats, Selected Poetry,* A.N. Jeffans (ed.), Macmillan, London
Yule G.U. (1902) 'Mendel's Laws and Their Probable Relations to Intra-racial Heredity', *New Phytologist, 2,* 235-42

Zirkle C. (1946) 'The Early History of the Idea of the Inheritance of Acquired Characters and of Pangenesis', *Trans. Am. Phil. Soc.,* N.S. *35,* 91-104
————(1949) *Death of a Science in Russia,* University of Pennsylvania Press, Philadelphia
Zuckerkandl E. (1975) 'The Appearance of New Structures and Functions in Proteins During Evolution', *J. Mol. Evol., 7,* 1-57

INDEX

DATE DUE